Mercury and the Everglades. A Synthesis and Model for Complex Ecosystem Restoration

Darren G. Rumbold • Curtis D. Pollman •
Donald M. Axelrad
Editors

Mercury and the Everglades. A Synthesis and Model for Complex Ecosystem Restoration

Volume II – Aquatic Mercury Cycling and Bioaccumulation in the Everglades

Editors
Darren G. Rumbold
Coastal Watershed Institute
Florida Gulf Coast University
Fort Myers, Florida, USA

Curtis D. Pollman
Aqua Lux Lucis, Inc
Gainesville, Florida, USA

Donald M. Axelrad
Institute of Public Health
Florida A&M University
Tallahassee, Florida, USA

ISBN 978-3-030-32059-1 ISBN 978-3-030-32057-7 (eBook)
https://doi.org/10.1007/978-3-030-32057-7

© Springer Nature Switzerland AG 2019
This work is subject to copyright. All rights are reserved by the Publisher, whether the whole or part of the material is concerned, specifically the rights of translation, reprinting, reuse of illustrations, recitation, broadcasting, reproduction on microfilms or in any other physical way, and transmission or information storage and retrieval, electronic adaptation, computer software, or by similar or dissimilar methodology now known or hereafter developed.
The use of general descriptive names, registered names, trademarks, service marks, etc. in this publication does not imply, even in the absence of a specific statement, that such names are exempt from the relevant protective laws and regulations and therefore free for general use.
The publisher, the authors, and the editors are safe to assume that the advice and information in this book are believed to be true and accurate at the date of publication. Neither the publisher nor the authors or the editors give a warranty, expressed or implied, with respect to the material contained herein or for any errors or omissions that may have been made. The publisher remains neutral with regard to jurisdictional claims in published maps and institutional affiliations.

This Springer imprint is published by the registered company Springer Nature Switzerland AG.
The registered company address is: Gewerbestrasse 11, 6330 Cham, Switzerland

Florida panther (credit Mark Cunningham).

Preface Volume II

As stated in the Preface to Volume I, the objective of this book is to provide a synthesis of the findings from the extensive research that has been conducted on the Everglades' mercury (Hg) problem. The goal is that this synthesis will serve as a framework for policy makers in making informed decisions. Additionally, it is hoped that this synthesis will serve as a reference for future researchers both within the Everglades and in other regions plagued with Hg problems.

Volume I of the book focused on the evolution of the Everglades as a perturbed ecosystem and the significance of atmospheric deposition of Hg to this system. This volume focuses on aquatic Hg cycling, biomagnification, and ecological and human health risks from exposure to Hg in the Everglades and south Florida. Volume III will focus on temporal trends in both Hg deposition and Hg residues in Everglades biota. It will also offer lessons learned and possible management strategies.

Hg is a metal that is mined from the earth. It has a long history of medical and industrial uses and, as a consequence, a long history of toxic incidents (for a review, see Hightower 2008; Rice et al. 2014). One of the earliest suspected cases of Hg toxicity was Qin Shi Huang, an early emperor of China, who reportedly died in 210 BCE after swallowing Hg pills meant to make him immortal (Rice et al. 2014 and references therein). His tomb was said to have a river of liquid Hg. One of the most well-known incidents of Hg poisoning was the "mad hatter disease" observed in hat makers who used Hg nitrate to soften fur pelts during the late nineteenth and early twentieth century. But these and other early incidents were the result of exposure to inorganic Hg.

The earliest known incident of toxicity to organic Hg was the deaths of two researchers exposed while studying the chemistry of dimethyl Hg in 1863. The incident was not widely reported until a publication in 1940 by Hunter et al. (1940). That publication also presented results of experiments exposing animals to organic Hg. Subsequently, Hunter and Russell (1954 as cited by Grandjean et al. 2010) reported cerebral and cerebellar atrophy in a human patient exposed to organic Hg (later termed Hunter–Russell syndrome). Organic Hg, principally monomethylmercury (MeHg), is the more toxic form of Hg that biomagnifies through food webs. Readers of this volume may be more familiar with the tragedy stemming from organic Hg exposure of residents of a fishing village in Minamata, Japan. Early in the 1950s, large numbers of fish began dying in the Minamata Bay. Birds were then found on the ground unable to fly. Later cats, which ate large quantities of fish in this fishing village, started behaving strangely—running around in circles or leaping up in the air only to die later. This was the so-called dancing cat disease (Yorifuji et al. 2013; Zillioux 2015). In 1955 and 1956, Minamata residents began showing symptoms: first a fisherman and then two young girls having difficulties walking and speaking were admitted to the local hospital (George 2001; Yorifuji et al. 2013). Readers may be surprised to learn of the challenges faced by researchers at the Kumamoto University in trying to identify the poison. In his book titled *Minamata Pollution and the Struggle for Democracy in Postwar Japan*, Timothy George (2001) describes the efforts by local and federal governments in Japan along with the chemical company, Shin Nihon Chisso Corporation, to evade blame. He describes what he terms the "indirect approaches" taken by Chisso to discredit the Hg toxicity theory put forward by the university, including promoting numerous "counter-theories." An early clue linking the problem to organic Hg came when a British neurologist, Douglas McAlpine, examined several patients while visiting the area and compared their symptoms to the Hunter–Russell syndrome (McAlpine and Araki 1958, as cited in George 2001). Yet, the identity of the poison was not completely resolved until much later when Takeuchi et al. (1962) published a paper that provided a clear causal analysis that included nine separate lines of evidence supporting organic Hg as the poison. Hachiya (2012) reports that while only 2271 patients were certified as having Minamata disease, more than 40,000 area residents were thought to exhibit some form of the disease. Although the levels of Hg were much greater in the fish of Minamata Bay than the Everglades, there are parallels between what happened in Minamata and what has happened in the Everglades. One such parallel is the value that animals may serve as sentinels of advance warning of environmental contamination, if properly recognized (for more details, see eco-epidemiology section of Chap. 10, this volume). It is important to note that in the Everglades, the death of panther No. 27 in 1989 from complications due to Hg poisoning (Roelke et al. 1991) prompted action by state and federal agencies.

One of the biggest hurdles the researchers from Kumamoto University faced in identifying the causative agent was to show how inorganic Hg, known to be used in the Chisso factory, could be methylated in the bay; however, as it turned out, this was a moot point in this particular case because the organic Hg was being formed through catalytic reactions inside the factory. It was not until 1967 that this critical connection was made when a study documented that microorganisms in aquatic sediments collected from a lake in Sweden were capable of methylating inorganic Hg (Jensen and Jernelov 1967, as cited by Jernelöv et al. 1975). The study was carried out because elevated levels of MeHg were being found in fish from lakes and rivers of Sweden in the early 1960s (Jernelöv et al. 1975). MeHg as a new environmental problem was not widely recognized in the USA until the early

1970s with a series of publications in the *New York Times* (with headlines such as "Mercury Hazard Found Nationwide," "Mercury in Tuna Leads to Recall," and "Rise in Mercury is Found in Eaters of Certain Fish") and the publication of Eugene Smith's photo essay on the Minamata tragedy in *Life* magazine. National attention turned to Florida on March 14, 1989, when the *New York Times* headline was "Mercury Levels Unsafe in Everglades Fish."

It took several more years before protocols were refined to allow for the collection of scrupulously clean samples (Fitzgerald and Watras 1989) and for the development of analytical methods to measure various forms of Hg at ultra-trace levels (Horvat et al. 1993; USEPA 1996, 1998). The South Florida Mercury Science Program (SFMSP) formed in 1995 as a partnership of federal, state, and local interests including the then Florida Department of Environmental Regulation (later to become the Florida Department of Environmental Protection, FDEP), the South Florida Water Management District (SFWMD), the U.S. Environmental Protection Agency (USEPA), the Florida Fish and Wildlife Conservation Commission (FFWCC), and the U.S. Geological Survey (USGS). Researchers from these agencies, particularly those at the state level, had to be trained in the new protocols to allow them to monitor Hg at ultra-trace levels. While SFWMD and FFWCC conducted some independent research, their primary role was to monitor Hg in various media along transects upstream and downstream of water control structures, i.e., taking a judgmental sampling approach, of stormwater treatment areas (which were being constructed to remove phosphorus from agricultural stormwater before being discharged to the Everglades) and within the Everglades Protection Area. The first permit-mandated requirement for extensive sampling of sediments, water, and biological tissues for Hg analysis came in the form of a permit in 1994 issued from FDEP to SFWMD to operate the Everglades Nutrient Removal Project (the experimental stormwater treatment area). By 2004, when the program was at its height, SFWMD was collecting Hg data under 33 different projects (including contracting with FFWCC to collect fish) at an annual cost of over $1.2 million making it one of the largest ultra-trace monitoring programs in the world (e.g., Rumbold et al. 2007a, b). Unlike the state agencies, USEPA's Regional Environmental Monitoring and Assessment Program (R-EMAP) took a probabilistic approach to randomly sample at hundreds of sites across the Everglades over a short period of time. The objective was to provide a synoptic look at the environmental conditions over a large area that would allow for inferential statistical tests. This required a large team and extensive helicopter time at great expense. To accomplish this, the USEPA assembled a team in the mid-1990s from within the agency and from Florida International University and FTN Associates. To investigate the biogeochemical processes leading to Hg methylation and accumulation, the USGS assembled an interdisciplinary team of experts from across the country from both within the survey and researchers at the University of Wisconsin-Madison, the Wisconsin Department of Natural Resources, the Academy of Natural Science, the University of Colorado, Texas A&M, and the University of Maryland. The effort was named the "Aquatic Cycling of Mercury in the Everglades (ACME) Project."

The experimental design employed by the ACME team in this early investigation of Hg biogeochemistry, which combined diurnal sampling from platforms in the Everglades, dosing mesocosms, and laboratory exposures of media collected from the field with radiotracers and stable isotopes, was transformative and has been duplicated by others in subsequent Hg studies. These different groups working together each revealed a different aspect of the Hg cycle.

The first section of Volume II contains six chapters that summarize information from studies both within the Everglades and elsewhere on Hg biogeochemistry:

Chapter 1. Aquatic Cycling of Mercury
Chapter 2. Sulfur Contamination in the Everglades, a Major Control on Mercury Methylation
Chapter 3. A Causal Analysis for the Dominant Factor in the Extreme Geographic and Temporal Variability in Mercury Biomagnification in the Everglades
Chapter 4. Dissolved Organic Matter Interactions with Mercury in the Florida Everglades
Chapter 5. Phosphorus and Redox Effects on Mercury Cycling
Chapter 6. Major Drivers of Mercury Methylation and Cycling in the Everglades: A Synthesis

These chapters reveal that following the inevitable early debate about methodology (e.g., sampling and analytical), a consensus emerged among the different groups of the SFMSP that while the principal cause of the problem was the high rate of atmospheric Hg deposition (see Volume I), the susceptibility was due to sulfate stimulation of MeHg production across the Everglades. However, as described in several of these chapters, beginning around 2007 managers at both SFWMD and FDEP began to question the interpretation of USGS and USEPA and even their own staff. Many of the questions they were raising centered around the source of the sulfate and its role in the Hg problem (for a review, see Chaps. 2, 3, and 6). Regrettably, while the debate on sulfate continues as this book is published, the level of state funding for monitoring or research has been greatly reduced and the SFMSP ended. And yet the Everglades' Hg problem remains.

In addition to assembling experts to develop new protocols for studying Hg deposition (see Volume I) and biogeochemistry, the SFMSP also assembled experts in Hg biomagnification and effects. Researchers from the University of Wisconsin-Madison and Florida International University along with staff from FFWCC carried out comprehensive food web studies to better understand Hg biomagnification (for details, see Chap. 8, this volume). Researchers from the University of Florida were funded to carry out both field studies and controlled, manipulative toxicological studies on wading birds. Up until that time most avian toxicological work had focused on temperate species (mallards or loons, in particular). Additionally, the SFMSP organized a peer review workshop by nationally recognized experts to review findings of these toxicological studies and resulting probabilistic ecological risk assessments (October 6–7, 1999; River Ranch, Florida; Suter 2000).

The second section of Volume II contains six chapters that summarize information from studies both within the Everglades and elsewhere on biomagnification and health risks from Hg exposure:

These chapters reveal general agreement among state and federal agencies that Hg levels remain high in the Everglades compared to other regions and as compared to toxicological benchmarks. Chapter 9 presents data that show Hg levels are significantly higher in Everglades' bass compared to the rest of the USA. Both ecological (Chap. 10) and human health risk assessments (Chap. 11) reveal continued moderate to high risk to both wildlife and Floridians who eat local fish. This may come as a surprise to some readers. The general public and even some resource managers appear to have the misconception that Florida's Hg problem has been resolved. This may be due in part to generalizations that were made in a few highly publicized reports in the early 2000s that Hg had declined across the system. However, as described in this volume, while biomagnification has lessened in some areas, hotspots remain. Additionally, the media coverage of Hg risks that was significant in the early 1990s has largely waned. The perception that the Hg problem has been resolved may also be due in part to the lack of attention given to the problem by state agencies.

This volume attempts to address remaining uncertainties and misconceptions regarding risks to humans and wildlife from Hg exposure from consuming fish and wildlife from the Everglades and south Florida. It explains why we should not expect to see acute death resulting from exposure to this neurotoxicant in either humans or wildlife. Instead, it describes how sublethal effects could impact humans (e.g., reduced I.Q., deficits in language, attention, motor function, memory; for a review, see Chap. 11, this volume) and leave wildlife at a competitive disadvantage (e.g., unable to find food causing malnutrition, increased susceptibility to disease, depressed reproduction) or at an increased risk of predation. In wildlife, the latter would be equivalent to ecological mortality (for a review, see Chap. 10, this volume).

It is our hope that with a better understanding of Hg cycling, biomagnification, and ecological and human health risks, resource managers will take actions to mitigate and reduce the Everglades' Hg problem.

Fort Myers, FL — Darren G. Rumbold
Gainesville, FL — Curtis D. Pollman
Tallahassee, FL — Donald M. Axelrad

References

Fitzgerald WF, Watras CJ (1989) Mercury in surficial waters of rural Wisconsin lakes. Sci Total Environ 87:223–232

George ST (2001) Minamata: pollution and the struggle for democracy in postwar Japan. Harvard University Press. ISBN 0-674-00785-9

Grandjean P, Satoh H, Murata K, Eto K (2010) Adverse effects of methylmercury: environmental health research implications. Environ Health Perspect 118:1137–1145

Hachiya N (2012) Epidemiological update of methylmercury and Minamata disease. In: Ceccatelli S, Aschner M (eds) Methylmercury and neurotoxicity, current topics in neurotoxicity. Springer, New York, pp 1–11

Hightower JM (2008) Diagnosis: mercury: money, politics, and poison. Island Press

Horvat M, Liang L, Bloom NS (1993) Comparison of distillation with other current isolation methods for the determination of methyl mercury compounds in low level environmental samples. Anal Chim Acta 282:153–168

Hunter D, Bomford RR, Russell DS (1940) Poisoning by methyl mercury compounds. QJM 9(3):193–226

Jernelöv A, Landner L, Larsson T (1975) Swedish perspectives on mercury pollution. Water Poll Control Fed 47(4):810–822

Rice KM, Walker Jr EM, Wu M, Gillette C, Blough ER (2014) Environmental mercury and its toxic effects. J Prev Med Public Health 47(2):74

Roelke M, Schultz D, Facemire C, Sundlof S, Royals H (1991) Mercury contamination in Florida panthers. Prepared by the Technical Subcommittee of the Florida Panther Interagency Committee

Rumbold D, Niemeyer N, Matson F, Atkins S, Jean-Jacques J, Nicholas K, Owens C, Strayer K, Warner B (2007a) Appendix 3B-1: Annual permit compliance monitoring report for mercury in downstream receiving waters of the Everglades Protection Area. In: 2007 South Florida Environmental Report, vol 1. South Florida Water Management District, West Palm Beach, FL. http://www.sfwmd.gov/sfer/SFER_2007/

Rumbold D, Niemeyer N, Matson F, Atkins S, Jean-Jacques J, Nicholas K, Owens C, Strayer K, Warner B (2007b) Appendix 5-6: Annual permit compliance monitoring report for mercury in Stormwater Treatment Areas. In: 2007

South Florida Environmental Report, vol 1. South Florida Water Management District, West Palm Beach, FL. http://www.sfwmd.gov/sfer/SFER_2007/

Suter GW II (2000) Summary of Comments by the Review Panel for the Workshop, Effects assessment for mercury in fish-eating birds: setting an avian wildlife criterion. Report to the Florida Department of Environmental Protection, 10 January 2000

Takeuchi T, Morikawa N, Matsumoto H, Shiraishi Y (1962) A pathological study of Minamata disease in Japan. Acta Neuropathologica 2(1):40–57

USEPA (1996) Method 1631: Mercury in water by oxidation, purge and trap, and cold vapor atomic fluorescence (CVAFS). Draft method EPA 821-R-96-001

USEPA (1998) Method 1630: Methyl mercury in water by distillation, aqueous ethylation, purge and trap, and cold vapor atomic fluorescence (CVAFS). Draft method

Yorifuji T, Tsuda T, Harada M (2013) Minamata disease: a challenge for democracy and justice (Chapter 5). In: Late lessons from early warnings: science, precaution, innovation. European Environment Agency Report No 1/2013, Copenhagen, Denmark, p 92. http://www.eea.europa.eu/publications/late-lessons-2. Accessed 22 Apr 2019

Zillioux EJ (2015) Mercury in fish: history, sources, pathways, effects, and indicator usage. Environmental indicators. Springer, Berlin, pp 743–766

Contents

Contributors

George Aiken (deceased)

Donald M. Axelrad Florida A&M University, Tallahassee, FL, USA

Alan Becker Florida A&M University, Tallahassee, FL, USA

Mark Cunningham Florida Fish and Wildlife Conservation Commission, Gainesville, FL, USA

Peter C. Frederick University of Florida, Gainesville, FL, USA

Andrew M. Graham Grinnell College, Grinnell, IA, USA

Charles Jagoe Florida A&M University, Tallahassee, FL, USA

David Krabbenhoft U.S. Geological Survey, Middleton, WI, USA

Ted Lange Florida Fish and Wildlife Conservation Commission, Eustis, FL, USA

William F. Loftus Aquatic Research & Communication, LLC, Vero Beach, FL, USA

William Orem U.S. Geological Survey, Reston, VA, USA

Todd Z. Osborne Wetland Biogeochemistry Laboratory, University of Florida, Gainesville, FL, USA
Whitney Laboratory for Marine Bioscience, University of Florida, St. Augustine, FL, USA

Sara A. Phelps Wetland Biogeochemistry Laboratory, University of Florida, Gainesville, FL, USA
Whitney Laboratory for Marine Bioscience, University of Florida, St. Augustine, FL, USA

Curtis D. Pollman Aqua Lux Lucis, Inc., Gainesville, FL, USA

Brett Poulin U.S. Geological Survey, Boulder, CO, USA

Darren G. Rumbold Florida Gulf Coast University, Fort Myers, FL, USA

Chapter 1
Aquatic Cycling of Mercury

William H. Orem, David P. Krabbenhoft, Brett A. Poulin, and George R. Aiken

Abstract This chapter examines crucial processes in the aquatic cycling of mercury (Hg) that may lead to microbial production of neurotoxic and bioaccumulative methylmercury (MeHg), and highlights environmental conditions in the Everglades that make it ideal for MeHg production and bioaccumulation. The role of complexation of Hg^{2+} in surface water, especially by dissolved organic matter (DOM), in the transport of mercury to sites of microbial methylation are discussed. Photochemical reactions important in Hg cycling in surface water are also discussed. A principal focus of the chapter is on the environmental conditions that promote MeHg production, especially the role of sulfide and DOM in transport of inorganic Hg into bacteria for methylation, and the types of bacteria that have the ability to methylate Hg. Finally, perturbations to the ecosystem (e.g., fire and drought) that have important effects on Hg cycling are discussed.

Keywords Mercury · Aquatic systems · Methylmercury · Redox · Sulfate reducing bacteria · Dissolved organic matter

---

Author "George R. Aiken" is deceased at the time of publication.

W. H. Orem (✉)
U.S. Geological Survey, Reston, VA, USA
e-mail: borem@usgs.gov

D. P. Krabbenhoft
U.S. Geological Survey, Middleton, WI, USA
e-mail: dpkrabbe@usgs.gov

B. A. Poulin
U.S. Geological Survey, Boulder, CO, USA
e-mail: bpoulin@usgs.gov

© Springer Nature Switzerland AG 2019

D. G. Rumbold et al. (eds.), *Mercury and the Everglades. A Synthesis and Model for Complex Ecosystem Restoration*, https://doi.org/10.1007/978-3-030-32057-7_1

1.1 Introduction

Although natural sources contribute mercury (Hg) to the environment, such as volcanism, forest fires, and runoff or groundwater weathering of certain Hg rich rocks, the large increase in Hg in the environment since the mid-1800s is largely from anthropogenic sources (Selin 2009; Driscoll et al. 2013). Although, in some instances Hg in water discharges is an important source, such as in industrial (Bloom et al. 1999) or mine discharges (May et al. 2000; Alpers et al. 2005), the primary source of Hg to most systems is atmosphere deposition from nearby and/or distant emission sources (for review of emission sources, see Chap. 5, Volume I).

The Hg biogeochemical cycle (Fig. 1.1) is complex due to the many oxidation states and chemical forms of Hg in these systems (Horvat 1996). The transformations of Hg can be purely chemical in nature, or facilitated by physical or biological processes, and Hg readily cycles between the atmosphere, surface water, and soil in aquatic environments (Schroeder et al. 1989). These many forms and transformations, and the generally low concentrations of Hg in the environment make the study of aquatic Hg biogeochemistry a challenge to researchers (Science for Environment Policy 2017).

The primary driver for undertaking biogeochemical studies of Hg in aquatic ecosystems is that it can lead to the production of methylmercury (MeHg), a more toxic form that can biomagnify through food webs (Morel et al. 1998; Grandjean et al. 2010; Driscoll et al. 2013; for review of the process of biomagnification, see Chap. 7, this volume). MeHg is a neurotoxin and can pass both the placental and blood/brain barriers in humans to cause neurological affects (Kerper et al. 1992; Kajiwara et al. 1996). The most well documented of these neurologic affects is Minamata Disease, which causes ataxia, dysarthria, constriction of visual fields, impaired hearing, sensory disturbance, and even death in extreme cases (Harada 1995). Children exposed *in utero* to toxic levels of MeHg can experience developmental impacts, including lowered IQ, and attention deficit disorder (Myers and Davidson 1998; Axelrad et al. 2007; Grandjean and Herz 2011; Pichery et al. 2012; Sakamoto et al. 2017). MeHg has also been linked to cardiovascular disease (Roman et al. 2011), and to autoimmune effects in some sensitive individuals (Motts et al. 2014; Maqbool et al. 2017; for review of human health risks, see Chap. 11, this volume). MeHg is also an endocrine disruptor, with known effects on reproductive success in piscivorous fauna in aquatic ecosystems, such as wading birds (Jayasenaa et al. 2011; for review of risks to fish and wildlife, see Chap. 10, this volume), and potential but largely undescribed endocrine effects in humans (Tan et al. 2009). Thus, for reasons related to environmental and human health, an understanding of the biogeochemical controls on the production of MeHg and the cycling of Hg is needed.

In this chapter we examine the major processes involved in the biogeochemical cycling of Hg, the resulting variety of forms of Hg, and the various Hg reservoirs in aquatic ecosystems with a focus on the Everglades.

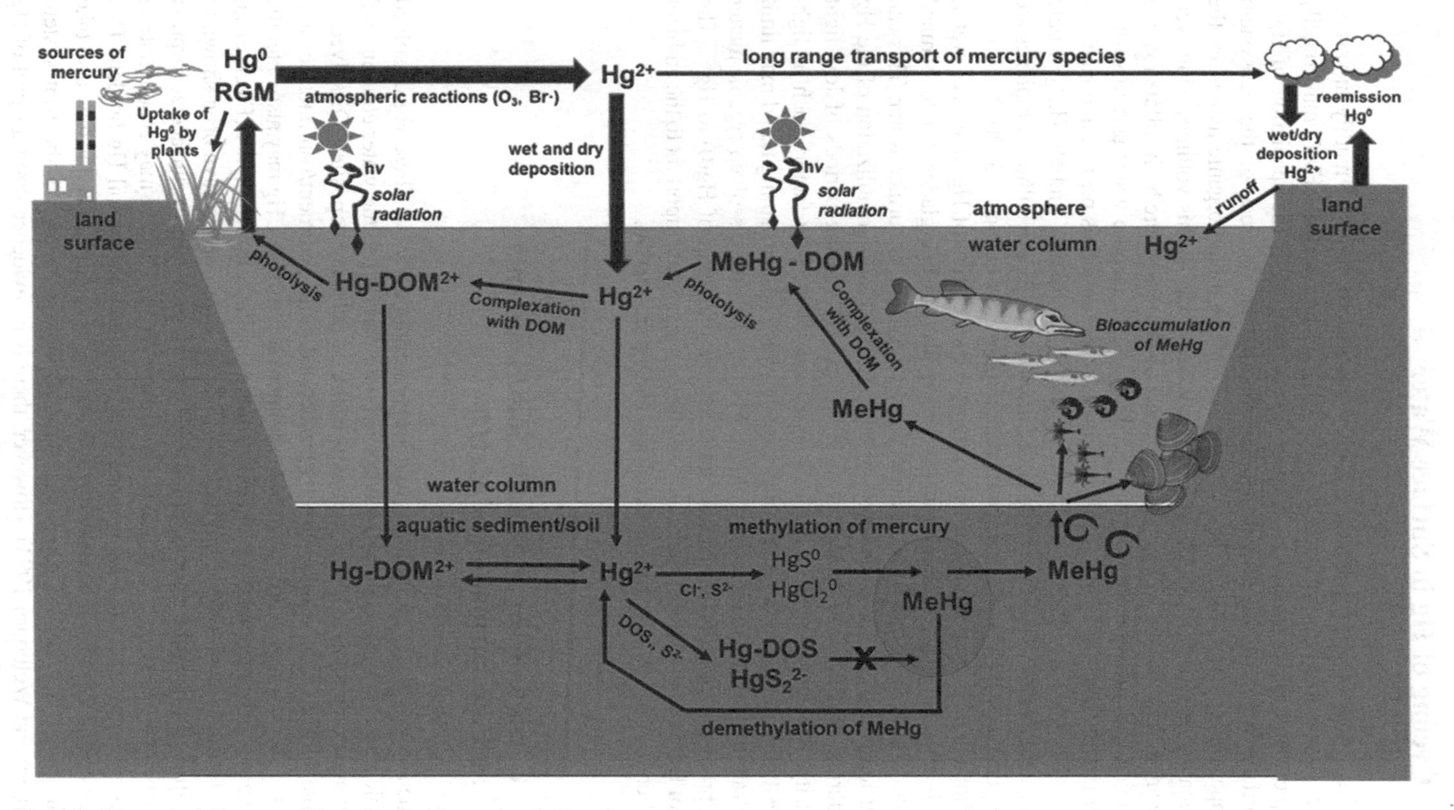

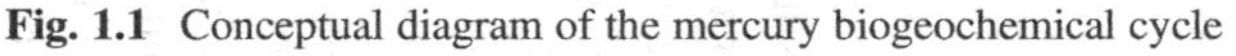
Fig. 1.1 Conceptual diagram of the mercury biogeochemical cycle

1.2 Cycling of Hg in Surface Water

The Hg^{2+} deposited on wetlands is highly soluble and dissolves readily in surface water. If dissolved organic matter or other ions are present, Hg^{2+} may readily complex with these substances. Geochemical modeling of the speciation of Hg^{2+} in surface water of the Everglades has shown that, in areas of the ecosystem unimpacted by canal water, Hg^{2+} complexes with dissolved organic matter are the dominant form (Jiang et al. 2018; for review, see Chap. 4, this volume). However, there are portions of the Everglades near canal discharges where high sulfate loading and microbial sulfate reduction may produce sulfide in pore water and in surface water (for review, see Chap. 2, this volume). Where sulfide is present at concentrations >20 ppb, Hg-sulfide complexes (e.g. HgS_2^{2-}, $HgHS_2^-$, and $Hg(HS)_2$) would dominate. Even at very low sulfide concentrations (0.32 ppb) sulfide complexes are as important as Hg-dissolved organic matter complexes (Jiang et al. 2018).

Photochemical reactions may also impact the speciation of Hg in surface water. This is especially true of subtropical wetlands like the Everglades. Photochemical reduction of Hg^{2+} to Hg(0) is an important process in near surface water. This can result in diurnal cycles of reduction of Hg^{2+} to Hg(0) and volatilization of the Hg (0) during daytime, and oxidation of Hg(0) to Hg^{2+} during periods of low light intensity. Reduction rates of Hg^{2+} to Hg(0) are affected by factors such as light intensity and dissolved organic matter content which adsorbs sunlight and may limit penetration of light into the water column. Photolysis of the dissolved organic matter may produce free radicals that also influence the oxidation of Hg(0) to Hg^{2+}. The loss of Hg(0) from surface waters by volatilization can be as important to the ambient Hg^{2+} concentration as sediment burial and MeHg production.

1.3 Cycling of Hg in Aquatic Soils/Sediments and the Formation of MeHg

As mentioned, Hg^{2+} readily binds to dissolved organic matter. These Hg-dissolved organic matter complexes may be adsorbed onto particles in the water column, and the complex transported to the sediments as the particles are deposited. Dissolved or complexed Hg^{2+} may also be directly transported to the sediment/soil interface and diffuse into pore water and adsorb onto sediment/soil surfaces. Hg may also enter the sediment/soil from uptake by plants, including aquatic macrophytes (Fleck et al. 1999; Ericksen et al. 2003; Lindberg et al. 2007). Macrophytes have been shown to take up Hg(0) from the atmosphere through the stomata and by surface adsorption (nonstomatal pathway; Stamenkovic and Gustin 2009). This Hg may accumulate in the plant tissue, and upon plant senescence contribute to the soil Hg burden. This source can be significant in some environments. Rooted macrophytes may also take up dissolved Hg species from the sediments or pore water via the root system (Coquery and Welbourn 1994). However, there is no evidence that transport of Hg from the roots to the leaves occurs. Thus, Hg taken up by the roots of macrophytes tends to stay in the roots and upon the death of the plant is incorporated into the sediment.

Once in the sediment/soil, Hg may partition between the solid phase and the pore water within the sediment/soil. The Hg^{2+} in the pore water may bind to dissolved organic matter, dissolved sulfide, and associate with other dissolved chemical species (e.g., Cl^- and OH^-). The partitioning of Hg between the solid phase and pore water is determined by binding sites on the sediment/soil surface, and the amounts and type of dissolved organic matter, and dissolved sulfide in pore water. The presence of sulfur functional groups on both sediment/soil organic matter and dissolved organic matter are particularly important in binding Hg and determining partitioning. Soil water partition coefficients (K_{Dsoil}) for Everglades peat have been measured as $10^{21.8}$–$10^{22.0}$ M^{-1}, while the K_D for dissolved organic matter was determined to have a range of $10^{22.8}$–$10^{23.2}$ M^{-1} (Drexel et al. 2002). This suggests that binding to dissolved organic matter may be stronger than that for peat surfaces. Functional groups in organic matter that bind Hg most effectively are reduced sulfur species (thiols, disulfides, disulfanes), with carboxyl and phenolic functional groups of secondary importance as binding sites. The magnitude of the binding sites on the peat allows organic matter (solid and dissolved) to compete with free dissolved sulfide for dissolved Hg^{2+}. In sulfidic aquatic environments, such as a large portion of the Everglades, much of the reduced sulfur functional groups on organic matter may arise from reaction of free sulfide with organic matter (for review, see Chap. 4, this volume). It seems likely that little dissolved free Hg^{2+} is present in aquatic pore water due to its strong affinity for complexing with organic matter and sulfide. The formation of complexes with dissolved organic matter (especially dissolved organic sulfur species) and dissolved sulfide has implications for the transport of Hg into microbial cells (across cell membranes) where production of MeHg occurs (Graham et al. 2017; Poulin et al. 2017). Various forms of Hg in pore water (e.g., Hg^{2+}, MeHg, etc.) are often present in higher concentration than in surface water, and therefore diffusion and/or advection of Hg forms from pore water into surface water may occur (Gilmour et al. 1992). Perhaps the most important form of Hg in wetland ecosystems is MeHg, due to its high degree of bioaccumulation and toxic effects. Concentrations of MeHg in surface water of wetlands can be dynamic since biotic and abiotic processes impact both production and destruction of this form of Hg.

Many different types of bacteria have the capacity to methylate inorganic Hg. The recent discovery of the genes controlling Hg methylation has expanded the number of organisms that have at least the genomic capacity for Hg methylation (Gilmour et al. 2013). Groups of microbes thought to have the genetic capacity to methylate Hg include sulfate reducing bacteria, iron reducing bacteria, manganese reducing bacteria, and methanogens. However, to date, Hg methylation in iron reducers and methanogens seems to be more restricted than among sulfate reducers (Gilmour et al. 2013; Christensen et al. 2016). Whether this capacity to methylate manifests into actual production of MeHg depends on a host of environmental factors, including temperature, redox conditions, pH, and the presence of suitable electron donors (e.g., organic carbon) and acceptors. All these groups of microorganisms are anaerobes, and it is generally considered that MeHg production occurs under anoxic conditions. The extensive wetland soils of the Everglades provide anoxic conditions for the microbes responsible for MeHg production to flourish (for review, see Chap. 5, this volume). The Everglades also has neutral pH, high levels of dissolved organic matter (which can complex, transport, and stabilize both inorganic Hg^{2+} and MeHg), high Hg deposition, high year-round temperatures promoting microbial activity including methylation, and an appropriate electron donor (organic matter) to drive microbial methylation.

For sulfate reducers to be the favored pathway requires a bioavailable form of oxidized sulfur to serve as an electron acceptor. However, sulfate reduction will be favored only after other electron accepters higher on the redox ladder (particularly oxygen) are exhausted from the environment. Yet, when sulfide, i.e., the product of this reduction, reaches a concentration above a variable threshold MeHg production from Hg^{2+} is inhibited (Benoit et al. 1999; Jay et al. 2000; for review of the threshold and the Goldilocks region where sulfate/sulfide is just right, see Chap. 2, this volume). It is thought that under highly sulfidic environments charged Hg-S complexes form that are not taken up as readily as neutral complexes by passive diffusion across the cell membrane. Studies also indicate that the more recently deposited Hg^{2+} is more readily converted to MeHg (Gilmour et al. 2004). This may be due to the complexation of "old" Hg^{2+} to dissolved organic matter, particles in the water column, or sediment/soil particles. This complexation may also reduce the flux of Hg^{2+} into microbial cells within which methylation occurs. Photochemical processes can also affect Hg bioavailability by transforming Hg^{2+} to HgS, which precipitates (Luo et al. 2017).

Once formed in anoxic environments in the water column or soils/sediments, MeHg may bind to dissolved organic matter. This may serve to stabilize MeHg and allow it to remain in the water column for periods of time sufficient for uptake by aquatic organisms; however, it can also affect its bioavailability (for review of DOM's influence, see Chap. 4, this volume).

It should be understood that the amount of MeHg available for subsequent biomagnification is dependent on the balance between rates of methylation and demethylation. Demethylation may occur biotically or abiotically. Many of the same organisms that methylate Hg can also demethylate it (Lu et al. 2016). However, in most systems (including the Everglades) studies to date indicate that microbial methylation greatly exceeds microbial demethylation leading to net MeHg production (Marvin-DiPasquale and Oremland 1998; Marvin-DiPasquale et al. 2000). The principal abiotic demethylation process is photochemical degradation (photolysis), which occurs in surface water (Seller et al. 1996). Complexation of MeHg by dissolved organic matter can enhance photolytic demethylation (Qian et al. 2014). This occurs as binding between the MeHg and thiolates in the dissolved organic matter allows directed energy transfer from aromatics excited by photons from sunlight sufficient enough to break carbon-Hg bonds in MeHg. The role that DOM plays in photodegradation is two-edged because, as mentioned previously, increasing DOM concentration (particularly the prevalent colored-DOM) may reduce light penetration and, thus, the rate of photodegradation.

1.4 Perturbations Affecting Hg Cycling

Aquatic soils/sediments represent a major sink for Hg^{2+} and MeHg. The affinity of Hg in binding to organic matter and sulfur stabilizes Hg species in the soils/sediments where Hg species remain retained and generally unavailable unless disturbances occur. However, aquatic ecosystems are subject to disturbances of both natural and anthropogenic origin that can change the environmental conditions in the soil/sediment leading to mobilization of Hg.

One type of disturbance which is common to the Everglades is the combination of drought and fire (Fig. 1.2). Drought is a natural part of the seasonal cycle in south Florida, which has a wet season and a dry season; typically, the dry season is October to May and the wet season occurs during the rest of the year. While the dry season does not always produce droughts in the Everglades, it is not an uncommon occurrence. Lightning strikes under drought conditions can lead to fires that can race through dry vegetation and under more severe drought cause burning of the peat soil. Both drought and fire cause oxidation of the peat soil and release of bound metals, including Hg upon rewetting (usually from rainfall and/or release of canal water into the ecosystem). The Hg remobilized from the oxidized peat soil is primarily present as Hg^{2+} released to surface and pore water and is reinjected into the biogeochemical cycling of Hg (Fig. 1.2). In the Everglades, drought and fire may also lead to the release of dissolved organic matter (DOM) and sulfur from the soil

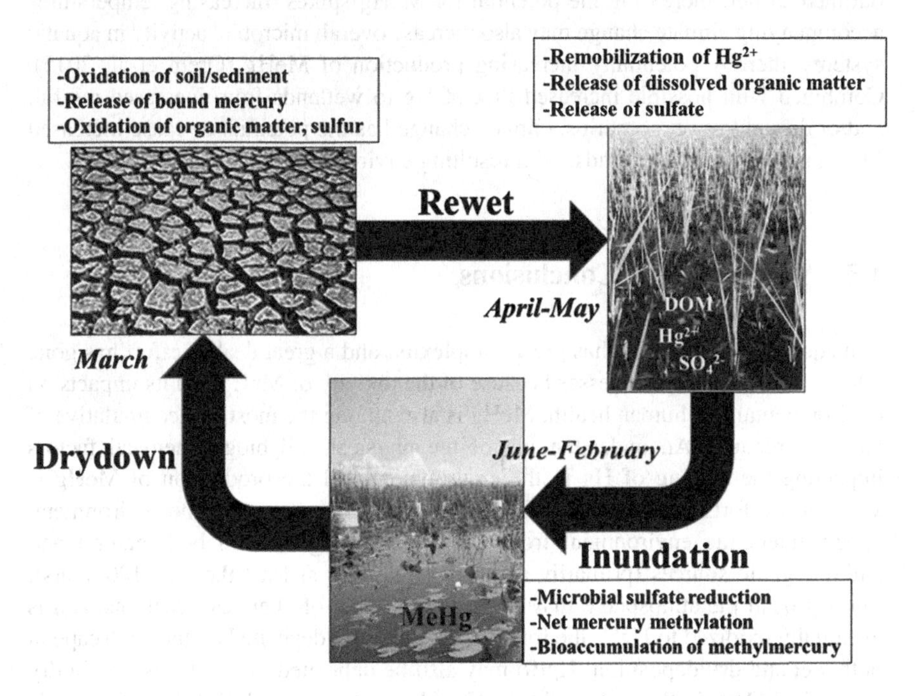

Fig. 1.2 Conceptual diagram of the impact that drought has on the mercury cycle

(oxidized to sulfate) and remobilization into surface and pore water upon rewetting. The presence of high concentrations of Hg^{2+}, sulfate, and DOM during further inundation of the dried or burned area produces conditions conducive to MeHg production. Indeed, large spikes of MeHg have been observed following rewetting and inundation of dried and burned peat soils in the Everglades (Krabbenhoft and Fink 2001).

Climate change with higher temperatures, altered precipitation, and the resulting sea level rise may also perturb Hg biogeochemical cycling in aquatic environments. However, little is known about the effects of climate change on Hg cycling. A study by Wängberg et al. (2010) using a Hg model assuming a 5 °K warming scenario suggested increased flux of mercury (Hg^0) from the ocean to the atmosphere, but decreased oxidation of gaseous elemental mercury (GEM) in the atmosphere. The decreased atmospheric oxidation of GEM reflected decreased concentrations of other reactive atmospheric species (i.e., lower Br concentrations) involved in GEM oxidation. Thus, according to this model, a + 5 °K temperature increase would result in accumulation of GEM in the atmosphere, potentially increasing the importance of long-range transport of GEM. Oceanic levels of Hg would decline in concert with the increase in atmospheric GEM. Krabbenhoft and Sunderland (2013) suggest that climate change may alter rainfall patterns, potentially resulting in increased Hg deposition on some aquatic ecosystems. However, other aquatic ecosystems may experience drought, and/or wild swings between excess precipitation and drought. These drought conditions may remobilize Hg sequestered in the soil by oxidation, as outlined earlier, increasing the potential for MeHg spikes. Increasing temperatures accompanying climate change may also increase overall microbial activity in aquatic systems, thereby potentially increasing production of MeHg (Orem et al. 2015). Combined with possible increased flux of Hg to wetlands from increased rainfall and/or drought rewet scenarios, climate change has the potential to cause increased MeHg production in wetlands, with resulting environmental effects.

1.5 Summary and Conclusions

The aquatic cycling of Hg has great complexity, and a great deal of effort has gone into unravelling these processes because of the toxicity of MeHg and its impacts on environmental and human health. MeHg is also among the most bioaccumulative of all contaminants. An understanding of the physical and biogeochemical factors impacting the cycling of Hg in the environment and the production of MeHg is key to any efforts aimed at mitigating the levels of this toxicant in the environment.

Hg enters the environment from emissions as Hg(0) from both natural and anthropogenic sources (primarily anthropogenic over at least the past 100 years). The Hg(0) in the atmosphere may undergo a number of chemical reactions, and is eventually oxidized to Hg^{2+}, the form that is primarily deposited on the landscape in both wet and dry deposition; Hg(0) may also be deposited on the landscape in dry deposition. Once in the water column, Hg^{2+} binds readily with DOM and inorganic anions to form complexes, or may be bound to particulates (Ortiz et al. 2015). Hg complexes are important in transport of Hg^{2+}, especially to sites where methylation can occur. Hg can also undergo photochemical reactions in surface water that can disrupt Hg-DOM complexes, as well as catalyzing the conversion of Hg^{2+} back to Hg(0), which can then evade from the water column into the atmosphere.

MeHg is an important form of Hg in aquatic systems because of its toxicity and ability to bioaccumulate. It is formed from the methylation of Hg^{2+} within different types of microorganisms (Eubacteria and Archaea) with genes that code for methylation of Hg. Experimental and field studies suggest that in the Everglades the most important groups of microorganisms for Hg methylation are the sulfate reducing bacteria. Sulfur biogeochemistry plays an important role in the methylation of Hg through stimulation of microbial sulfate reduction by sulfate (production of MeHg), and complexation of Hg^{2+} by sulfide and dissolved organic sulfur that may inhibit transport of Hg^{2+} into the cell where methylation occurs (see Chap. 2, this volume for detailed discussion). Once produced, MeHg may bind with DOM, stabilizing it, and be incorporated into the food chain. In the Everglades, MeHg is primarily produced in anoxic soils, and it is thought that zooplankton (hiding in soil from predators during daylight hours but emerging at night) may incorporate MeHg and be important vectors for this toxicant into the food chain.

Perturbations to the ecosystem, such as agricultural amendments (see Chapter 2, this volume), fire, drought, and climate change can have large effects on the cycling of Hg in aquatic systems. Factors such as soil oxidation may release Hg bound to soil organic matter, thereby increasing Hg^{2+} and its bioavailability for methylation. Climate change in the Everglades may result in more extreme events, such as extended periods of drought and fire or more rainfall during other periods. These extreme events can have significant impacts on Hg cycling and MeHg production in ecosystems, and an understanding of these effects is necessary for land and water managers to prepare for the impacts.

Acknowledgments This work was supported by the USGS Priority Ecosystems Studies Program for South Florida—Nick Aumen, Program Executive. Any use of trade, firm, or product names in this report is for descriptive purposes only and does not imply endorsement by the USGS or the U.S. Government.

References

Alpers CN, Hunerlach MP, May JT, Hothem RL, Taylor HE, Antweiler RC, De Wild JF, Lawler DA (2005) Geochemical characterization of water, sediment, and biota affected by mercury contamination and acidic drainage from historical gold mining, Greenhorn Creek, Nevada County, CA, 1999–2001. U.S. Geological Survey Scientific Investigations Report 2004-5251, 278 p. https://pubs.usgs.gov/sir/2004-5251/

Axelrad DA, Bellinger DC, Ryan LM, Woodruff TJ (2007) Dose-response relationship of prenatal mercury exposure and IQ: an integrative analysis of epidemiologic data. Environ Health Perspect 115:609–615

Benoit JM, Mason RP, Gilmour CC (1999) Estimation of mercury-sulfide speciation in sediment pore waters using octanol-water partitioning and implications for availability to methylating bacteria. Environ Toxicol Chem 18:2138–2141

Bloom NS, Gill GA, Cappellino S, Dobbs C, McShea L, Driscoll C, Mason R, Rudd J (1999) Speciation and cycling of mercury in Lavaca Bay, Texas, sediments. Environ Sci Technol 33:7–13

Christensen GA, Wymore AM, King AJ, Podar M, Hurt R Jr, Santillan EU, Soren A, Brandt CC, Brown SD, Palumbo AV, Wall JD, Gilmour CC, Elias DA (2016) Development and validation of broad-range qualitative and clade-specific quantitative molecular probes for assessing mercury methylation in the environment. Appl Environ Microbiol. https://doi.org/10.1128/AEM.01271-16

Coquery M, Welbourn PM (1994) Mercury uptake from contaminated water and sediment by the rooted and submerged aquatic macrophyte *Eriocaulon septangulare*. Arch Environ Contam Toxicol 26:335–341

Drexel RT, Ushaitzer M, Ryan J, Aiken G, Nagy K (2002) Mercury(II) sorption to two Florida Everglades peats: evidence for strong and weak binding and competition by dissolved organic matter released from the peat. Environ Sci Technol 36:4058–4064

Driscoll CT, Mason RP, Chan HM, Jacob DJ, Pirrone N (2013) Mercury as a global pollutant: sources, pathways, and effects. Environ Sci Technol 47(10):4967–4983. https://doi.org/10.1021/es305071v

Ericksen JA, Gustin MS, Schorran DE, Johnson DW, Lindberg SE, Coleman JS (2003) Accumulation of atmospheric mercury in forest foliage. Atmos Environ 37:1613–1622

Fleck JA, Grigal DF, Nater EA (1999) Mercury uptake by trees: an observational experiment. Water Air Soil Pollut 115:513–523

Gilmour CC, Henry EA, Mitchell R (1992) Sulfate stimulation of mercury methylation in freshwater sediments. Environ Sci Technol 26:2281–2287

Gilmour CC, Krabbenhoft DP, Orem WO (2004) Appendix 2B-3: mesocosm studies to quantify how methylmercury in the Everglades responds to changes in mercury, sulfur, and nutrient loading. In: Redfield G (ed) 2004 Everglades consolidated report. South Florida Water Management District, West Palm Beach, FL

Gilmour CC, Podar M, Bullock AL, Graham AM, Brown SD, Somenahally AC, Johs A, Hurt RA Jr, Bailey KL, Elias DA (2013) Mercury methylation by novel microorganisms from new environments. Environ Sci Technol 47:11810–11820

Graham AM, Cameron-Burr KT, Hajic HA, Lee C, Msekela D, Gilmour CC (2017) Sulfurization of dissolved organic matter increases Hg–sulfide–dissolved organic matter bioavailability to a Hg-methylating bacterium. Environ Sci Technol 51(16):9080–9088. https://doi.org/10.1021/acs.est.7b02781

Grandjean P, Herz KT (2011) Methylmercury and brain development: imprecision and underestimation of developmental neurotoxicity in humans. Mt Sinai J Med: J Transl Pers Med 78 (1):107–118

Grandjean P, Satoh H, Murata K, Eto K (2010) Adverse effects of methylmercury: environmental health research implications. Environ Health Perspect 118(8):1137–1145

Harada M (1995) Minamata disease: methylmercury poisoning in Japan caused by environmental pollution. Crit Rev Toxicol 25(1):1–24. https://doi.org/10.3109/10408449509089885

Horvat M (1996) Mercury analysis and speciation in environmental samples. In: Baeyens W, Ebinghaus R, Vasiliev O (eds) Global and regional mercury cycles: sources, fluxes and mass balances. Kluwer Academic, Dordrecht. In cooperation with NATO Scientific Affairs Division, NATO ASI series (ASEN2, vol 21), pp 1–31

Jay JA, Morel FM, Hemond HF (2000) Mercury speciation in the presence of polysulfides. Environ Sci Technol 34:2196–2200

Jayasenaa N, Frederick PC, Larkinb IL (2011) Endocrine disruption in white ibises (*Eudocimus albus*) caused by exposure to environmentally relevant levels of methylmercury. Aquat Toxicol 105:321–327

Jiang P, Liu G, Cui W, Cai Y (2018) Geochemical modeling of mercury speciation in surface water and implications on mercury cycling in the everglades wetland. Sci Total Environ 640–641:454–465

Kajiwara Y, Yasutake A, Adachi T, Hirayama K (1996) Methylmercury transport across the placenta via neutral amino acid carrier. Arch Toxicol 70(5):310–314

Kerper IE, Ballatori N, Clarkson TW (1992) Methylmercury transport across the blood-brain barrier by an amino acid carrier. Am J Phys 262(5 Pt 2):R761–R765

Krabbenhoft DP, Fink L (2001) Appendix 7–8: The effect of dry down and natural fires on mercury methylation in the Florida Everglades. In: Redfield G (ed) 2001 Everglades consolidated report. South Florida Water Management District, West Palm Beach, FL, 14 p

Krabbenhoft DP, Sunderland EM (2013) Global change and mercury. Science 341:1457–1458. https://doi.org/10.1126/science.1242838

Lindberg S, Bullock R, Ebinghaus R, Engstrom D, Feng X, Fitzgerald W, Pirrone N, Prestbo E, Seigneur CA (2007) Synthesis of progress and uncertainties in attributing the sources of mercury in deposition. Ambio 36:19–32

Lu X, Liu Y, Johs A, Zhao L, Wang T, Yang Z, Lin H, Elias DA, Pierce EM, Liang L, Barkay T, Gu B (2016) Anaerobic mercury methylation and demethylation by *Geobacter bemidjiensis* Bem. Environ Sci Technol 50(8):4366–4373. https://doi.org/10.1021/acs.est.6b00401

Luo H-W, Yin X, Jubb AM et al (2017) Photochemical reactions between mercury (Hg) and dissolved organic matter decrease Hg bioavailability and methylation. Environ Pollut. 220: 1359-1365. https://doi.org/10.1016/j.envpol.2016.10.099

Maqbool F, Niaz K, Hassan FI, Khan F, Abdollahi M (2017) Immunotoxicity of mercury: pathological and toxicological effects. J Environ Sci Health C 35(1):29–46. https://doi.org/10.1080/10590501.2016.1278299

Marvin-DiPasquale MC, Oremland RS (1998) Bacterial methylmercury degradation in Florida Everglades peat sediment. Environ Sci Technol 32(17):2556–2563. https://doi.org/10.1021/es971099l

Marvin-DiPasquale M, Agee J, McGowan C, Oremland RS, Thomas M, Krabbenhoft D, Gilmour CC (2000) Methyl-mercury degradation pathways: a comparison among three mercury-impacted ecosystems. Environ Sci Technol 34(23):4908–4916. https://doi.org/10.1021/es0013125

May JT, Hothem RL, Alpers CN, Law MA (2000) Mercury bioaccumulation in fish in a region affected by historic gold mining: the South Yuba River, Deer Creek, and Bear River watersheds, California, 1999. U.S. Geological Survey Open-File Report 00-367, 30 p. https://pubs.water.usgs.gov/ofr00-367/

Morel FMM, Kraepiel AML, Amyot M (1998) The chemical cycle and bioaccumulation of mercury. Annu Rev Ecol Syst 29:543–566

Motts JA, Shirley DL, Silbergeld EK, Nyland JF (2014) Novel biomarkers of mercury-induced autoimmune dysfunction: a cross-sectional study in Amazonian Brazil. Environ Res 132:12–18. https://doi.org/10.1016/j.envres.2014.03.024

Myers GJ, Davidson PW (1998) Prenatal methylmercury exposure and children: neurologic, developmental, and behavioral research. Environ Health Perspect 106(Suppl 3):841–847

Orem W, Newman S, Osborne TZ, Reddy KR (2015) Projecting changes in Everglades soil biogeochemistry for carbon and other key elements, to possible 2060 climate and hydrologic scenarios. Environ Manag 55:776–798

Ortiz VL, Mason RP, Ward JE (2015) An examination of the factors influencing mercury and methylmercury particulate distributions, methylation and demethylation rates in laboratory-generated marine snow. Mar Chem 177:753–762

Pichery C, Bellanger M, Zmirou-Navier D, Fréry N, Cordier S, Roue-LeGall A, Hartemann P, Grandjean P (2012) Economic evaluation of health consequences of prenatal methylmercury exposure in France. Environ Health (Open Access) 11(1):53. https://doi.org/10.1186/1476-069X-11-53

Poulin BA, Ryan JN, Nagy KI, Stubbins A, Dittmar T, Orem W, Krabbenhoft DP, Aiken GR (2017) Spatial dependence of reduced sulfur in everglades dissolved organic matter controlled by sulfate enrichment. Environ Sci Technol 51:3630–3639

Qian Y, Yin X, Lin H, Rao B, Brooks SC, Liang L, Gu B (2014) Why dissolved organic matter enhances photodegradation of methylmercury. Environ Sci Technol Lett 1(10):426–431

Roman HA, Walsh TL, Coull BA, Dewailly E, Guallar E, Hattis D, Mariën K, Schwartz J, Stern AH, Virtanen JK, Rice G (2011) Evaluation of the cardiovascular effects of methylmercury exposures: current evidence supports development of a dose–response function for regulatory benefits analysis. Environ Health Perspect 119(5):607–614. https://doi.org/10.1289/ehp.1003012

Sakamoto M, Itai T, Murata K (2017) Effects of prenatal methylmercury exposure: from Minamata disease to environmental health studies. Nihon Eiseigaku Zasshi 72(3):140–148. https://doi.org/10.1265/jjh.72.140

Schroeder WH, Munthe J, Lindqvist O (1989) Cycling of mercury between water, air, and soil compartments of the environment. Water Air Soil Pollut 48(3–4):337–347

Science for Environment Policy (2017) Tackling mercury pollution in the EU and worldwide. In-depth report 15 produced for the European Commission, DG Environment by the Science Communication Unit, UWE, Bristol. http://ec.europa.eu/science-environment-policy

Selin N (2009) Global biogeochemical cycling of mercury: a review. Annu Rev Environ Resour 34:43–63

Seller P, Kelly CA, Rudd JWM, MacHutchon AR (1996) Photodegradation of methylmercury in lakes. Nature 380:694–697

Stamenkovic J, Gustin MS (2009) Nonstomatal versus stomatal uptake of atmospheric mercury. Environ Sci Technol 43(5):1367–1372. https://doi.org/10.1021/es801583a

Tan SW, Meiller JC, Mahaffey KR (2009) The endocrine effects of mercury in humans and wildlife. Crit Rev Toxicol 39(3):228–269. https://doi.org/10.1080/10408440802233259

Wängberg I, Moldanová J, Munthe J (2010) Mercury cycling in the environment—effects of climate change. IVL Swedish Environmental Research Institute. Report B1921, 23 p. https://www.ivl.se/download/18.4b1c947d15125e72dda163c/1449751893676/B1921.pdf

Chapter 2
Sulfur Contamination in the Everglades, a Major Control on Mercury Methylation

William H. Orem, David P. Krabbenhoft, Brett A. Poulin, and George R. Aiken

Abstract In this chapter sulfur contamination of the Everglades and its role as a major control on methylmercury (MeHg) production is examined. Sulfate concentrations over large portions of the Everglades (60% of the ecosystem) are elevated or greatly elevated compared to background conditions of <1 mg/L. Land and water management practices in south Florida are the primary reason for the high levels of sulfate loading to the Everglades. Marshes in the northern Everglades that are highly enriched in sulfate have average concentrations of 60 mg/L, but water in canals in the Everglades Agricultural Area (EAA) contain the highest concentrations of sulfate averaging 60–70 mg/L. Studies that examined the mass balance of sulfur to the Everglades have determined that the primary sources of sulfate include: sulfur currently used in agriculture, and natural and legacy agricultural sulfur released by oxidation of organic soil within the EAA. The extensive loading of sulfate to the ecosystem increases microbial sulfate reduction, the dominant microbial process driving mercury methylation and MeHg production. The biogeochemical processes linking sulfate loading and MeHg production, however, are complex. MeHg production increases as sulfate levels rise from levels <1 mg/L up to about 20 mg/L. However, production of sulfide (a byproduct of microbial sulfate reduction) starts to inhibit MeHg production above 20 mg/L. Sulfate loading to canals in the EAA has impacted the northern Everglades the most, but the Everglades canal system can transport sulfate as far as Everglades National Park (ENP), 80 km further south.

Author "George R. Aiken" is deceased at the time of publication.

W. H. Orem (✉)
U.S. Geological Survey, Reston, VA, USA
e-mail: borem@usgs.gov

D. P. Krabbenhoft
U.S. Geological Survey, Middleton, WI, USA
e-mail: dpkrabbe@usgs.gov

B. A. Poulin
U.S. Geological Survey, Boulder, CO, USA
e-mail: bpoulin@usgs.gov

© Springer Nature Switzerland AG 2019
D. G. Rumbold et al. (eds.), *Mercury and the Everglades. A Synthesis and Model for Complex Ecosystem Restoration*, https://doi.org/10.1007/978-3-030-32057-7_2

Plans to deliver more water to ENP as part of restoration may increase overall sulfate loads to the southern Everglades.

Reduction of sulfate loading should be a major goal of Everglades restoration because of the many negative effects of sulfate on the ecosystem. The ecosystem has been shown to respond quickly to reductions in sulfate loading, and strategies for reducing sulfate loading may produce positive outcomes for the Everglades in the near-term. Strategies for reducing sulfate loading will need to include: best management practices for agricultural use of sulfate, approaches to minimize soil oxidation in the EAA, and modifications to stormwater treatment areas to improve sulfate retention.

Keywords Everglades · Sulfur · Mercury · Water quality · Sulfate reduction

Abbreviations

BCNP Big Cypress National Preserve
EAA Everglades Agricultural Area
ENP Everglades National Park
STA Stormwater Treatment Area
WCA Water Conservation Area

2.1 Introduction

There are few ecosystems worldwide like the subtropical Florida Everglades. Its great extent, variety of habitats, and unique mix of plants and animals set it apart. It is also an urban ecosystem, bordered on both east and west by large cities, and with an intensively cultivated agricultural area to the north. As summarized in Chap. 1 (Volume I), for many years starting in the late 1800s the Everglades was viewed primarily as an area for development, with agriculture displacing pond apple swamps south of Lake Okeechobee, canals dug to drain the agricultural lands and protect against flooding, and cities that began on the coast gradually displacing wetlands as they expanded inland. Highways were constructed across the Everglades to provide access to the cities. Non-native plant and animal species brought into south Florida for various purposes often found an exploitable environment in the Everglades wetlands. All these factors had an impact on the Everglades ecosystem, mostly detrimental. In 1947, Marjory Stoneham Douglas published "The Everglades: River of Grass" that first called attention to the damage being done to this unique ecosystem. In the 1980s, the discovery of eutrophication in Lake Okeechobee and parts of the Everglades by phosphorus from agricultural runoff triggered a lawsuit by the Federal Government against the State of Florida. Settlement of the lawsuit produced the Everglades Forever Act passed by the Florida Legislature in 1994 and a combined Federal, State, and private effort to restore the Everglades.

The Everglades suffers from a myriad of problems, including: inadequate and unreliable water supply, distribution of water (some areas get too much and some areas too little), impaired water quality, and ecosystem disruption from invasive species (for review, see Chaps. 1 and 2, Volume I). The eutrophication of the ecosystem resulting from phosphorus contamination from agricultural runoff was an issue that initiated the restoration. Phosphorus has been the principal focus of restoration efforts addressing water quality, and these efforts have been largely successful in reducing phosphorus contamination. However, phosphorus is not the only contamination issue facing the ecosystem. The linked issues of sulfur and methylmercury (MeHg) are serious water quality issues that pose a threat to Everglades wildlife and ecosystem value. MeHg, a neurotoxin and endocrine disruptor that is bioaccumulated, has been found in high concentrations in freshwater fish from the Everglades, and poses a potential threat to piscivorous wildlife (Jurczyk 1993; Rumbold et al. 2008; for review see Chap. 10, this volume) and to human health (especially developing fetuses; for review, see Chap. 11, this volume) through fish consumption. Elevated levels of MeHg in gamefish caused Florida to issue fish consumption advisories for all the Everglades (Florida Department of Health 2003). High levels of MeHg in Everglades fish were first reported from Everglades National Park (ENP) as early as 1974 (Ogden et al. 1974), and current levels of MeHg in top predator fish in ENP are among the highest in the nation for freshwater fish (Axelrad et al. 2009; for review, see Chap. 9, this volume), MeHg continues to impact vast areas of the Everglades, especially ENP. The Everglades has very high rates of deposition of inorganic mercury (Guentzel et al. 1995, 2001; Dvonch et al. 2005; for review, see Chap. 4, Volume I), but total mercury concentrations and spatial patterns in soil and water in the Everglades do not explain the bioaccumulation and distribution of mercury in prey fish (Stober et al. 2001; Scheidt and Kalla 2007). Other factors must be at work.

Sulfur in the form of sulfate is linked to MeHg production through the stimulation of microbial sulfate reduction (Fig. 2.1; Chap. 1, this volume). This anaerobic microbial process involves the oxidation of organic matter (or H_2) to CO_2 and the reduction of sulfate to sulfide, thereby producing energy for cellular processes. Sulfate reducing microorganisms also methylate mercury if it is available inside the cell to produce MeHg. While other microorganisms have also been shown to methylate mercury, microbial sulfate reduction appears to be the dominant process producing MeHg in the Everglades. Although many freshwater wetlands have low sulfate concentrations, the Everglades has sulfate concentrations approaching brackish water levels in some areas. This is not a natural condition of the ecosystem, but represents a water quality issue directly tied to MeHg contamination. The high levels of sulfate entering the Everglades may also have other effects on the ecosystem, including: changes to microbial ecology in the Everglades, buildup of toxic H_2S and lowered redox conditions in soil, changes in trace metal speciation and solubility, production of dissolved and solid phase organic sulfur compounds, and mobilization of nutrients from peat soil. In this chapter we review the extent of sulfur contamination in the Everglades, the sources of the contamination, the complex biogeochemistry of sulfur and MeHg production, other impacts of sulfur on the ecosystem, modeling used to look at how restoration may change sulfur and MeHg

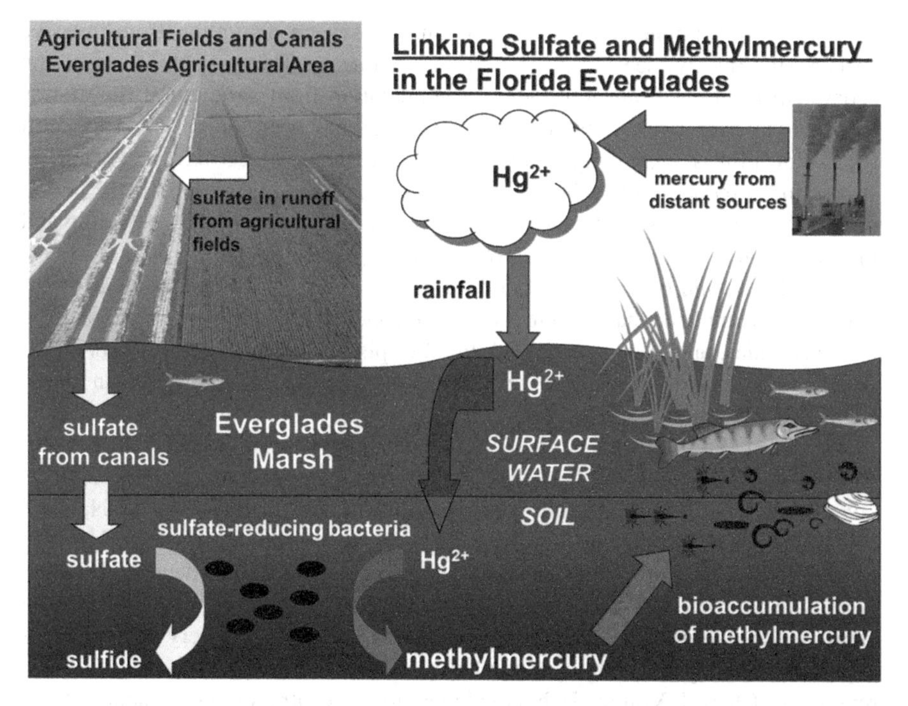

Fig. 2.1 Schematic representation of the link between sulfur and methylmercury (MeHg) production for the Everglades, Florida

contamination in the ecosystem, and ecosystem management approaches for minimizing and mitigating sulfur/MeHg contamination in the Everglades.

2.2 Historical and Background Levels of Sulfur in the Everglades

There is little reliable data on levels of sulfur species in the pre-drainage Everglades (e.g. prior to development in the 1920s). Thus, discussion of specific sulfur levels (especially sulfate levels in surface water) in the pre-drainage ecosystem are speculative, and based on (1) an understanding of historic water sources to the ecosystem (Harvey and McCormick 2009), (2) current levels of sulfate (and other sulfur species) in surface water in areas of the ecosystem far from canal influences (Orem et al. 2011), (3) current levels of sulfate in rain water, and (4) models of the pre-drainage ecosystem (SFWMD 2006). All these sources of information support the hypothesis that the pre-drainage Everglades was a mineral-poor and oligotrophic system, with low levels of dissolved ions and nutrients in surface water (Harvey and McCormick 2009).

Harvey and McCormick (2009) estimate that the pre-drainage ecosystem received 81% of its water from rainfall and 8% from overflow from Lake Okeechobee (the rest from direct runoff from other areas, such as the eastern ridge). Rainfall chemical data from south Florida in the 1990s show very low levels of dissolved ions: median specific conductance of <20 μS/cm and concentrations of most major ions (including sulfate) of <1 mg/L (SFWMD 2009). Recent measurements may overestimate historic levels of ions in rainwater, as anthropogenic activities and dust from cleared land may increase levels of dissolved ions in rainwater. The 8% of water from Lake Okeechobee in the pre-development period may have had higher levels of dissolved ions (due to evaporative concentration in the lake, and groundwater or surface runoff from the Kissimmee River Basin) than rainwater, but is still expected to be low. Runoff to Lake Okeechobee would have passed over soils that are largely sand with low ionic content (Harvey and McCormick 2009), and therefore would contribute few ions to the water entering the lake. This analysis is supported by data from the late 1930s and early 1940s (Parker et al. 1955) which indicate low ion content in tributaries of the lake (e.g., Kissimmee River, Fisheating Creek, Taylor Creek). Parker et al. (1955) report sulfate levels of 1–2 mg/L in Fisheating Creek and 4–9 mg/L in the Kissimmee River, with higher concentrations in Lake Okeechobee (ranging from 13 to 34 mg/L, average of about 24 mg/L) resulting from evapoconcentration in the shallow water (maximum depth of about 5 m) of the lake. The data of Parker et al. (1955) were collected during an early stage of development in south Florida, and may, therefore, be somewhat elevated compared to pre-development levels. Nevertheless, it seems clear that Lake Okeechobee had higher levels of sulfate and other ions compared to other water sources to the Everglades (rainfall and runoff from eastern upland areas). Overflow from the lake (mostly during the summer rainy season) probably represented the dominant source of dissolved ions (including sulfate) to the Everglades prior to development; though precise concentration ranges remain elusive.

There is little evidence of significant groundwater discharge into the ecosystem prior to development in the 1920s. Modeling suggests that groundwater accounted for only 1% of total inflows to the pre-development ecosystem (SFWMD 2006). Estimated vertical fluxes of about 0.01 cm/day between the upper 8 m of groundwater and surface water suggests that any input of sulfate and other inorganic ions from groundwater to the Everglades was minimal in the pre-development period, at least for most of the ecosystem where thick peat separated the mineral sources in the bedrock from surface water. In the southern part of the Everglades, thinner peat, limestone outcrop (rocky Glades area), and greater transmissivity in the bedrock may have allowed somewhat greater groundwater exchange with surface water (Harvey et al. 2006). This may have supported some greater flux of sulfate and other dissolved ions to surface water in the southern Everglades compared to the greater part of the northern and central Everglades. However, the historical sulfate flux is still likely to be much smaller than present conditions (Price and Swart 2006).

Current water quality and sulfate levels in areas of the freshwater Everglades far from canals and other anthropogenic influences may approximate what pre-development conditions were. In general, these areas have very low levels of

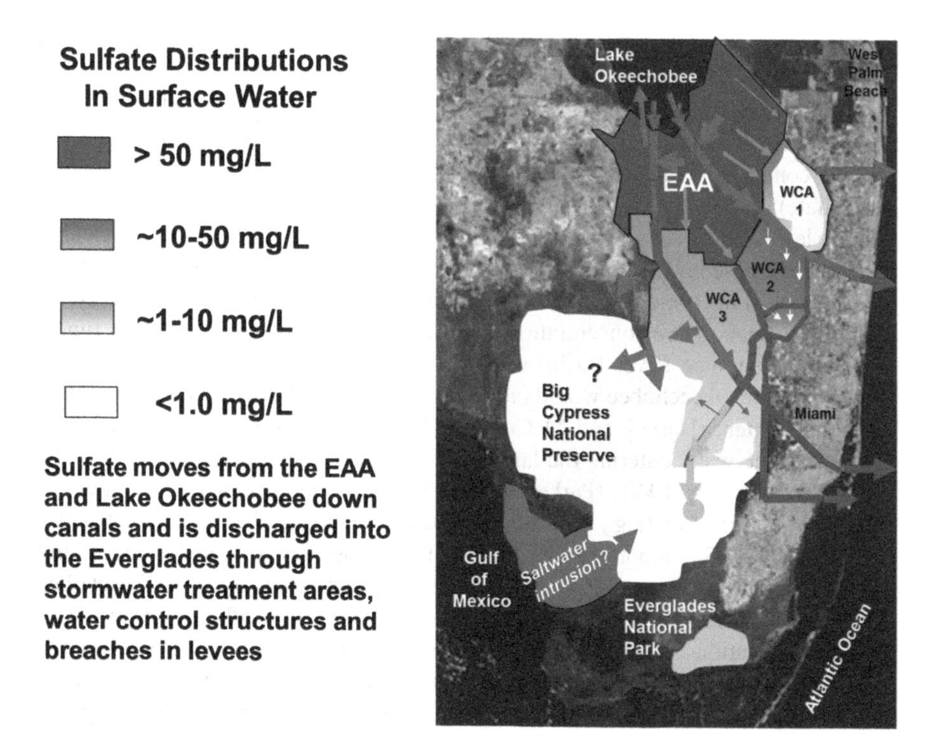

Fig. 2.2 Current sulfate distributions in the Everglades and canal systems (Orem et al. 2011). Color coding reflects concentration of sulfate and arrows show direction of discharges from canals into the Everglades

sulfate in surface water, typically <1 mg/L, and often <0.1 mg/L (Gilmour et al. 2007b; Scheidt and Kalla 2007; Orem et al. 2011). For example, Water Conservation Area 1 (WCA 1; Loxahatchee National Wildlife Refuge) is somewhat protected from canal water inflow and is thus a rainfall dominated system, especially toward the center area away from canal water intrusion under and through the levee (locations of the WCAs are shown in Fig. 2.2). Levels of sulfate in central WCA 1 are <1 mg/L, but increase towards the canal levees at the edge of the area (McCormick and Harvey 2011). The main portion of Big Cypress National Preserve (BCNP) and the portion of Everglades National Park (ENP) west of Shark River Slough also have sulfate levels generally ≪1 mg/L (Orem et al. 2011). These areas are also far removed from canal discharge. Flora and Rosendahl (1981, 1982a, b) observed trends of increasing major ion concentrations in Shark River Slough related to completion of the S-12 structures in the early 1960s. Exceptions to these low concentrations include marshes just north of BCNP (2–3 mg/L) and canals such as the L28 (10 mg/L sulfate) that surround BCNP, areas influenced by anthropogenic activities (e.g., agriculture, residential development, energy exploration activities). Central WCA 3A also currently has very low sulfate concentrations (<0.1 mg/L, Orem et al. 2011). This area had sulfate concentrations of 6–10 mg/L during the late

1990s, but changes in water flow resulting from Everglades restoration activities has closed off canal water previously migrating to this part of the ecosystem. This area currently appears to be rainfall dominated with resulting very low sulfate levels.

Overall, the available evidence suggests that the pre-development Everglades was a low sulfate, low total dissolved solids, oligotrophic system. The principal source of sulfate in this pre-development era was likely rainfall and possibly some contributions from overflow from Lake Okeechobee, quickly diluted as it entered the Everglades. It is possible that sulfate levels higher than 1 mg/L were present in confined areas of the region immediately south of the lake, primarily pond apple swamp in the pre-development era. The vast majority of the sawgrass ridge/slough Everglades likely had sulfate levels $\ll$1 mg/L prior to development in the 1920s.

2.3 Present Levels of Sulfate in the Everglades

Sulfate and total dissolved solids (TDS) levels in Lake Okeechobee and the Everglades increased beginning in the 1920s, largely from agricultural development in the Everglades Agricultural Area (EAA), located directly south of the Lake, and from canal construction (Parker et al. 1955; Joyner 1974; Harvey and McCormick 2009; Orem et al. 2011). Currently, the Everglades has concentrations of sulfate greatly elevated over historical levels (Fig. 2.2). Although some areas of the ecosystem located far from any canal water discharge still have concentrations of sulfate consistent with estimated historical levels as discussed above, large portions of the Everglades have elevated concentrations. This is especially true of the northern part of the Everglades. Scheidt and Kalla (2007) estimate that about 57% of the remaining Everglades has sulfate concentrations >1 mg/L. This 1 mg/L level represents the estimated maximum pre-development concentration of sulfate in the ecosystem.

The highest concentrations of sulfate in the Everglades are in canals within the EAA (Fig. 2.2; Orem et al. 2011). Long term monitoring of the major canals within the EAA (mid 1990s to present) shows average concentrations of between 60 and 70 mg/L, with concentrations ranging from 25 to over 200 mg/L (Fig. 2.3). The low end of this range represents the approximate current concentration of sulfate in Lake Okeechobee water, with water from the lake dominating the canal water during long dry periods. Sulfate concentrations on the high end of this range primarily occur during the periods of heavy rainfall, especially rainfall following long dry periods. Parker et al. (1955) and Gleason (1974) observed that canal water had higher total dissolved solids following heavy rainfall. These results suggest that rainfall is washing soluble sulfate (and other ions) from agricultural fields into EAA canals. High levels of sulfate and other dissolved ions are also observed in small farm canals within the EAA (Parker et al. 1955; Chen et al. 2006). The observation of highest sulfate concentrations in canals within the EAA, and lower levels in Lake Okeechobee and in canal water moving downflow and away from the EAA suggests that the source of much of the excess sulfate loading originates from within the EAA.

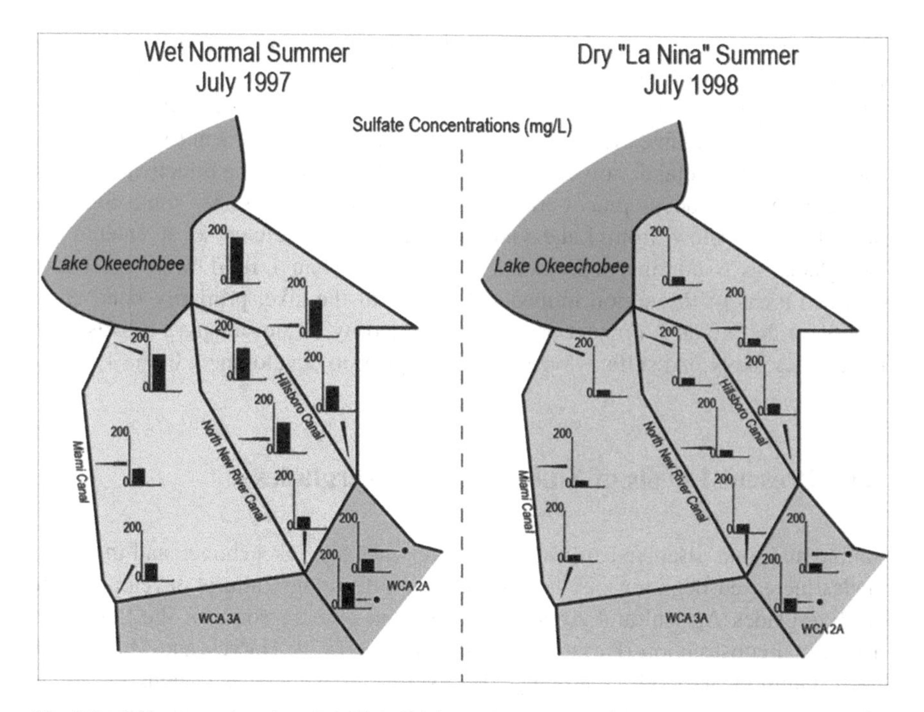

Fig. 2.3 Sulfate concentrations (mg/L) in EAA canal water during a normal wet summer following an extended dry period (left), and during a dry summer affected by the "La Nina" atmospheric effect. During the wet period following a dry spell, rainfall washes sulfate from agricultural fields into the canals, resulting in very high sulfate concentrations in canal water. In contrast, during dry periods (no rainfall) the sulfate concentrations in canals drops to that observed in Lake Okeechobee; with canal water essentially representing the lake water under these conditions (Axelrad et al. 2008)

Sulfate concentrations in Lake Okeechobee seem to be largely controlled by discharges to the lake from the EAA and rivers entering the lake from the north (McCormick and James 2008; James and McCormick 2012), and from evapoconcentration in this broad shallow lake (James et al. 1995). Sulfate concentrations were about 25 mg/L in 1940–1941 (Love 1955), 28 mg/L in 1950–1952 (Brown and Crooks 1955), and 22 mg/L in 2005 (Scheidt and Kalla 2007). Sulfate concentrations in the lake were >60 mg/L in the 1970s due to backpumping of EAA runoff water. However, backpumping of water to the lake was greatly reduced in the 1980s to reduce nutrient loading (mostly phosphorus) and resulting eutrophication, and sulfate concentrations also dropped to between 20 and 25 mg/L. Overall, the lake acts as a reservoir that evapoconcentrates sulfate entering the lake from surface drainage. There are no major sources of sulfate from Lake Okeechobee (Orem et al. 2011), though the lake is a small sink for sulfate (James and McCormick 2012) as is true for most lakes in Florida (Pollman and Canfield 1991). Lake Okeechobee exports a modest amount of sulfate (about 4500 mt/year) to the EAA canals, most of which began at the lake (Orem et al. 2011). The sulfate in the canal water is either discharged into the Everglades, or transported all the way down the canals to the east where they terminate at the Atlantic Ocean. The apparent source of sulfate from canals within the EAA drives the general north to south concentration gradient for sulfate observed in Everglades marshes (Scheidt and Kalla 2007; Orem et al. 2011). In marshes, sulfate concentrations range from <0.05 to 100 mg/L (Payne et al. 2009; Orem et al. 2011). There is also a seasonal trend in sulfate distributions in Everglades marshes, with the highest concentrations occurring during the wet season due to the pumping of stormwater from the EAA and canals into the Everglades for flood control (Scheidt et al. 2000). Concentrations of sulfate <1 mg/L occur in areas far from canal discharge points and generally rainfall dominated. The low sulfate levels thus reflect the low concentrations of sulfate in rain water. Surveys of sulfate concentration in the ecosystem indicate that the most sulfate-enriched parts of the Everglades are in WCA 2A, and northern WCA 3A (Gilmour et al. 2007b; Scheidt and Kalla 2007; Orem et al. 2011). Sulfate concentrations are >40 mg/L throughout the entire area of WCA 2A. Stormwater Treatment Area 2 (STA 2) discharges water with sulfate concentrations as high as 100 mg/L during the wet season into northwestern WCA 2A; these sulfate concentrations are the highest of any of the STAs (Scheidt and Kalla 2007). In WCA 3A, sulfate concentrations are highest in the north and east, and at sites near the Miami and L67 Canals, and decrease toward the south and west. Sulfate concentrations range up to 100 mg/L in WCA 3A near points of canal discharge, but average concentrations typically range from 5 to 20 mg/L in northern WCA 3A. Sulfate concentrations in central, southern, and western WCA 3A are generally $\ll$1 mg/L, except near canal discharge points. BCNP also has sulfate concentrations $\ll$1 mg/L, except for areas near canals.

Superimposed on the general north to south gradient in sulfate, many areas of marsh also exhibit sulfate gradients. These gradients are established from points of high sulfate canal water discharge into the marsh. The discharge points for canal water were previously (1990s and earlier) mostly pumping stations along canals, but the principal sources of discharge into the Everglades since the 2000s are STAs. In many marsh areas there is an overall gradient in sulfate concentration moving away from an STA and into the interior of the marsh. For example, in WCA 1 (A.R.-M. Loxahatchee National Wildlife Refuge), Gilmour et al. (2007b) observed sulfate concentrations rapidly decrease from about 15 to <0.2 mg/L along a transect from inside the levee of the Hillsboro Canal to the center of WCA 1. Sites near the canal in WCA 1 appear to receive some discharge of canal water across or from beneath the levee as well as input of high sulfate canal water from discharge from the West Palm Beach Canal through STAs 1E and 1W. WCA 1 is somewhat protected from these canal inputs by a rim canal that directs most flows around this wetland. However, Wang et al. (2009) have suggested from modeling studies that during the wet season, when water discharge from the STAs is highest, sulfate from canal water may penetrate the interior of WCA 1 and raise concentrations. WCA 2 also exhibits a sulfate concentration gradient from points of discharge of canal water to the interior of the marsh. STA 2 discharges canal water (Hillsboro Canal) into the northwestern portion of WCA 2 through a series of culverts. This discharge has sulfate concentrations ranging from 60 to 100 mg/L, which gradually diminish to 40–50 mg/L in the center of WCA 2.

Elevated sulfate concentrations can be found near major canals throughout the ecosystem, even areas to the south such as along the L-67 canal in southern WCA 3A near breaks in the levee, and where the L-67 terminates in ENP. Indeed, canals are the major conduit for sulfate contamination throughout the Everglades. Thus, sulfate levels in the freshwater part of ENP are highest in the northern part of Shark River Slough where the L67 canal terminates, with concentrations of 5–10 mg/L extending well south (Stober et al. 2001; Scheidt and Kalla 2007; Orem et al. 2011). It is unclear what effect restoration efforts to increase water flow to ENP by constructing a series of bridges along Tamiami Trail (Route 41), that previously blocked sheet flow of water from WCA 3, will have on sulfate levels in ENP. Indeed, the impacts of Everglades restoration efforts, changes in the EAA, and climate change on sulfur distributions in the ecosystem are unknown and areas of active research (Orem et al. 2014, 2015).

2.4 Other Forms of Sulfur in the Everglades

There are other forms of sulfur in the Everglades besides sulfate. Most of these other forms are a direct result of the biological utilization of sulfate as a terminal electron acceptor in anaerobic metabolism by microbial sulfate reduction. In this well described process (Fauque et al. 1991; Shen and Buick 2004; many others), various groups of sulfate-reducing bacteria reduce sulfate (sulfur in +6 oxidation state) to sulfide (sulfur in −2 oxidation state) while oxidizing organic matter (or H_2) to CO_2 to generate energy for cellular processes. Intermediates in the process of microbial sulfate reduction include thiosulfate (+6 oxidation state) and sulfite (+4 oxidation state). These species have been observed in the Everglades (Fig. 2.4).

Sulfate is the overwhelming form of sulfur in canals and Everglades surface water. Canal water contains up to 1% particulate sulfur, but reduced forms of sulfur are not typically present (except when canals are blocked and go stagnant during repair operations). Sulfide and other forms of reduced sulfur (thiosulfate and sulfite) are typically only observed in anoxic wetland soils in the Everglades (Orem et al. 2011). Sulfide concentrations in Everglades soil porewater generally follow surface water sulfate concentrations, and range from >10 mg/L in heavily sulfate enriched areas, to below detection (<1 μg/L at sites with sulfate levels <1 mg/L). Vertical profiles of sulfide in soil porewater often exhibit a maximum concentration within the upper 20 cm of soil. Since sulfide is absent or has much lower concentrations in surface water than in porewater, a concentration gradient occurs that drives diffusion of sulfide toward the surface. This upward diffusing sulfide may be oxidized to sulfate, often by sulfur oxidizing bacteria at the surface. Sulfate concentrations often decrease with depth in Everglades soil porewater as microbial sulfate reduction occurs. This sulfate concentration gradient (higher concentrations in surface water) drives diffusion of sulfate into soil porewater. Heavily sulfate enriched sites in the

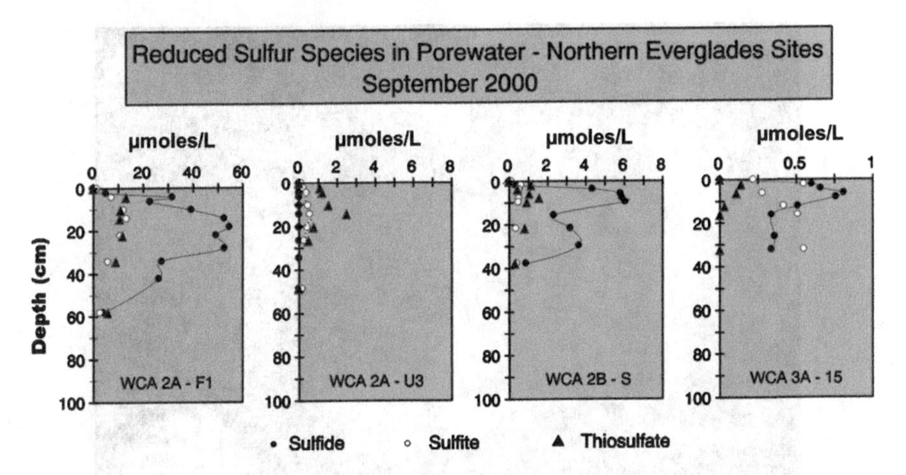

Fig. 2.4 Reduced forms of sulfur (sulfide, sulfite, thiosulfate) in pore water from Everglades soil in WCA 2A, 2B, and 3A collected in 2000 (Orem et al. 2011)

Everglades may have measurable concentrations of sulfide in surface water; >100 μg/L sulfide concentrations have been observed (Orem et al. 1997). Thus, heavily sulfate impacted areas of the Everglades may have sulfide concentrations that exceed EPA standards of 2 μg/L for sulfide in surface water (USEPA 2006).

Sulfide is highly reactive with trace metals and organic matter (Casagrande et al. 1979; Morse and Luther 1999; Heitmann and Blodau 2006). Recent work by Poulin et al. (2017) has demonstrated the reaction of sulfide with dissolved organic matter in the Everglades, and the importance of these dissolved organic sulfur species in trace metal complexation in the ecosystem. While metal sulfides (e.g., pyrite, and monosulfides) are important sinks for sulfide in clastic marine sediments (Vairavamurthy et al. 1995), metal disulfides (pyrite) and monosulfides account for only 10–30%, and 0–2% of the sulfur in Everglades soil, respectively (Bates et al. 1998). This is likely due to iron limitation as the Everglades lies on a carbonate platform with soil dominated by organic peat. However, pyrite framboids can be observed in some parts of the ecosystem (Fig. 2.5). The major form of sulfur in Everglades soil is organic sulfur accounting for 50–85% of the total sulfur (Casagrande et al. 1977; Altschuler et al. 1983; Bates et al. 1998). The organic sulfur originates from sulfur present in the original plant material, and from reaction of sulfide with organic matter. Sequestration of reduced sulfur (organic sulfur and metal sulfides) in Everglades peat soil is the major sink for sulfur in the ecosystem. This reduced sulfur remains in place in the soil unless disturbed or oxidized. Drought and fire, which are frequent events in the Everglades, may result in oxidation of this reduced sulfur back to sulfate (Gilmour et al. 2004; and discussion below).

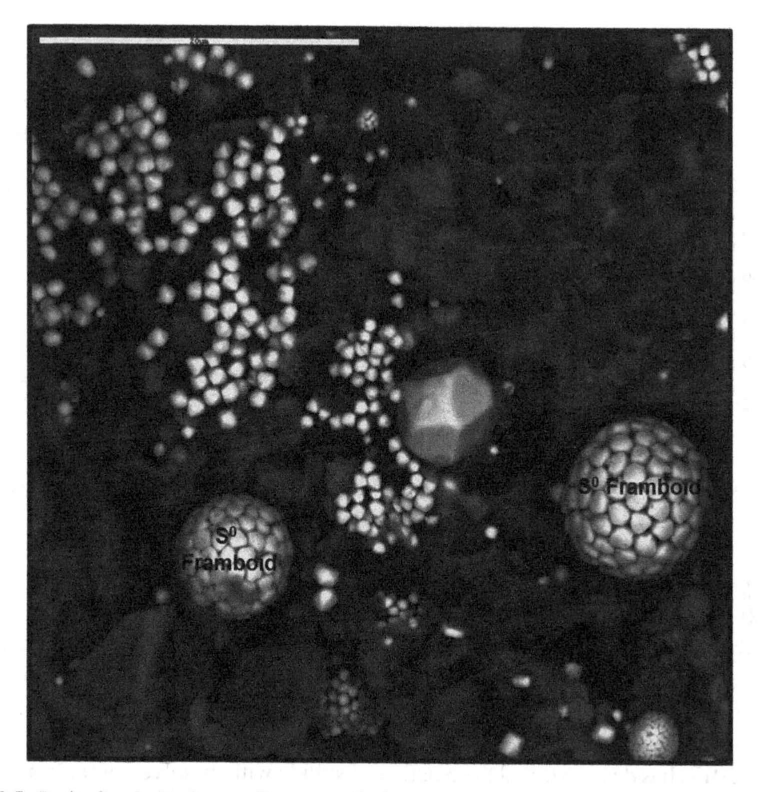

Fig. 2.5 Pyrite framboids from sediments on the bottom of Taylor Creek prior to its discharge into Lake Okeechobee. Photograph courtesy of Robert Zielinski, U.S. Geological Survey

2.5 Sources of Sulfate to the Everglades

Identifying the sources of sulfate to the Everglades is a complex problem both scientifically and from the standpoint of policy/politics (Orem et al. 2011). Proposed policies to reduce anthropogenic sulfate loading (Axelrad et al. 2008) are contentious, due to implications for agricultural practices. The contentious nature of the issue, and the possible economic implications of implementing best management practices for sulfur use in agriculture, make it important to get the science right.

The extent of sulfate contamination in the ecosystem and its link to the production of neurotoxic and bioaccumulative MeHg in the Everglades makes it a major contamination issue. The distributions of sulfate within the greater Everglades point to the EAA as the major source of sulfate to the ecosystem. As previously discussed, canals within the EAA have the highest levels of sulfate within the ecosystem, and the north to south gradient in sulfate concentration in Everglades marshes points to proximity to the EAA as a major factor in predicting sulfate concentration (an exception is WCA 1 which is somewhat protected from canal water discharge).

Sulfate and Phosphorus Phosphorus, another major Everglades water quality issue (which is discussed in Chap. 5 of this volume, including its effects on Hg biogeochemical cycling), is also tied to sources in the EAA and shows a north to south concentration gradient as well (Koch and Reddy 1992; Davis 1994). Phosphorus originates from fertilizer use in the EAA that washes into canals during rain events and is discharged into Everglades marshes. Levels of phosphorus (principally in the form of phosphate ion, PO_4^{3-}) in water discharged into the Everglades were as high as 150 μg/L in the northern Everglades in the 1970s, compared to 4–10 μg/L prior to development in the early 1900s (McCormick et al. 2002). These levels of phosphorus caused eutrophication of an originally oligotrophic ecosystem, changes in macrophyte distributions (e.g., cattail displacing sawgrass), and impacts on other aquatic organisms (McCormick et al. 1996, 2002; Richardson et al. 1999). Concerns over these changes triggered a lawsuit that began the Everglades restoration effort, perhaps the most extensive and expensive ecosystem restoration ever undertaken (for review, see Preface of Volume I). The implementation of best management practices (BMPs) for phosphorus use in the EAA, and the construction of stormwater treatment areas (STAs) have reduced phosphorus in water discharged to the Everglades to current levels of 30–50 μg/L, producing many positive benefits for the ecosystem (Reddy et al. 2006). However, phosphorus is not the only water quality issue facing the Everglades (Bates et al. 2002; Pfeuffer and Rand 2004; Scheidt and Kalla 2007; Perry 2008), and these efforts to reduce phosphorus contamination have had little impact on sulfate loading to the ecosystem.

Although phosphorus and sulfate are both Everglades contaminants, and both primarily originate from within the EAA, there are some significant differences between these two contaminants. Perhaps the biggest difference is that while phosphorus entered the ecosystem at levels of 150 μg/L prior to restoration efforts to reduce this loading, sulfate is currently entering the Everglades in discharge as high as 100 mg/L; levels nearly 1000 times higher compared to phosphorus (Gilmour et al. 2007b). While phosphorus levels entering the Everglades at their peak were about 15 times background levels from wet deposition (the principal source), sulfate levels are currently at least 100 times background. Sulfur is a required nutrient for plant growth at about the same levels as phosphorus (Hawkesford and DeKok 2007) and sulfate is actively taken up by plants. However, the high anthropogenic load of sulfate greatly exceeds the capacity of native vegetation to assimilate this nutrient. Thus, sulfate penetrates much farther into marshes from points of discharge than phosphorus. This is clearly demonstrated in WCA 2, where phosphorus concentrations fall off sharply with distance from the canal discharge point, eventually reaching background levels. While sulfate levels do decrease with distance they never come close to reaching background levels. Consequently, virtually all of WCA 2 is heavily contaminated with sulfate. The STAs which are very effective at removing phosphorus (at μg/L levels) remove little sulfate as presently constructed because the levels of sulfate (mg/L) are simply too great for removal by the plants. As discussed earlier, sulfate is also removed from surface waters as it is reduced to sulfide in peat soil and sequestered as reduced sulfur. However, this process is limited by rates of diffusion of sulfate into soil, rates of microbial sulfate reduction, and rates of reaction of sulfide with metals or soil organic matter.

Sources of Sulfate Most of the sulfate discharged into the Everglades originates from within the EAA. Sulfate concentrations in rainfall are too low to account for the high levels of sulfate in EAA canals, with rainwater from the northern Everglades having sulfate concentrations ranging from <1 to 2.5 mg/L (Bates et al. 2002; McCormick and Harvey 2011), and rainwater from ENP having annual volume-weighted mean sulfate concentrations of 0.5–0.7 mg/L sulfate (NADP 2008). Furthermore, sulfate in rainfall has sulfur isotopic ($\delta^{34}S$) compositions lighter (+2 to +6‰) than canal water (+15 to +23‰; Katz et al. 1995; Bates et al. 2001, 2002). Sulfur isotopes are often used to distinguish sources of sulfur (Mitchell et al. 2001), and the distinct differences between rainfall values and that of canal water indicates that rainfall is not an important source of the sulfate in the canals. Rainfall may be the dominant source of sulfate to areas of the ecosystem far from canal discharges but cannot account for the high levels of sulfate in canals or impacted marshes. Direct measurements of dry deposition of sulfate are somewhat uncertain (Pollman and Canfield 1991), but dry deposition is considered to be equal to or less than sulfate from wet deposition and, therefore, is not a significant source (Orem et al. 2011).

Groundwater underlying the Everglades, especially groundwater >9 m depth below the surface could represent a significant source of sulfate to the ecosystem. Miller (1988) measured sulfate concentrations in deep groundwater underlying the EAA (45 feet depth) ranging from 25 to 580 mg/L, and another study (CH2MHILL 1978) reported deep groundwater sulfate concentrations under the EAA of 20–490 mg/L. Shallow groundwater (<9 m) appears to have generally lower sulfate concentrations. Bates et al. (2002) reported sulfate concentrations of <2 mg/L in shallow groundwater (<9 m) under WCA 2 and the southern EAA. Biscayne Aquifer, a shallow aquifer under the southeastern portion of WCA 3 and ENP, also has relatively low sulfate levels (median concentration of 17 mg/L; Radell and Katz 1991). The Biscayne aquifer is susceptible to contamination from surface anthropogenic sources, including agriculture (Klein and Hull 1978). Harvey and McCormick (2009) have reported that a significant quantity of the fertilizer and agricultural additives used in the EAA may migrate to and be temporarily stored in shallow groundwater before being discharged into EAA canals.

Pollman (2012) used concentrations of major cations and anions (Na^+ and Cl^-, SO_4^{2-} and Cl^-, K+ and Na+) in connate seawater and canal water in the EAA along with a principal components analysis approach to identify and estimate possible groundwater contributions to sulfate in EAA canals that ranges from 27 to 35%. Exactly how or where this deep groundwater would enter canals is unclear, but presumably this would happen through cracks in the canal bottoms. Transects downstream along canals from the EAA show that sulfate levels gradually decrease with distance from the EAA (Orem et al. 2011). Thus, if groundwater is flowing upward through canal bottoms this is occurring primarily within the EAA. There is no evidence of groundwater contributing sulfate to Everglades marshes as marshes with high surface water sulfate concentrations have porewater sulfate profiles that exponentially decrease with depth, indicating no advective flux of sulfate-rich groundwater (Orem et al. 2011). Orem et al. (2011) point to three lines of geochemical evidence that contravene the theory that groundwater significantly contributes sulfate to canal water, including: (1) deep groundwater has a sulfur isotopic composition ($\delta^{34}S$) of about +12‰, distinct from values (+18 to +25‰) typical of sulfate in canals and enriched marsh sites, (2) sulfate/chloride ratios are 0.2 in deep groundwater compared to 0.5 in canal water and high-sulfate marsh sites, and (3) the uranium concentration and $^{234}U/^{238}U$ activity ratio (AR) of groundwater (AR = 1.30) is distinct from that of canal water (AR = 0.98) and surface water of Everglades marshes (AR = 0.97). Landing (2015) evaluated these apparently conflicting data and concluded that, assuming all the chloride in the system is coming from the groundwater (connate seawater), the uranium and sulfur isotopic data is not in total conflict with the elemental data analysis conducted by Pollman (2012). Landing (2015) concludes that perhaps as much as 25% of the sulfate in EAA canals may be coming from upwelling deep groundwater. However, if other sources of chloride exist in the EAA, then the groundwater contribution would decrease proportionately (Landing 2015). Macro- and micro-nutrients added to EAA soils include K, Fe, Mg, and Mn (Bottcher and Izuno 1994), which can be added as either chloride or sulfate salts. Chloride (and sulfate) are also the dominant anions in EAA shallow farm canals (Chen et al. 2006), suggesting that chloride (and sulfate) are washing off farm fields. This evidence points to a significant chloride source from EAA agricultural fields supporting the idea that it is agricultural fields and not groundwater that are the source of chloride and sulfate to the canals.

Bates et al. (2001, 2002) were the first to point to sulfur used in agriculture as a potential source of sulfate to EAA canals. Stable isotopic analysis ($\delta^{34}S$) of sulfate in surface water marshes and canals was used to examine sources of sulfate to the Everglades. Canal water from the EAA had the highest sulfate concentrations and the lowest $\delta^{34}S$ values, while concentrations decreased and $\delta^{34}S$ values increased moving downstream along the canals and out into the Everglades because of progressive microbial sulfate reduction in canals and marsh soils and dilution by low sulfate water (Bates et al. 2002; Orem 2004; Orem et al. 2011). When these authors plotted sulfate concentration versus sulfur isotopic composition ($\delta^{34}S$) they found wide scatter at low sulfate concentration, reflecting the different sources and redox changes that can contribute to the sulfate pool at unimpacted areas. At higher sulfate concentrations a distinct trend emerged, suggesting a single source of sulfate with $\delta^{34}S$ values approaching +16‰ produced the highest sulfate concentrations (Bates et al. 2002). Sulfur isotopic analyses ($\delta^{34}S$) of elemental sulfur (agricultural sulfur) used in the EAA as a soil amendment and fungicide (Bottcher and Izuno 1994) had a range of values (15–20‰) consistent with the isotopic composition ($\delta^{34}S$) of sulfate in EAA canals (Bates et al. 2001, 2002). In addition sulfate extracted from the top 10 cm of soil in an active sugarcane field in the EAA had a $\delta^{34}S$ value of 15.6‰ (Bates et al. 2001, 2002), consistent with that of agricultural sulfur. These data are consistent with a scenario wherein: (1) sulfur applications in the EAA are oxidized to sulfate in the largely aerobic soils, (2) remobilized from the soils by rainfall and/or irrigation, and (3) transported as sulfate in runoff to the canals in the EAA and then discharged in canal water into the Everglades marshes.

Elemental sulfur (which readily oxidizes to form sulfate) is used commonly in agriculture as a nutrient (Tabatabai 1984), soil amendment for pH adjustment (Boswell and Friesen 1993; Bottcher and Izuno 1994), and as a fungicide (Meyer 1977; Michaud and Grant 2003; McCoy et al. 2003). Sulfur may also be added to EAA fields as gypsum ($CaSO_4$) to reduce erosion, as copper sulfate ($CuSO_4$) for use as a fungicide, and as a counter ion (e.g. ammonium sulfate, iron sulfate) or trace constituent of other fertilizers or soil amendments. However, the total amount of sulfur added annually in all these forms is unknown. Also unknown is the total sulfate entering canals as runoff from EAA fields. Published recommendations to farmers suggest 500–1700 and 560 kg/ha-year of elemental sulfur should be applied in the EAA for vegetables and for multiyear sugarcane production, respectively (Schueneman and Sanchez 1994). This recommendation is currently under reevaluation (Landing 2015). Schueneman (2000) estimates that actual current elemental sulfur use in the EAA (elemental sulfur and sulfur in phosphorus fertilizer) is about 37 kg/ha-year (111 kg/ha-year converted from sulfur to sulfate). Clearly, better information on total sulfur applications in the EAA, the amount that is mobilized in runoff, and the legacy sulfur in the soil is needed to improve mass balance estimates.

Sulfur Mass Balance Based on the available, admittedly incomplete datasets available on sulfur sources in the EAA, two significant attempts at constructing a sulfur mass balance for the system have been published: Gabriel (2010), and Corrales et al. (2011). Results of the two mass balance studies are summarized in Table 2.1. Both mass balance studies are attempts to quantify sulfur sources to the Everglades, but both also suffer from a paucity of available information. For example, Gabriel (2010) ignores groundwater contributions entirely, and Corrales et al. (2011) assign a minimal value to groundwater. As discussed earlier, the groundwater contribution to sulfate loading to the Everglades is problematic at present, with some evidence pointing to contributions as high as 37%, while other evidence suggests only a small contribution. Both models confirm that atmospheric contributions (wet and dry) are very small (<5%). The biggest contributors in these mass balance studies include: Lake Okeechobee (31–50%) and sulfur released by soil oxidation in the EAA (38–44%). However, much of the sulfate released from Lake Okeechobee (about 36%) originates from the EAA (James and McCormick 2012). Furthermore, the Lake Okeechobee release estimates of both Gabriel (2010) and Corrales et al. (2011) differ from those of James and McCormick (2012), and may have misinterpreted older data as on a sulfur basis when it was actually on a sulfate basis (Landing 2015). Landing (2015) reevaluated the various mass balance estimates and suggests that the actual sulfur contribution from the lake to the EAA is closer to 10%. A similar problem of unit misinterpretation in older data (i.e., concentrations originally expressed as sulfate but mistakenly assumed to be expressed as sulfur) may also be present in the estimate for infiltration under and through the levee (Corrales et al. 2011) as pointed out by Landing (2015). The

Table 2.1 Sulfur mass balance comparison of results from Gabriel (2010) to Corrales et al. (2011)

	Gabriel (2010)						Corrales et al. (2011)	
	Moderate year		Wet year		Dry year		Average year	
	(metric tons/ year)	(% of total)	(metric tons/ year)	(% of total)	(metric tons/ year)	(% of total)	(metric tons/ year)	(% of total)
Total sulfur exported from EAA (to canals and the Everglades)	72,382	xx	80,906	xx	69,229	xx	110,303	xx
Sources								
Atmospheric wet deposition	3295	4.6	3864	4.8	2861	4.1	4229	3.8
Atmospheric dry deposition	529	0.7	487	0.6	508	0.7	nd	nd
Lake Okeechobee	31,057	42.9	40,626	50.2	28,494	41.2	35,217	31.9
Input through/under levees	nd	nd	nd	nd	nd	nd	5858	5.3
Groundwater	nd	nd	nd	nd	nd	nd	4055	3.7
Agricultural sulfur applications	6286	8.7	6286	7.8	6286	9.1	11,775	10.7
Soil oxidation in the EAA	30,646	42.3	30,646	37.9	30,646	44.3	49,169	44.6

Relevant information	Gabriel (2010)	Corrales et al. (2011)
EAA area (hectares)	280,466	290,600
Agricultural sulfur application rate ((kg/ha)/year)	22.4	37.0
Soil oxidation rate in EAA (cm/year)	1.3	1.4
Sulfur in EAA soil (%)	0.37	0.35
Sulfur loss from EAA by crop removal (metric tons/ year)	25,500	23,182

reevaluation by Landing (2015) makes clear that soil oxidation in the EAA is by far the biggest source of sulfur, and together with the newly applied agricultural sulfur accounts for more than 75% of the total sulfur to the Everglades (Table 2.1). If the current estimates are correct, current applications of agricultural sulfur appear to be roughly balanced by removal of sulfur in crops (Landing 2015). However, there is still considerable uncertainty of the total sulfur utilized within the EAA for all purposes. A considerable amount of the sulfur released by soil oxidation in the EAA likely represents legacy sulfur from historical applications.

Clearly, existing mass balances still suffer from many shortcomings, mostly related to the absence of information. A mass balance that includes detailed information on sulfur sources to the canals and Everglades marshes will be crucial for any attempts to mitigate this serious contamination problem. A more detailed accounting of all sulfur used in the EAA, a more thorough investigation of the actual contributions of groundwater and where groundwater comes to the surface, and a better understanding of release of sulfur from soil oxidation appear to be the key factors of concern. The contributions to sulfur loading from atmospheric sources and Lake Okeechobee appear to be better constrained at this time.

2.6 Remobilization of Sulfur from Drought and Fire

A source of sulfate to Everglades marshes that does not largely originate from canal discharge is remobilization of sulfur following drought and/or fire. Drought and fire have historically occurred in the Everglades and are a natural part of the ecosystem dynamics, especially during the dry season (October to May; Gunderson and Snyder 1994; see also Chap. 1, Volume I). Frequency of lightning strikes also plays a role. However, the frequency and severity of drought and fire has increased as a consequence of development (construction of canals, levees, and pumping stations) and water management practices in response to water demand from agriculture and urban areas (Wu et al. 1996; Lockwood et al. 2003; see also Chap. 2, Volume I). During extended drought or fire (fire that actually burns into the soil) organic soil is oxidized and reducing conditions in the soil are removed. Sequestered forms of reduced sulfur (metal sulfides and organic sulfur) that are stable under reducing conditions are oxidized to sulfate during fires; some sulfur also may be volatilized to the atmosphere. When rainfall returns, the rewetting of the surface mobilizes the sulfate formed by the oxidation process, producing high levels of sulfate in surface water. Accumulating surface water will also limit transfer of oxygen to marsh soil; anoxic conditions are then re-reestablished in the soil, and microbial sulfate reduction is stimulated by the remobilized sulfate.

One example of this occurred in the Everglades in May/June 1999 when drought during an extended dry season produced a fire that affected most of northern WCA 3A. Before the fire, surface water sulfate concentrations averaged about 7 mg/L, while following the fire and rewetting of the area sulfate levels in the burned area averaged 58 mg/L, with sulfate concentrations well exceeding 100 mg/L in some locations (Fig. 2.6). A year later (September 2000), surface water sulfate levels had decreased to pre-burn levels (average of 5 mg/L), and vertical profiles of sulfate in porewater showed normal exponentially decreasing concentrations with depth, indicating that equilibrium microbial sulfate reduction had been reestablished. Gilmour et al. (2007b) conducted controlled laboratory experiments using Everglades soil cores to further evaluate sulfur release from Everglades soils following drying and later rewetting (40 days under simulated natural lighting conditions). Dried peat soils exhibited high concentrations of sulfate in overlying water (>200 mg/L) after rewetting, while controls (not dried) had very low overlying water sulfate concentrations (<1 mg/L). Sulfide was detected in the rewetted soil cores, indicating that rewetting and remobilization of sulfate stimulated microbial sulfate reduction. Results of this lab experiment confirmed field studies that drying Everglades organic soil: (1) oxidizes reduced sulfur species to sulfate, (2) mobilizes sulfate following rewetting, and (3) the excess sulfate stimulates microbial sulfate reduction once anoxic conditions are reestablished in the soil. MeHg production was also stimulated in the soil cores for several months after rewettimg (Gilmour et al. 2007a, b).

2.7 Sulfur Controls on Mercury Methylation

The Everglades has nearly the ideal conditions to promote the production of bioaccumulative and toxic MeHg (see Chap. 1, this volume). The two key factors that produce high levels of MeHg in the Everglades are high levels of mercury deposition (as Hg^{2+} primarily in rainfall) that provides mercury for methylation, and high sulfate loading that drives the engine of microbial methylation of mercury via sulfate reduction. Because both factors are necessary to produce MeHg, any significant reduction in either would mitigate MeHg levels in the Everglades. Indeed, the Florida Department of Environmental Protection (DEP) attempted such an approach through reductions of local Hg emissions (Atkeson et al. 2003, 2005). At the time, local emissions of mercury in south Florida were considered the dominant source of mercury deposited on the Everglades (Dvonch et al. 1999), so reducing local emissions seemed a reasonable approach to the problem. This program was highly successful in reducing local emissions Chap. 5, Volume I). Unfortunately, this effort had little impact on mercury deposition on the ecosystem, and it is now known that most of the mercury deposited on the Everglades comes from outside of south Florida (Chap. 5, Volume I) and not amenable to local regulations. Reducing sulfate loading, is amenable to local controls and, thus, has a greater likelihood of success in mitigating the mercury problem.

Sulfate has been known for some time to be important in the production of MeHg in aquatic ecosystems (Gilmour et al. 1992, 2011; Benoit et al. 2003; Jeremiason et al. 2006; Munthe et al. 2007; Mitchell et al. 2008; Marvin-DiPasquale et al. 2014), including the Everglades (Gilmour et al. 1998). Available evidence from the Everglades suggests that sulfate reducers are likely the most important (but not the only) microorganisms capable of methylating mercury to MeHg.

MeHg Maximum Production and Goldilocks In wetland ecosystems like the Everglades where production of MeHg is thought to be primarily a result of microbial sulfate reduction, the addition of sulfate is expected to increase methylation rates. However, as first pointed out by Gilmour et al. (1992), the relationship between sulfate loading and mercury methylation is complex and non-linear. This non-linearity arises from two major factors: (1) sulfate reduction produces highly reactive sulfide as a byproduct that may influence the bioavailability of inorganic Hg^{2+}, and (2) methylation occurs inside of the cell, requiring transport of Hg^{2+} into the cell for MeHg production. Thus, mercury methylation by sulfate reduction reflects a balance between sulfate stimulation, and buildup of sulfide that appears to inhibit the bioavailability of Hg^{2+} for methylation (Benoit et al. 2003). Plots of sulfate concentration versus MeHg concentration or production rates exhibit a maximum in MeHg production at about 10–20 mg/L sulfate (exact maximum

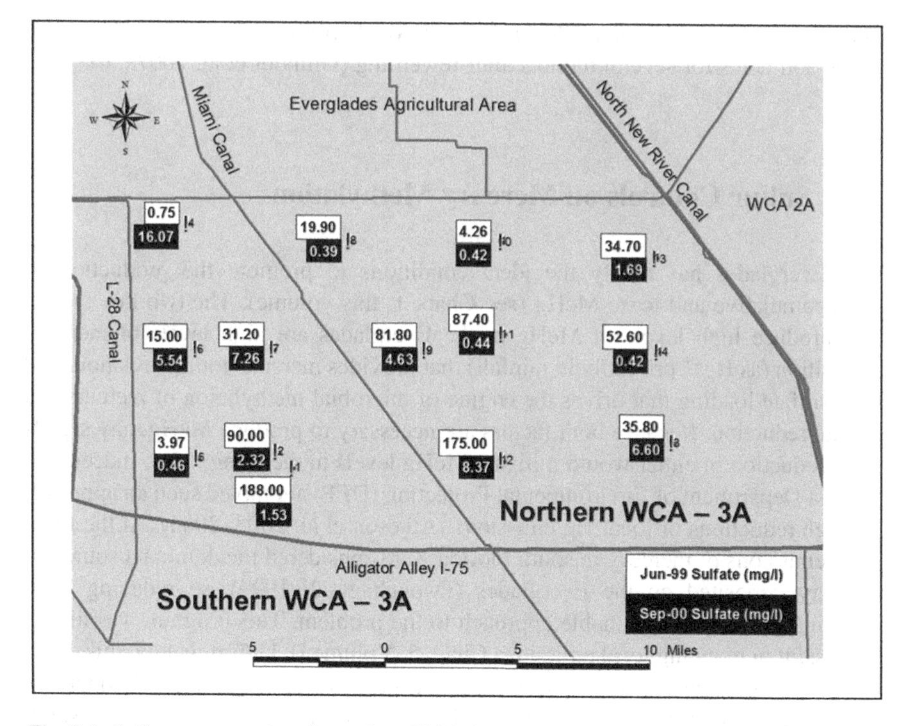

Fig. 2.6 Sulfate concentrations in northern WCA-3A following rewet immediately after a drought and fire in 1999 (clear boxes), and 1 year later (black boxes); from Gilmour et al. (2004)

dependent on location) followed by a decline in MeHg at higher sulfate concentrations (Fig. 2.7). This maximum is often referred to as the Goldilocks zone, where things are "just right" to promote methylation (a balance of sulfate stimulation and sulfide inhibition that maximizes MeHg production).

Field observations, laboratory experiments, and mesocosm studies have all demonstrated this Goldilocks phenomenon in the Everglades (Gilmour et al. 2007a; Orem et al. 2011; for summary, also see Chap. 3, this volume). In much of the WCAs, positive correlations are observed between net MeHg production and surface water sulfate concentrations over the range of 0.5–20 mg/L sulfate (Gilmour et al. 2007a). Yet, Scheidt and Kalla (2007) found that while sulfate correlates with MeHg in Everglades surface water and benthic periphyton, the positive correlation breaks down at pore water sulfide concentrations >1 mg/L due to the buildup of sulfide which becomes inhibitory to MeHg production. Across the Everglades, porewater sulfide is negatively correlated with MeHg concentration and bioaccumulation (Scheidt and Kalla 2007). This Goldilocks relationship has also been demonstrated for MeHg in mosquitofish from the Everglades, demonstrating the linkage between biogeochemical processes influencing MeHg production and bioaccumulation of MeHg in fish (Pollman 2014; Gabriel et al. 2014).

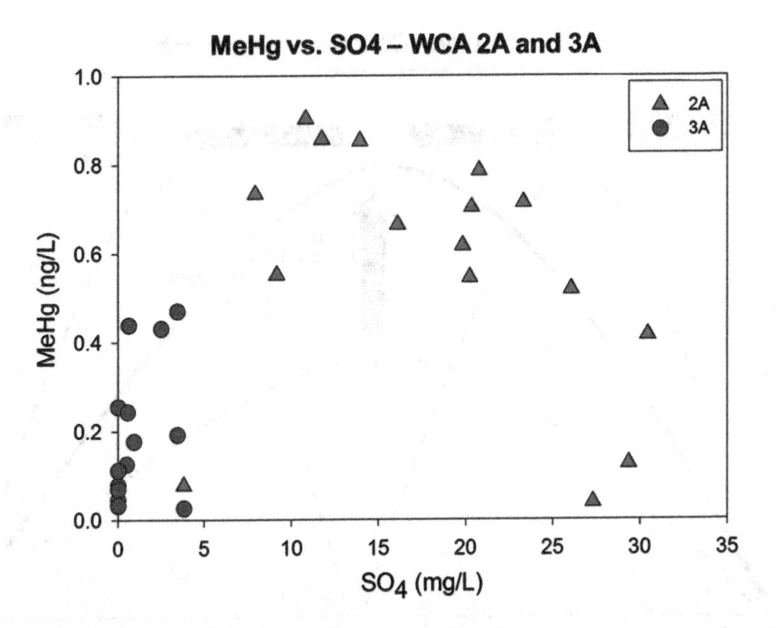

Fig. 2.7 Sulfate (mg/L) and MeHg (ng/L) concentrations in surface water across the central Everglades Water Conservation Areas (WCA) illustrating the "Goldilocks" maximum in MeHg at about 12–14 mg/L sulfate (Orem et al. 2011)

The biogeochemistry behind the Goldilocks effect on MeHg production is not completely understood at this time and is a subject of ongoing research. A conceptual model of the Goldilocks effect is illustrated in Fig. 2.8. Benoit et al. (2001) proposed that sulfide inhibits the bioavailability of Hg^{2+} for methylation by forming sulfide—Hg^{2+} complexes. At lower sulfide concentrations the complex is mostly a mercury monosulfide that is zero charged and may pass the cell membrane unhindered. At higher sulfide concentrations, more and more mercury disulfide complexes occur, and these negatively charged complexes have difficulty transiting the cell membrane. This would result in lower amounts of Hg^{2+} inside the cell for methylation. Others have postulated that the reaction of sulfide with dissolved organic matter creates dissolved organic sulfur compounds that can complex Hg^{2+} and influence transport into the cell and availability for methylation (Aiken et al. 2011; Poulin et al. 2017). It may be that both processes and others are involved in the effects of sulfide buildup on methylation of mercury and Goldilocks.

Two other factors that influence the observed Goldilocks effect are: (1) the degree to which to which increasing concentrations of sulfide are bound or complexed by other metals; and (2) organic matter. As previously discussed, sulfide is highly reactive with both organic matter and metals. High organic matter soils (peats) or soils that contain high concentrations of trace metals (especially iron) can impact the amount of sulfide that is present in pore water and thereby alter the observed Goldilocks curve. For example, while the Goldilocks maximum in MeHg concentrations in the organic peat soils of the WCAs usually occurs at sulfate

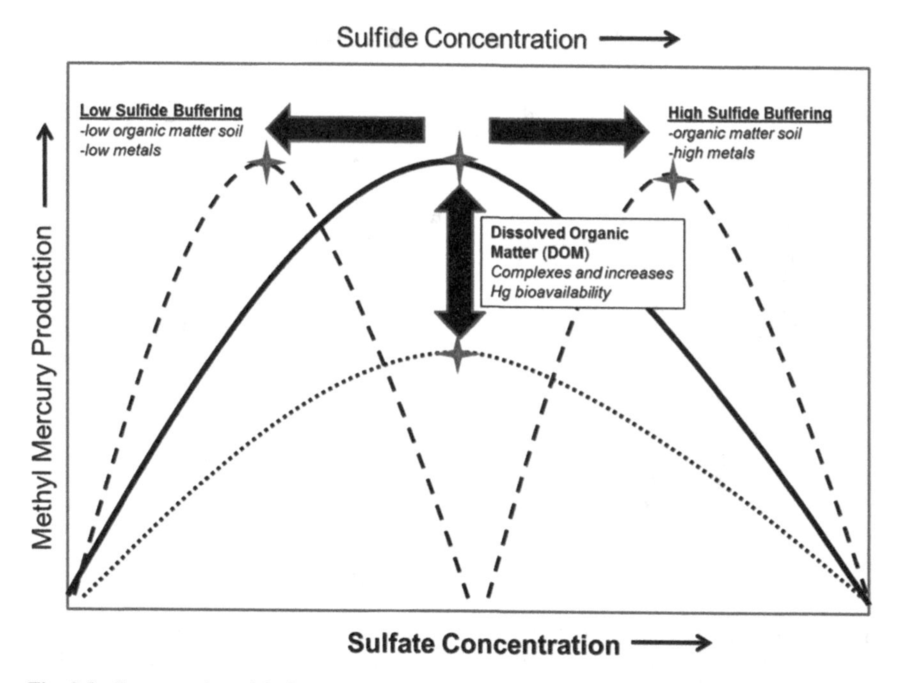

Fig. 2.8 Conceptual model of sulfate stimulation and sulfide inhibition of MeHg production, and the additional effects of dissolved organic matter (DOM) and soil composition (organics and metals)

concentrations of between 10 and 20 mg/L, in the much lower organic content marl soils of ENP the Goldilocks maximum is closer to 2–3 mg/L sulfate. In the WCAs, the high organic matter content of the soil essentially buffers the free sulfide levels in the pore water that affect the bioavailability of Hg^{2+} for methylation. Thus, the point at which maximum MeHg occurs is at a higher sulfate level compared to a low organic matter/low trace metal system (metals like iron can also reduce free sulfide by forming insoluble metal sulfides) like ENP. This effect is illustrated in Fig. 2.8 by left or right movement of the curve.

Dissolved organic matter may also influence the Goldilocks curve by movement in a vertical direction (Fig. 2.8). Dissolved organic matter causes this effect by complexing dissolved Hg^{2+} and facilitating transport across the cell membrane for methylation (for review, see Chap. 4, this volume). This change in bioavailability of Hg^{2+} for methylation may affect the amount of MeHg produced at a given level of sulfate but does not affect the sulfate concentration producing maximal MeHg (except in the event of formation of organic sulfur compounds, as previously discussed).

Sulfur and MeHg Across the Everglades The Goldilocks effect on mercury methylation is the primary driver of where in the ecosystem MeHg production hot spots occur. This is especially true as mercury deposition is evenly distributed over the Everglades (Keeler et al. 2001). Pristine areas of the ecosystem with sulfate <1 mg/L have generally low levels of MeHg because of low sulfate availability that limits microbial sulfate reduction. In sulfate-contaminated areas (>20 mg/L sulfate in the WCAs), mostly in the northern Everglades or at locations near canal discharge, sulfide inhibition limits MeHg production. Areas with intermediate concentrations of sulfate (10–20 mg/L) have sulfate and sulfide levels that are "just right" and maximize MeHg production.

During the 1990s, the hot spot for MeHg levels in soil (Gilmour et al. 1998), fish (Stober et al. 1996, 2001), and wading birds (Frederick et al. 1997) in the Everglades was located in central WCA 3A. High mercury deposition (Chap. 4, Volume I) and sulfate levels of 6–10 mg/L (pore water sulfide concentrations of only 5–150 μg/L) promoted high MeHg concentrations that entered the food chain (Orem et al. 1997; Stober et al. 1996, 2001). However, since about 2000 this area exhibited a sharp decline in levels of MeHg in surface water (<0.1 ng/L), and levels of MeHg in fish and wading birds have also declined sharply. This area is no longer the Everglades MeHg hot spot. Sulfate (and sulfide) levels in central WCA 3A dropped at the same time that MeHg production dropped at this site. Thus, the observed reduction in sulfate loading to levels $\ll$1 mg/L virtually extinguished microbial sulfate reduction in this area and may also be contributing to the dramatic decline in MeHg production (Gilmour et al. 2007a; Axelrad et al. 2008; Orem et al. 2011). Sources of sulfate to central WCA 3A include Miami and L-67 canal water distributed through levee breaks. The observed decline in sulfate in central WCA 3A during the late 1990s may reflect changes in water management accompanying the Everglades restoration effort. However, overall loading of sulfate to the ecosystem has not declined, only the location of hot spots of mercury methylation resulting from the changes in water distribution accompanying the restoration. The decline of MeHg levels in central WCA 3A is balanced by increasing levels in fish and wading birds in ENP, possibly driven by increased delivery of canal water down the L67 canal as part of the restoration effort (Maglio et al. 2015). Axelrad et al. (2008) reported that deposition of total Hg on the central Everglades (WCA 3A) has stayed relatively constant over the past 15 years. However, more recent analyses (based on mercury in sediment cores from Lake Annie, north of Lake Okeechobee) suggest mercury deposition may have declined over the period from 1994 to 2013 by 10–15% (see Chap. 1, Volume III for a discussion of this). Lake Annie is some distance from the central Everglades (approximately 150 km) but may be reflective of deposition on the central Everglades. The decline in sulfate and a possible decline in mercury deposition together may be contributing to the decline in MeHg production and presence in biota in central WCA 3A.

2.8 Other Impacts of Sulfur on the Everglades

Sulfide Toxicity Although the most important impact of sulfur contamination on the Everglades is as a driver of mercury methylation, there are also other effects. Sulfide, one of the major endproducts of microbial sulfate reduction, is toxic to plants

(Mendelssohn and McKee 1988; Koch and Mendelssohn 1989; Koch et al. 1990; Armstrong et al. 1996) and animals (National Research Council 1979; Wang and Chapman 1999). In plants, sulfide affects cell metabolism and reduces nutrient uptake by roots (Dobermann and Fairhurst 2000; Gao et al. 2003). Aquatic macrophytes show varying sensitivities to sulfide concentrations: Rice at >70 μg/L porewater sulfide (Allam and Hollis 1972), the freshwater macrophyte *Nitella flexilis* at >1700 μg/L sulfide (van der Welle et al. 2006), and *Phragmites australis* at 6800–10,200 μg/L (Hotes et al. 2005). Li et al. (2009) demonstrated that *Cladium jamaicense* (sawgrass) is more susceptible to sulfide toxicity than *Typha domingensis* (cattails); *Cladium* impacted at 7480 μg/L sulfide, and *Typha* unaffected up to 23,460 μg/L sulfide. In portions of the Everglades where *Typha* has displaced *Cladium*, porewater levels of sulfide approaching 13,000 μg/L occur (Gilmour et al. 2007b). These areas tend to also be heavily enriched with phosphorus from canal discharge, and eutrophication may explain this change in macrophyte dominance (Miao et al. 2000; Childers et al. 2003). However, sulfide toxicity may also play a role, since sulfide levels near canal discharge may often exceed those toxic to *Cladium* (Li et al. 2009). The toxicity of sulfide to other Everglades flora has not been examined.

Sulfide toxicity for fish ranges from 47 to 1000 μg/L in fresh water (Adelman and Smith 1970; USEPA 1976; Thurston et al. 1979), and a marine shrimp (*Crangon crangon*) showed 50% mortality at 680 μg/L sulfide (Vismann 1996). A freshwater oligochaete (*Ophidonais serpentine*) showed sulfide toxicity at 1700 μg/L (van der Welle et al. 2006). Sulfide levels in surface water in Everglades marshes are generally $\ll 1$ μg/L, but in heavily sulfate-enriched areas may attain levels of 100 s of μg/L (Orem et al. 2011). These heavily impacted areas have surface water sulfide levels that exceed EPA guidelines for sulfide in freshwater of 2 μg/L (Smith et al. 1976). Much higher levels of up to 13,000 μg/L sulfide are present in soil porewater from sulfate-contaminated areas of the Everglades (Orem et al. 2011), and aquatic organisms that spend part/all of their life cycle in soils could experience sulfide toxicity, but studies of this have not been conducted. These levels of sulfide in surface water and soil porewater are an unnatural condition for the freshwater Everglades.

Internal Eutrophication Sulfate loading has been demonstrated to increase mobilization of N and P from wetland soils via several redox-related mechanisms referred to as "internal eutrophication" (Lamers et al. 1998; Smolders et al. 2006). Gilmour et al. (2007b) reported on mesocosm studies conducted in the central Everglades that examined sulfate loading and nutrient from soils. These studies showed release of ammonium and phosphate to both surface water and porewater from Everglades soils at sulfate levels >20 mg/L. Phosphate and ammonium were enhanced by up to 50× and 20×, respectively, relative to controls, and increased with increased sulfate loading, though in a non-linear manner. Garrett and Ivanoff (2008) observed increased sulfate loading from the release of STA 2 water to northwestern WCA 2A, and also observed high levels of phosphate in porewater here even though surface water phosphate levels decreased and phosphate concentrations in STA

2 are <30 μg/L. Garrett and Ivanoff hypothesized that increased anaerobic microbial activity due to increased water depth resulted in the higher porewater phosphate levels. However, given the demonstrated releases of phosphate in sulfate-contaminated waters (e.g., Gilmour et al. 2007b), it is likely that increased sulfate loading and microbial sulfate reduction contributed to phosphate remobilization in STA2.

Suppression of Redox Conditions and Effects on Metal Speciation Redox conditions in aquatic soils/sediments are typically controlled by microbial metabolism and the biodegradation of organic matter (Berner 1980). Highly reducing conditions are often associated with large amounts of bioavailable organic matter and electron acceptors such as sulfate, nitrate and iron for microbial redox reactions. Excess sulfate entering the Everglades has lowered redox conditions in soils over large areas of the ecosystem. Areas near to canal or STA discharge have lower redox potentials compared to sites remote from canals (Drake et al. 1996). A sulfate addition experiment (sulfate added at 10–100 mg/L, with controls having no sulfate addition) in the central Everglades (WCA-3) using mesocosms demonstrates the effect of buildup of sulfide resulting from sulfate loading to the ecosystem on redox potential (Fig. 2.9). Lissner et al. (2003) have shown that nutrient uptake and growth of aquatic macrophytes is reduced under lower soil redox conditions.

Although little is known about how sulfur loading affects metal speciation in the Everglades, sulfide is reactive with many transition metals. Sulfide may change both the speciation and solubility of metals through the formation of highly insoluble metal sulfides. The formation of insoluble iron mono- and disulfides has been shown to cause iron deficiency in some species of rooted macrophytes in other wetlands (Smolders et al. 1995; Van der Welle et al. 2006). Smolders et al. (2003) implicated the combination of iron deficiency from iron sulfide precipitation, sulfide toxicity, and ammonium toxicity (ammonium from internal eutrophication) in the decline of *Stratiotes aloides* L., a formerly abundant and keystone freshwater macrophyte in

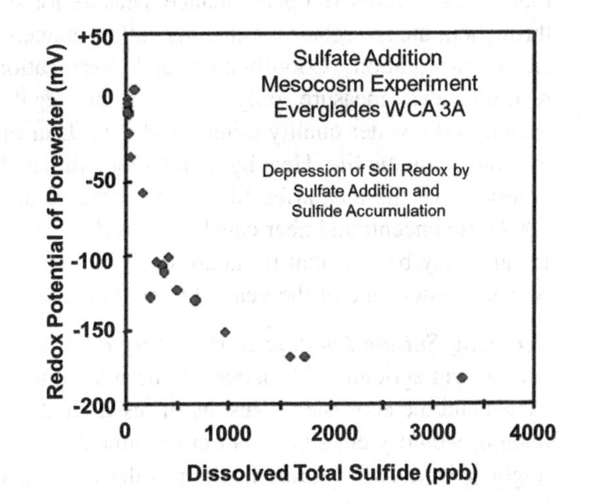

Fig. 2.9 Plot of total sulfide versus redox potential (mV) in porewater from a sulfate addition mesocosm experiment. Sulfide buildup in porewater dramatically depresses redox potential (Axelrad et al. 2009)

the Netherlands. The impacts of sulfur loading on metal chemistry in the Everglades has not been studied extensively, except for the case of mercury.

2.9 Conclusions

Sulfur is a major contaminant in the Everglades. The high loading of sulfate to Everglades marshes from canal discharge has greatly altered the ecosystem, fundamentally changing the natural microbial ecology over large areas from a primarily methanogenic soil microbiome to one dominated by microbial sulfate reduction. This change has adversely impacted the Everglades, with the biggest impact being the stimulation of mercury methylation by microbial sulfate reduction. Without the extensive sulfate contamination observed, it is probable that the Everglades would have a greatly reduced MeHg problem and one that would be manageable. This is demonstrated by MeHg levels of <0.1 ng/L in water in areas of the ecosystem where there is no sulfate contamination. Of particular interest is the reduction of MeHg in water and biota in central WCA 3A discussed earlier. This former MeHg hot spot saw a decline in surface water sulfate levels from 10 mg/L to current levels of <0.1 mg/L as well as possible declines in mercury deposition, and resulting major drops in MeHg in surface water from 0.8 to <0.1 ng/L (Orem et al. 2011). This change in central WCA 3A occurred abruptly over a 5-year period, indicating that changes to the biogeochemical drivers of MeHg production can reduce levels of MeHg quickly. Shutting off sulfate loading to the ecosystem and returning to more natural levels of sulfate in the ecosystem (e.g. <1 mg/L) may be an effective management tool for controlling MeHg levels in the Everglades.

Currently, no USEPA or Florida numeric water quality criteria for sulfate concentration in the context of ecosystem protection are in place as a guide to management. On the recommendation of EPA, the Comprehensive Everglades Restoration Plan (CERP) adopted a performance measure for surface water sulfate of <1 mg/L throughout the Everglades, reflecting sulfate concentrations found in the unimpacted areas (Restoration, Coordination and Verification 2007). However, this is a recommended measure only and has no legal weight. The USEPA (2006) established a water quality criterion of 2 μg/L in place for sulfide in surface water to protect aquatic life. Heavily sulfate-contaminated areas often greatly exceed this criterion, but the area affected may be <5% of the Everglades (Scheidt and Kalla 2007), is concentrated near canal or STA discharge (surface water sulfate >40 mg/L), and may be seasonal in nature (e.g. exceeding EPA standards only during the warmer/wetter time of the year) (Orem et al. 2011).

Reducing Sulfate Loading to the Everglades Research on sulfur shows that past and present agricultural practices in the EAA introduce most of the sulfate entering canals and the ecosystem. Results of research also show that a reduction in sulfate loading would yield significant environmental benefits to the Everglades. So how might reductions in sulfate loading to the Everglades be achieved, and how costly would this be? First, better mass balance estimates of sulfur are needed in the EAA. Present mass balance models were developed based on incomplete information on total use of sulfur in the EAA, groundwater sources of sulfate, and rates of sulfur release from soil oxidation. However, even with our current understanding of sulfur sources from the EAA, some effort at mitigation can begin.

Reductions in sulfur currently used in EAA agriculture could be achieved via implementation of best management practices (BMPs), or agricultural practices that allow sustainable/economic crop production and agricultural land use, but with minimal adverse environmental impacts. The use of BMPs to mitigate phosphorus pollution of the Everglades from the EAA has a long history and has been highly successful (Bottcher and Izuno 1994). Indeed, the Florida Legislature passed the "Florida Everglades Forever Act" in 1994, which mandated both STAs and BMPs for management of EAA phosphorus loading to the Everglades. This could serve as a model for BMPs applied to sulfur use. However, sulfur use in the EAA differs from that of phosphorus. While phosphorus is added as a nutrient to stimulate plant growth, crops grown in organic EAA soils are unlikely to be deficient in sulfur as a plant nutrient (Rice et al. 2006). Sulfur is used more as a soil amendment for soil pH adjustment and is present as an ingredient in other agricultural chemicals. Ye et al. (2010) have suggested that, because of soil oxidation and $CaCO_3$ buildup in EAA soils, the effect of elemental sulfur added for pH control may be minimal. The acidity produced by the oxidation of elemental sulfur to sulfate instead of promoting the bioavailability of added phosphorus fertilizer as intended reacts with the $CaCO_3$ concentrated in the remaining soil. Development of BMPs for reduction of sulfur use may require information on: (1) sulfur contents of soil and plant tissue to achieve minimum sulfur application to maintain crop sulfur nutrient needs (this is possibly zero application needs), (2) methods for adjusting soil pH without sulfur, (3) removing sulfur from other agricultural chemicals where it is an unnecessary ingredient (e.g. replacing $MgSO_4$ and K_2SO_4 with $MgCl_2$ and KCl), and (4) examining alternatives for sulfur-based fungicides.

Another approach to reducing sulfate loading to the Everglades would be redesigning the existing STAs for the removal of other contaminants (e.g., sulfate) in addition to phosphorus. The STAs are designed to remove phosphorus through uptake by emergent, submerged, and floating aquatic vegetation. The STAs have reduced total phosphorus (TP) loads to the ecosystem by 70% (>1000 metric tons of TP retained) over the past 15 years (Pietro et al. 2009). However, the current STAs remove relatively little sulfate (ca. 70% TP removal vs. 11% sulfate removal). A big factor in the poor performance of STAs in removing sulfate compared to TP that the mass inflow of sulfate exceeds that of TP by a factor of over 1000. Plant requirements for sulfur and phosphorus are roughly equivalent (Tabatabai 1984), but the tidal wave of sulfate entering the STAs overwhelms the capacity of the plants to remove sulfate. Sequestration of sulfur in STA anoxic soils through microbial reduction of sulfate to sulfide and formation of solid phase organic sulfur and metal sulfides may be effective but requires a substantial residence time (Morgan and Good 1988; Morgan 1990). The major factor limiting this process is the slow rate of diffusion of sulfate into aquatic soils where microbial reduction to sulfide occurs (Krom and Berner 1980; Ullman and Aller 1982). Reducing STA flow-through rate and increasing residence time, and possibly adding an iron source may improve sulfate removal performance of STAs (Orem 2007).

Permeable reactive barriers (PRBs) are a relatively new approach for removal of contaminants from water through in situ passive treatment (Scherer et al. 2008). PRBs may be constructed using materials that remove contaminants by chemical sequestration, or chemical or microbial breakdown. Creating areas in the Everglades STAs that incorporate PRBs might be effective in increasing sulfate removal, but testing in pilot studies would need to be conducted. Materials to utilize in PRBs that may prove effective in sulfate removal include minerals such as feldspar (Priyantha and Perera 2000), and modified synthetic zeolites (Haggerty and Bowman 2002).

Orem (2007) and Orem et al. (2011) have reviewed additional methods for sulfate removal from surface water, and describe the effectiveness and cost of these approaches. Some approaches described in these reviews (e.g., nanofiltration and ion exchange) are likely to be problematic for the large volumes of water discharged into the Everglades and may also be prohibitively expensive.

Management of Sulfur for Restoration Everglades restoration efforts aim to reestablish natural sheet flow of water in the ecosystem by removing man made barriers, and to funnel more water to areas (mostly in the south) that currently lack sufficient water. It is hoped that this will move the ecosystem to a condition approximating its pre-development state. However, this will require using water originating in Lake Okeechobee and passing through the EAA, and that is contaminated with a chemicals harmful to the ecosystem, such as sulfate. Sulfate is of particular concern because of its high concentration, links to MeHg production, and other adverse effects. Land and water resource managers have a need to understand that actions taken to improve water distribution may produce water quality concerns with unintended harmful ecological impacts, and costs and benefits need to be carefully weighed.

Reduction of sulfate loading to the Everglades will require an integrated approach involving BMPs for sulfur use in agriculture, reducing soil oxidation in the EAA by modifying farming practices, modifying the existing STAs for more efficient sulfate removal, shutting off groundwater leakage high in sulfate to the surface, and other mitigation strategies outlined elsewhere (Orem et al. 2011).

Reducing sulfate loading to the CERP performance measure of <1 mg/L is a desirable goal, but any significant reduction in overall sulfate loads to the Everglades will benefit the ecosystem. The number of sulfur sources and the high level of sulfate loading will make attaining the CERP sulfate performance measure difficult. However, research in the Everglades has demonstrated that a significant reduction in sulfate levels benefits the ecosystem by reducing MeHg production and contamination of the aquatic food web. Because the vast majority of the mercury currently deposited on the Everglades appears to originate from distant sources (outside the reach of State and Federal regulators), reductions in sulfate loading to the Everglades may be the most viable approach for reducing MeHg production and bioaccumulation within an ecosystem that has some of the highest levels of MeHg in biota of any wetland in the USA. The Everglades is resilient, and the response of the ecosystem to reduced sulfur loading appears to be rapid.

Acknowledgments This work was supported by the USGS Priority Ecosystems Studies for South Florida Program—Nick Aumen, Program Executive. Any use of trade, firm, or product names in this report is for descriptive purposes only and does not imply endorsement by the USGS or the U.S. Government. All figures and tables used are original creations for this chapter. Thanks to Matthew Varonka, Anne Bates, Tiffani Schell, Cynthia Gilmour, John DeWild, and many others who contributed to the USGS Aquatic Cycling of Mercury in the Everglades (ACME) Project over the years.

References

Adelman IR, Smith LL (1970) Effect of hydrogen sulfide on northern pike eggs and sac fry. Trans Am Fish Soc 99:501–508

Aiken GR, Gilmour CC, Krabbenhoft DP, Orem W (2011) Dissolved organic matter in the Florida Everglades: implications for ecosystem restoration. Crit Rev Environ Sci Technol 41:217–248

Allam AI, Hollis JP (1972) Sulfide inhibition of oxidases in rice roots. Phytopathology 62:634–639

Altschuler ZS, Schnepfe MM, Silber CC, Simon FO (1983) Sulfur diagenesis in Everglades peat and origin of pyrite in coal. Science 221:221–227

Armstrong J, Armstrong W, Van der Putten WH (1996) Phragmites die-back: bud and root death, blockages within the aeration and vascular systems and the possible role of phytotoxins. New Phytol 133:399–414

Atkeson T, Axelrad D, Pollman C, Keeler G (2003) Integrating atmospheric mercury deposition and aquatic cycling in the Florida Everglades: an approach for conducting a total maximum daily load analysis for an atmospherically derived pollutant. Integrated summary. Final report. Florida Department of the Environment

Atkeson TD, Pollman CD, Axelrad DM (2005) Chapter 26: Recent trends in Hg emissions, deposition, and biota in the Florida Everglades: a monitoring and modeling analysis. In: Pirrone N, Mahaffey K (eds) Dynamics of mercury pollution on regional and global scales: atmospheric processes, human exposure around the world. Springer, Norwell, MA, pp 637–656

Axelrad DM, Lange T, Gabriel M, Atkeson TD, Pollman CD, Orem WH, Scheidt DJ, Kalla PI, Frederick PC, Gilmour CC (2008) Mercury and sulfur monitoring, research and environmental assessment in South Florida. South Florida environmental report, Chapter 3B, South Florida Water Management District, West Palm Beach, FL, 53 p

Axelrad DM, Lange T, Atkeson TD, Gabriel MC (2009) Mercury and sulfur monitoring research and environmental assessment in South Florida. South Florida environmental report, Chapter 3B, South Florida Water Management District, West Palm Beach, FL, 32 p

Bates AL, Spiker EC, Holmes CW (1998) Speciation and isotopic composition of sedimentary sulfur in the Everglades, Florida, USA. Chem Geol 146:155–170

Bates AL, Orem WH, Harvey JW, Spiker EC (2001) Geochemistry of sulfur in the Florida Everglades; 1994 through 1999. U.S. Geological Survey Open-File Report 01-0007, 54 p

Bates AL, Orem WH, Harvey JW, Spiker EC (2002) Tracing sources of sulfur in the Florida Everglades. J Environ Qual 31:287–299

Benoit JM, Gilmour CC, Mason RP (2001) The influence of sulfide on solid-phase mercury bioavailability for methylation by pure cultures of *Desulfobulbous propionicus* (1pr3). Environ Sci Technol 35:127–132

Benoit JM, Gilmour CC, Heyes A, Mason RP, Miller CL (2003) Chapter 19: Geochemical and biological controls over methylmercury production and degradation in aquatic ecosystems. In:

Cai Y, Braids OC (eds) Biogeochemistry of environmentally important trace elements, ACS symposium series, vol 835, pp 262–297
Berner RA (1980) Early diagenesis: a theoretical approach. Princeton University Press, Princeton, NJ, p 241
Boswell CC, Friesen DK (1993) Elemental sulfur fertilizers and their use on crops and pastures. Fert Res 35:127–149
Bottcher AB, Izuno FT (1994) Everglades Agricultural Area (EAA)—water, soil, crop, and environmental management. University Press of Florida, Gainesville, FL, p 318
Brown E, Crooks JW (1955) Chemical character of surface waters in the Central and Southern Florida Flood Control District. USGS Open File Report FL 55002, 13 p
Casagrande DJ, Siefert K, Berschinski C, Sutton N (1977) Sulfur in peat-forming systems of the Okefenokee Swamp and Florida Everglades: origins of sulfur in coal. Geochim Cosmochim Acta 41:161–167
Casagrande DJ, Idowu G, Friedman A, Rickert P, Siefert K, Schlenz D (1979) H_2S incorporation in coal precursors: origins of organic sulphur in coal. Nature 282:599–600
CH2MHILL (1978) Water quality studies in the Everglades Agricultural Area. Report submitted to the Florida Sugarcane League. Gainesville, FL, 136 p
Chen M, Daroub SH, Lang TA, Diaz OA (2006) Specific conductance and ionic characteristics of farm canals in the Everglades Agricultural Area. J Environ Qual 35:141–150
Childers DL, Doren RF, Jones R, Noe GB, Rugge M, Scinto LJ (2003) Decadal change in vegetation and soil phosphorus pattern across the Everglades landscape. J Environ Qual 32:344–362
Corrales J, Naja GM, Dziuba C, Rivero RG, Orem W (2011) Sulfate threshold target to control methylmercury levels in wetland ecosystems. Sci Total Environ 409:2156–2162
Davis SM (1994) Phosphorus inputs and vegetation sensitivity in the Everglades. In: Davis SM, Ogden JC (eds) Everglades: the ecosystem and its restoration. St. Lucie Press, Delray Beach, FL, pp 357–378
Dobermann A, Fairhurst TH (2000) Rice: nutrient disorders and nutrient management. Potash and Phosphate Institute, International Rice Research Institute, Singapore, Makati City, 254 p
Drake HL, Aumen NG, Kuhner C, Wagner C, Griesshammer A, Schmittroth M (1996) Anaerobic microflora of Everglades sediments: effects of nutrients on population profiles and activities. Appl Environ Microbiol 62:486–493
Dvonch JT, Graney JR, Keeler GJ, Stevens RK (1999) Use of elemental tracers to source apportion mercury in South Florida precipitation. Environ Sci Technol 33:4522–4527
Dvonch JT, Keeler GJ, Marsik FJ (2005) The influence of meteorological conditions on the wet deposition of mercury in Southern Florida. J Appl Meteorol 44:1421–1435
Fauque G, LeGall J, Barton LL (1991) Sulfate-reducing and sulfur-reducing bacteria. In: Shively JM, Barton LL (eds) Variations in autotrophic life. Academic, London, pp 271–337
Flora MD, Rosendahl PC (1981) Specific conductance and ionic characteristics of the Shark River Slough, Everglades National Park, Florida. National Park Service, Homestead, FL, Report T-615, 55 p
Flora MD, Rosendahl PC (1982a) The response of specific conductance to environmental conditions in the Everglades National Park, Florida. Water Air Soil Pollut 17:51–59
Flora MD, Rosendahl PC (1982b) Historical changes in the conductivity and ionic characteristics of the source water for the Shark River Slough, Everglades National Park, Florida, U.S.A. Hydrobiologia 97:249–254
Florida Department of Health (2003) Florida fish consumption advisories. ftp://ftp.dep.state.fl.us/pub/lab/assessment/mercury/fishadvisory/pdf. Accessed 29 June 2018
Frederick PC, Spalding MG, Sepulveda MS, William G, Bouton S, Lynch H, Arrecis J, Lorezel S, Hoffman D (1997) Effects of environmental mercury exposure on reproduction, health and survival of wading birds in the Florida Everglades. Final report for the Florida Department of Environmental Protection. Tallahassee, FL, 206 p

Gabriel MC, Axelrad DM, Lange T, Dirk L (2010) Mercury and sulfur monitoring, research and environmental assessment in South Florida. In: 2010 South Florida environmental report, Chapter 3B, South Florida Water Management District, West Palm Beach, FL, 49 p

Gabriel MC, Howard N, Osborne TZ (2014) Fish mercury and surface water sulfate relationships in the Everglades protection area. Environ Manag 53:583–593. https://doi.org/10.1007/s00267-013-0224-4

Gao S, Tanji KK, Scardaci SC (2003) Incorporating straw may induce sulfide toxicity in paddy rice. Calif Agric 57:55–59

Garrett B, Ivanoff D (2008) Hydropattern restoration in Water Conservation Area 2A. Prepared for the Florida Department of Environmental Protection in Fulfillment of Permit # 0126704-001-GL (STA-2), by the STA Management Division, South Florida Water Management District, 113 p

Gilmour C, Henry EA, Mitchell R (1992) Sulfate stimulation of mercury methylation in freshwater sediments. Environ Sci Technol 26:2287–2294

Gilmour C, Riedel GS, Ederington MC, Bell JT, Benoit JM, Gill GA, Stordal MC (1998) Methylmercury concentrations and production rates across a trophic gradient in the Northern Everglades. Biogeochemistry 40:327–345

Gilmour CC, Krabbenhoft D, Orem W, Aiken G (2004) Appendix 2B-1: influence of drying and rewetting on mercury and sulfur cycling in Everglades and STA soils. In: 2004 Everglades consolidated report, South Florida Water Management District, West Palm Beach, FL

Gilmour C, Krabbenhoft D, Orem W, Aiken G, Roden E (2007a) Status report on ACME studies on the control of Hg methylation and bioaccumulation in the Everglades. In: 2007 South Florida environmental report, Appendix 3B-2, South Florida Water Management District, West Palm Beach, FL, 37 p

Gilmour C, Orem W, Krabbenhoft D, Roy S, Mendelssohn I (2007b) Preliminary assessment of sulfur sources, trends and effects in the Everglades. In: 2007 South Florida environmental report, Appendix 3B-3, South Florida Water Management District, West Palm Beach, FL, 46 p

Gilmour CC et al (2011) Sulfate-reducing bacterium *Desulfovibrio desulfuricans* ND132 as a model for understanding bacterial mercury methylation. Appl Environ Microbiol 77:3938–3951

Gleason PJ (ed) (1974) Environments of South Florida: present and past. Miami Geological Society, Miami

Guentzel JL, Landing WM, Gill GA, Pollman CD (1995) Atmospheric deposition of mercury in Florida: the fams project (1992–1994). Water Air Soil Pollut 80:393–402

Guentzel JL, Landing WM, Gill GA, Pollman CD (2001) Processes influencing rainfall deposition of mercury in Florida. Environ Sci Technol 35:863–873

Gunderson LH, Snyder JR (1994) Fire patterns in the Southern Everglades. In: Davis SM, Ogden JC (eds) Everglades: the ecosystem and its restoration. St. Lucie Press, Delray Beach, FL, pp 291–305

Haggerty GM, Bowman RS (2002) Sorption of chromate and other inorganic anions by organo-zeolite. Environ Sci Technol 28(3):452–458

Harvey JW, McCormick PV (2009) Groundwater's significance to changing hydrology, water chemistry, and biological communities of a floodplain ecosystem, Everglades, South Florida, USA. Hydrogeol J 17:185–201

Harvey JW, Newlin JT, Krupa SL (2006) Modeling decadal timescale interactions between surface water and ground water in the central Everglades, Florida, USA. J Hydrol 320:400–420

Hawkesford MJ, DeKok LJ (2007) Sulfur in plants: an ecological perspective. Springer, Dordrecht, p 264

Heitmann T, Blodau C (2006) Oxidation and incorporation of hydrogen sulfide by dissolved organic matter. Chem Geol 235:12–20

Hotes S, Adema E, Grootjans A, Inoue T, Poschlod P (2005) Reed die-back related to increased sulfide concentration in a coastal mire in Eastern Hokkaido, Japan. Wetl Ecol Manag 13:83–91

James RT, McCormick P (2012) The sulfate budget of a shallow subtropical lake. Fundam Appl Limnol 181(4):253–269

James RT, Jones BL, Smith VH (1995) Historical trends in the Lake Okeechobee ecosystem II. Nutrient budgets. Arch Hydrobiol Suppl 107:25–47

Jeremiason JD, Engstrom DR, Swain EB, Nater EA, Johnson BM, Almendinger JE, Monson BA, Kolka RK (2006) Sulfate addition increases methylmercury production in an experimental wetland. Environ Sci Technol 40:3800–3806

Joyner BF (1974) Chemical and biological conditions of Lake Okeechobee, Florida, 1969–70. Open-File Report 71006. U.S. Geological Survey, Tallahassee, FL

Jurczyk NU (1993) An ecological risk assessment of the impact of mercury contamination in the Florida Everglades. MS thesis, University of Florida, Gainesville, FL

Katz BG, Plummer LN, Busenberg E, Revesz KM, Jones BF, Lee TM (1995) Chemical evolution of groundwater near a Sinkhole Lake, Northern Florida, 2. Chemical patterns, mass-transfer modeling, and rates of chemical reactions. Water Resour Res 31:1564–1584

Keeler GJ, Marsik FJ, Al-Wali KI, Dvonch JT (2001) Appendix 7–6: status of the atmospheric dispersion and deposition model. In: 2001 Everglades consolidated report. South Florida Water Management District and Florida Department of Environmental Protection, West Palm Beach, FL

Klein H, Hull JE (1978) Biscayne aquifer, Southeast Florida. U.S. Geological Survey, Water Resources Investigations Report 78-107, 52 p

Koch MS, Mendelssohn IA (1989) Sulfide as a soil phytotoxin: differential responses in two marsh species. J Ecol 77:565–578

Koch MS, Reddy KR (1992) Distribution of soil and plant nutrients along a trophic gradient in the Florida Everglades. Soil Sci Soc Am J 56:1492–1499

Koch MS, Mendelssohn IA, McKee KL (1990) Mechanism for the hydrogen sulfide-induced growth limitation in wetland macrophytes. Limnol Oceanogr 35:399–408

Lamers LM, Tomassen HM, Roelofs JM (1998) Sulfate-induced eutrophication and phytotoxicity in freshwater wetlands. Environ Sci Technol 32:199–205

Landing W (2015) Peer-review report on the Everglades Agricultural Area regional sulfur mass balance: technical webinar. In: 2015 South Florida environmental report, Appendix 3B-2, South Florida Water Management District, West Palm Beach, FL, 44 p

Li S, Mendelssohn IA, Chen H, Orem WH (2009) Does sulfate enrichment promote *Typha domingensis* (cattail) expansion into the *Cladium jamaicence* (sawgrass)-dominated Florida Everglades? Freshw Biol 54:1909–1823

Lissner J, Mendelssohn IA, Lorenzen B, Brix H, McKee KL, Miao S (2003) Interactive effects of redox intensity and phosphate availability on growth and nutrient relations of *Cladium jamaicense* (Cyperaceae). Am J Bot 90:736–748

Lockwood JL, Ross MS, Sah JP (2003) Smoke on the water: the interplay of fire and water flow on Everglades restoration. Front Ecol Environ 1(9):462–468

Love SK (1955) Quality of ground and surface waters. In: Parker G, Ferguson GE, Love SK, others (eds) Water resources of Southeastern Florida with special reference to the geology and ground water of the Miami area, U.S. Geological Survey Water-Supply Paper 1255, Washington, DC, pp 727–833

Maglio M, Krabbenhoft D, Tate M, DeWild J, Ogorek J, Thompson C, Aiken G, Orem W, Kline J, Castro J, Gilmour C (2015) Drivers of geospatial and temporal variability in the distribution of mercury and methylmercury in Everglades National Park. GEER meeting, Coral Springs, FL, April 2015. Program and Abstracts

Marvin-DiPasquale M, Windham-Myers L, Agee JL, Kakouros E, Kieu le H, Fleck JA, Alpers CN, Stricker CA (2014) Methylmercury production in sediment from agricultural and non-agricultural wetlands in the Yolo Bypass, California, USA. Sci Total Environ 484:288–299

McCormick PV, Harvey JW (2011) Influence of changing water sources and mineral chemistry on the Everglades ecosystem. Crit Rev Environ Sci Technol 41(S1):28–63

McCormick PV, James RT (2008) Lake Okeechobee: regional sulfate source, sink, or reservoir? Presented at the 19th Annual Florida Lake Management Society Conference and 2008 NALMS Southeast Regional Conference, June 3, 2008

McCormick PV, Rawlick PS, Lurding K, Smith EP, Sklar FH (1996) Periphyton–water quality relationships along a nutrient gradient in the Florida Everglades. J N Am Benthol Soc 15:433–449

McCormick PV, Newman S, Miao S, Gawlik DE, Marley D, Reddy KR, Fontaine TD (2002) Effects of anthropogenic phosphorus inputs on the Everglades. In: Porter JW, Porter KG (eds) The Everglades, Florida Bay, and coral reefs of the Florida keys, an ecosystem sourcebook. CRC, Boca Raton, FL, pp 83–126

McCoy CW, Nigg HN, Timmer LW, Futch SH (2003) Use of pesticides in citrus IPM. In: Timmer LW (ed) 2003 Florida citrus pest management guide. University of Florida Cooperative Extension Service, Institute of Food and Agricultural Services

Mendelssohn IA, McKee KL (1988) *Spartina alterniflora* dieback in Louisiana: time course investigation of soil waterlogging effects. J Ecol 76:509–521

Meyer B (1977) Sulfur, energy, and environment. Elsevier, Amsterdam, p 457

Miao S, Newman S, Sklar FH (2000) Effects of habitat nutrients and seed sources on growth and expansion of *Typha domingensis*. Aquat Bot 68:297–311

Michaud JP, Grant AKJ (2003) Sub-lethal effects of a copper sulfate fungicide on development and reproduction in three coccinellid species. J Insect Sci 3:16–22

Miller WL (1988) Description and evaluation of the effects of urban and agricultural development on the surficial aquifer system, Palm Beach County, Florida. U.S. Geological Survey Water-Resources Investigations Report 88-4056

Mitchell MJ, Mayer B, Bailey SW et al (2001) Use of stable isotope ratios for evaluating sulfur sources and losses at the Hubbard Brook Experimental Forest. In: Proceedings of acid rain 2000, Japan. Water Air Soil Pollut 130:75–86

Mitchell CPJ, Branfireun BA, Kolka RK (2008) Assessing sulfate and carbon controls on net methylmercury production in peatlands: an in situ mesocosm approach. Appl Geochem 23:503–518

Morgan MD (1990) Streams in the New Jersey pinelands directly reflect changes in atmospheric deposition chemistry. J Environ Qual 19(2):296

Morgan MD, Good RE (1988) Stream chemistry in the New Jersey pinelands: the influence of precipitation and watershed disturbance. Water Resourc Res 24:1091–1100

Morse JW, Luther GW III (1999) Chemical influences on trace metal-sulfide interactions in anoxic sediments. Geochim Cosmochim Acta 63:3373–3378

Munthe J, Bodaly R, Branfireun B, Driscoll C, Gilmour C, Harris R, Horvat M, Lucotte M, Malm O (2007) Recovery of mercury-contaminated fisheries. Ambio 36:33–44

NADP, National Atmospheric Deposition Program (2008) National Atmospheric Deposition Program data, site FL11, annual data summaries. http://nadp.sws.uiuc.edu/default.html

National Research Council (1979) Hydrogen sulfide. University Park Press, Baltimore, MD

Ogden JC, Robertson WB, Davis GE, Schmidt TW (1974) Pesticides, polychlorinated biphenyls and heavy metals in upper food chain levels, Everglades National Park and vicinity. U.S. Department of the Interior, National Technical Information Service, No. PB-235 359

Orem WH (2004) Impacts of sulfate contamination on the Florida Everglades ecosystem. USGS Fact Sheet FS 109-03, 4 p

Orem W (2007) Sulfur contamination in the Florida Everglades: initial examination of mitigation strategies. U.S. Geological Survey Open-File Report 2007-1374, 53 pp. http://sofia.usgs.gov/publications/ofr/2007-1374/

Orem WH, Lerch HE, Rawlik P (1997) Descriptive geochemistry of surface and pore water from USGS 1994 and 1995 coring sites in South Florida wetlands. USGS Open-File Report 97-454, 70 p

Orem W, Gilmour C, Axelrad D, Krabbenhoft D, Scheidt D, Kalla P, McCormick P, Gabriel M, Aiken G (2011) Sulfur in the South Florida ecosystem: distribution, sources, biogeochemistry, impacts, and management for restoration. Rev Environ Sci Technol 41(S1):249–288

Orem W, Newman S, Osborne TZ, Reddy KR (2014) Projecting changes in Everglades soil biogeochemistry for carbon and other key elements, to possible 2060 climate and hydrologic scenarios. Environ Manag 55:776–798

Orem WH, Fitz C, Krabbenhoft D, Tate M, Gilmour C, Shafer M (2015) Modeling sulfate transport and distribution and methylmercury production associated with aquifer storage and recovery implementation in the everglades protection area. Sustain Water Qual Ecol 3–4:33–46

Parker GG, Ferguson GE, Love SK et al (1955) Water resources of Southeastern Florida—with special reference to the geology and ground water of the interior Miami area. U.S. Geological Survey Water-Supply Paper 1255, pp 1–4, 157

Payne GG, Xue SK, Weaver KC (2009) Chapter 3A: Status of water quality in the Everglades protection area. In: 2009 South Florida environmental report—volume I. South Florida Water Management District, West Palm Beach, FL

Perry W (2008) Everglades restoration and water quality challenges in South Florida. Ecotoxicology 17:569–578

Pfeuffer R, Rand G (2004) South Florida ambient pesticide monitoring Program. Ecotoxicology 13:195–205

Pietro K, Bearzotti R, Germain G, Iricanin N (2009) Chapter 5: STA performance, compliance and optimization. In: 2009 South Florida Environmental Report, vol I. South Florida Water Management District, West Palm Beach, FL

Pollman CD (2012) Modeling sulfate and *Gambusia* mercury relationships in the Everglades. Final reported submitted to the Florida Department of Environmental Protection. Tallahassee, FL. Aqua Lux Lucis, Gainesville, FL

Pollman CD (2014) Mercury cycling and trophic state in aquatic ecosystems: implications from structural equation modeling. Sci Total Environ 499:62–73

Pollman CD, Canfield DE Jr (1991) Florida. In: Charles DF (ed) Acid deposition and aquatic ecosystems: regional case studies. Springer, New York, pp 367–416

Poulin BA, Ryan JN, Nagy KL, Stubbins A, Dittmar T, Orem W, Krabbenhoft DP, Aiken GR (2017) Spatial dependence of reduced sulfur in everglades dissolved organic matter controlled by sulfate enrichment. Environ Sci Technol 51:3630–3639

Price RM, Swart PK (2006) Geochemical indicators of groundwater recharge in the surficial aquifer system, Everglades National Park, Florida, USA. GSA Spec Pap 404:251–266. https://doi.org/10.1130/2006.2404(21)

Priyantha N, Perera S (2000) Water Resour Manag 14(6):417–434

Radell MJ, Katz BG (1991) Major-ion and selected trace metal chemistry of the Biscayne Aquifer, Southeast Florida. U.S. Geological Survey Water Resources Investigations Report 91-4009. Tallahassee, FL, 18 p

Reddy KR, Kadlec RH, Chimney MJ (2006) The Everglades nutrient removal project. Ecol Eng 27:265–267

Restoration, Coordination and Verification (2007) Comprehensive everglades restoration plan system-wide performance measures

Rice RW, Gilbert RA, Lentini RS (2006) Nutritional requirements for florida sugarcane. Document SS-AGR-228 of the Agronomy Department, Florida Cooperative Extension Service, Institute of Food and Agricultural Sciences, University of Florida. Online at http://edis.ifas.ufl.edu/SC028

Richardson CJ, Ferrell GM, Vaithiyanathan P (1999) Nutrient effects on stand structure, resorption efficiency, and secondary compounds in Everglades sawgrass. Ecology 80:2182–2192

Rumbold DG, Lange TR, Axelrad DM, Atkeson TD (2008) Ecological risk of methylmercury in Everglades National Park, Florida, USA. Ecotoxicology 17(7):632–641. https://doi.org/10.1007/s10646-008-0234-9

Scheidt DJ, Kalla PI (2007) Everglades ecosystem assessment: water management and quality, eutrophication, mercury contamination, soils and habitat: monitoring for adaptive management: a R-EMAP status report. USEPA Region 4. EPA 904-R-07-001. Athens, GA, 98 p. http://www.epa.gov/Region4/sesd/reports/epa904r07001/epa904r07001.pdf

Scheidt D, Stober J, Jones R, Thornton K (2000) South Florida ecosystem assessment: water management, soil loss, eutrophication and habitat. United States Environmental Protection Agency Report 904-R-00-003. Atlanta, GA, 46 p. http://www.epa.gov/region4/sesd/reports/epa904r00003/epa904r00003.pdf

Scherer MM, Richter S, Valentine RL, Alvarez PJJ (2008) Chemistry and microbiology of permeable reactive barriers for groundwater clean up. Crit Rev Microbiol 26(4):221–264

Schueneman TJ (2000) Characterization of sulfur sources in the EAA. Soil Crop Sci Soc Fla Proc 60:20–22

Schueneman TJ, Sanchez CA (1994) Vegetable production in the EAA. In: Bottcher AB, Izuno FT (eds) Everglades Agricultural Area (EAA): water, soil, crop, and environmental management. University Press of Florida, Gainesville, FL, pp 238–277

SFWMD (2006) Natural system model (NSM) version 4.5. https://my.sfwmd.gov/pls/portal/docs/page/pg_grp_sfwmd_hesm/portlet_nsm/portlet_subtab_nsm_documents/tab1354050/nsm45.pdf

SFWMD (2009) DBHYDRO. South Florida Water Management District, West Palm Beach, FL. http://my.sfwmd.gov/dbhydroplsql/show_dbkey_info.main_menu

Shen Y, Buick R (2004) The antiquity of microbial sulfate reduction. Earth Sci Rev 64:243–272

Smith LL Jr, Oseid DM, Adelman LR, Broderius SJ (1976) Effect of hydrogen sulfide on fish and invertebrates, part I. Acute and Chronic Toxicity Studies, United States Environmental Protection Agency, Washington D.C., USA (1976) EPA-600/3-76-062a

Smolders AJP, Nijboer RC, Roelofs JGM (1995) Prevention of sulfide accumulation and phosphate mobilization by the addition of iron(II) chloride to a reduced sediment: an enclosure experiment. Freshw Biol 34:559–568

Smolders AJP, Lamers LPM, den Hartog C, Roelofs JGM (2003) Mechanisms involved in the decline of Stratiotes aloides L. in The Netherlands: sulphate as a key variable. Hydrobiologia 506–509:603–610

Smolders AJP, Lamers LPM, Lucassen ECHET, Van der Velde G, Roelofs JGM (2006) Internal eutrophication: how it works and what to do about it—a review. Chem Ecol 22:93–111

Stober J, Scheidt D, Jones R, Thornton K, Ambrose R, France D (1996) South Florida ecosystem assessment. Monitoring for adaptive management: implications for ecosystem restoration. Interim report. United States Environmental Protection Agency EPA-904-R-96-008

Stober J, Thornton K, Jones R, Richards J, Ivey C, Welch R, Madden M, Trexler J, Gaiser E, Scheidt D, Rathbun S (2001) South Florida ecosystem assessment: phase I/II summary report. Everglades stressor interactions: hydropatterns, eutrophication, habitat alteration, and mercury contamination. EPA 904-R-01-002. USEPA Region 4 Science and Ecosystem Support Division. Athens, GA

Tabatabai MA (1984) Importance of sulphur in crop production. Biogeochemistry 1:45–62

Thurston RV, Russo RC, Fetterolf CM Jr, Edsall TA, Barber YM Jr (1979) A review of the EPA Red Book: quality criteria for water. Water Quality Section, American Fisheries Society, Bethesda, MD

Ullman WJ, Aller RC (1982) Diffusion coefficients in nearshore marine sediments limnol. Oceanography 27:552–556

USEPA (1976) Quality criteria for water. US Environmental Protection Agency, Washington, DC

USEPA (2006) Clean Air Status and Trends Network (CASTNET) 2005 Annual Report. U.S. Environmental Protection Agency, Office of Air and Radiation, Clean Air Markets Division, Washington, DC, 48 pp. plus references and appendices. http://www.epa.gov/castnet

Vairavamurthy MA, Schoonen MAA, Eglinton TI, Luther GW III, Manowitz B (1995) Geochemical transformations of sedimentary sulfur, American Chemical Society Symposium Series 612. American Chemical Society, Washington, DC

Van der Welle MEW, Cuppens M, Lamers LPM, Roelofs JGM (2006) Detoxifying toxicants: interactions between sulfide and iron toxicity in freshwater wetlands. Environ Toxicol Chem 25:1592–1597

Vismann B (1996) Sulfide species and total sulfide toxicity in the shrimp *Crangon crangon*. J Exp Mar Biol Ecol 204:141–154
Wang F, Chapman PM (1999) Biological implications of sulfide in sediment—a review focusing on sediment toxicity. Environ Toxicol Chem 18:2526–2532
Wang H, Waldon M, Meselhe E, Arceneaux J, Chen C, Harwell M (2009) Surface water sulfate dynamics in the Northern Florida Everglades. J Environ Qual 38:734–741
William O, Gilmour C, Axelrad D, Krabbenhoft D, Scheidt D, Kalla P, McCormick P, Gabriel M, Aiken G (2011) Sulfur in the South Florida ecosystem: distribution, sources, biogeochemistry, impacts, and management for restoration. Crit Rev Environ Sci Technol 41(Supp 1):249–288
Wu Y, Sklar FH, Gopu K, Rutchey K (1996) Fire simulations in the Everglades landscape using parallel programming. Ecol Model 93:113–124
Ye R, Wright AL, Orem WH, McCray JM (2010) Sulfur distribution and transformations in everglades agricultural area soil as influenced by sulfur amendment. Soil Sci 175(6):263–269

Chapter 3
A Causal Analysis for the Dominant Factor in the Extreme Geographic and Temporal Variability in Mercury Biomagnification in the Everglades

Darren G. Rumbold

Abstract The objective of this chapter is to assemble and evaluate existing evidence (both from within and outside the Everglades) in a formal, causal analysis using criteria-guided judgement to determine the cause of the observed susceptibility of the Everglades to high rates of Hg biomagnification and its extreme geographic and temporal variability.

Keywords Mechanistic plausibility · Laboratory tests · Mesocosms · Spatial gradients · Temporal gradients · Verified predictions

3.1 Introduction

Following the discovery of high mercury (Hg) levels in Everglades' biota (Ware et al. 1990; for history, see Preface of Volume 1) numerous theories were put forward to account for the system's susceptibility and for the observed geographic variation in the magnitude of biomagnification (Table 3.1; Roelke et al. 1991; Stober et al. 1996; Lange et al. 1999). Of concern were the biological Hg hotspots, defined as a localized area where biota exhibit elevated Hg levels as compared to the surrounding landscape and in excess of established criteria for protection of human or wildlife health. Most of the candidate causes focused on Hg deposition rates or factors controlling net methylation (for review of early theories, see Stober et al. 1998; Fink et al. 1999; Krabbenhoft et al. 2001). Studies were then designed to test these candidate causes (Table 3.1).

Based on the results of early studies (also see Table 3.2) and the contemporaneous discovery of widespread sulfate contamination (Orem et al. 1997; Bates et al. 1998),

D. G. Rumbold (✉)
Florida Gulf Coast University, Fort Myers, FL, USA
e-mail: drumbold@fgcu.edu

© Springer Nature Switzerland AG 2019

D. G. Rumbold et al. (eds.), *Mercury and the Everglades. A Synthesis and Model for Complex Ecosystem Restoration*, https://doi.org/10.1007/978-3-030-32057-7_3

Table 3.1 Early theories and attendant studies put forward to explain the observed susceptibility of the Florida Everglades to high rates of mercury biomagnification and its extreme geographical and temporal variability

Candidate causes	Studies conducted within the system to test theory
Unusually high atmospheric Hg deposition with high seasonal and geographical variation	Guentzel et al. (1995), Landing et al. (1995), Pollman et al. (1995) and Dvonch et al. (1998)
Hg release from sugar cane burning	Patrick et al. (1994)
Hg release from sediments due to the extreme dry-wet cycles or the historically high rates of peat oxidation within the Everglades Agricultural Area (EAA)	Rood et al. (1995) and Stober et al. (1998)
High rates of net methylation due to the influence of Everglades eutrophication	Gilmour et al. (1998a, b)
High rates of net methylation due to the influence of sulfate contamination; later modified to include sulfide inhibition of methylation in some areas	Gilmour et al. (1998a, b) (see Table 3.3.2 ref.)
Net methylation influenced by iron—initially focused on Fe scavenging of sulfide, then re-focused on iron-reducing bacteria	Fink et al. (1999) and Gilmour et al. (2007a) on 2005 mesocosms
Biomagnification being strongly influenced by biodilution; arguments that "cold spots" were a function of eutrophication	Exponent (formerly PTI) (1998)
Biomagnification was being strongly influenced by quantity and quality of DOM (resulting from complexation and altered bioavailability of inorganic Hg, MeHg or both)	Ravichandran et al. (1998) and Aiken et al. (2003)
Biomagnification strongly influenced by food web structure	Cleckner et al. (1998), Lange et al. (1999) and Loftus (2000)

a consensus emerged that while the principal cause of the problem was the high rate of atmospheric Hg deposition (see Chaps. 3 and 4, Volume I), the susceptibility was due to sulfate stimulation of methylmercury (MeHg) production across the Everglades (Gilmour et al. 1998b; Stober et al. 2001). Moreover, it was the balance between sulfate and sulfide that was the driver of the observed geographic and temporal variability. This became the prevailing paradigm of federal and state agencies, including the United States Geological Survey (USGS; Krabbenhoft and Fink 2001), the U.S. Environmental Protection Agency (USEPA; Stober et al. 1996, 1998, 2001), the Florida Department of Environmental Protection (FDEP; Atkeson and Parks 2001), and the South Florida Water Management District (SFWMD; Fink and Rawlik 2000; Rumbold and Fink 2003a; Rumbold et al. 2006). This paradigm was incorporated into monitoring plans and mitigation strategies developed by the SFWMD (Rumbold 2005a; Rumbold and Fink 2006; Rumbold et al. 2007). It was also the basis of a restoration target developed by USGS and USEPA for the Comprehensive Everglades Restoration Plan (i.e., "*maintain or reduce sulfate concentrations to one ppm or less*", for reviews, see Scheidt and Kalla 2007; Corrales et al. 2011).

Table 3.2 Evidence supporting or refuting a causative link between sulfur cycle and Hg methylation in the Everglades, generated from within the system

Types of evidence	Description	References	Score
Spatial gradient, i.e., concentration-response in the field	"Methylmercury concentrations in surficial Everglades sediments and MeHg as a percent of HgT (%MeHg) increased dramatically from north to south, opposite the gradients in nutrient, sulfate and sulfide concentrations"	Gilmour et al. (1998b, p. 341)	++
	"Factor and principal component analyses of canal and marsh data partitioned THg in fish and MeHg in water as two independent components, with total phosphorus (TP), total organic carbon (TOC), and total ionic sulfate (TSO4) aggregated as a third component accounting for the variance in THg in fish"	Stober et al. (1998, p. ii)	+
	"In addition, there were interactions among the inorganic ligands, TOC, sulfide, and soil organic content (AFDW) and total mercury, and methylmercury in water and soil. The interactions among hydropattern and nutrient, organic carbon, and sulfate loadings from the EAA, with mercury contamination change from north to south in the South Florida ecosystem"	Stober et al. (2001, pp. 7–13)	+
	". . .sulfate gradient may account for the spatial difference in MeHg concentration between P33 and P34"	Rumbold and Fink (2003a, p. 4)	+
	". . .data in hand show a positive relationship between sulfate and MeHg south of 3A15, and experimental studies show sulfate stimulation of methylation in surface sediments. This information supports the idea that the very low sulfate concentrations found in Taylor Slough and central LNWR limit the activity of SRB and therefore limit methylation rates"	Gilmour and Krabbenhoft (2001, p. A7-4-7)	++
	"For the ENP dataset, low sulfate concentrations (less than 1 mg/L) were associated with a substrate limitation response, which means Hg methylation by SRB was limited by the amount of available sulfate. High sulfate concentrations (greater than about 5 mg/L) were associated with an inhibition effect, presumably due to sulfide accumulation in porewater causing reduced Hg bioavailability. At mid-level sulfate concentrations (1–5 mg/L), MeHg production appears to be maximal in the ENP"	Krabbenhoft et al. (2010) as referenced by Gu et al. (2012); also see USGS data from ENP presented in Corrales et al. (2011)	++
	"Parameters most highly correlated with BAFm (bioaccumulation factor for mercury) were surface water DOC ($r = -0.65$, $p < 0.001$), porewater sulfide ($r = -0.63$, $p < 0.001$), porewater sulfate ($r = -0.54$, $p < 0.001$) and MeHg in floating periphyton ($r = -0.47$, $p = 0.047$)"	Scheidt and Kalla (2007, p. 80)	+

(continued)

Table 3.2 (continued)

Types of evidence	Description	References	Score
Spatial gradient-cont.	Using data on Gambusia Hg and water column SO_4^{2-} concentrations collected from marsh sites throughout the EPA as part the USEPA's Regional Monitoring and Assessment Program (REMAP): "study results indicate a unimodal response similar to that originally hypothesized by Gilmour and Henry (1991)"	Axelrad et al. (2013, p. 3B-30) and Pollman (2012, referenced therein)	++
	"The relationship between sulfate and THg in each fish type is nonlinear and resembles a skewed trend (Fig. 3). Fish THg in each type abruptly increases up to ~1 mg/L sulfate (Fig. 3), displays peak THg levels between 1 and 12 mg/L sulfate, has a downward sloping trend between 12 and 25 mg/L sulfate then a slight downward sloping to zero-slope trend for sulfate concentrations ≥25 mg/L"	Gabriel et al. (2014, p. 588)	++
	There is a high level of variability between the S-Hg unimodal relationship during each monitoring phase. Qualitatively, some of the data seem to conform to a unimodal distribution with what appears to be peaks at key locations. However, based on a quartic regression, these relationships are not statistically significant ($p > 0.05$; Fig. 3B-19) and explain very little of the variation in the proportion of MeHg	Julian et al. (2014, p. 3B-47)	0
Laboratory tests of site media	Based on incubations of intact sediment cores from Everglades sites with addition of $^{203}Hg^{(II)}$ or various inhibitors: "Methylation rates generally increased to the south. Methylation rates appeared to follow the same trend as MeHg, with methylation rates increasing toward the south…" … "the addition of molybdate, an inhibitor of sulfate reduction, inhibited methylation; and addition of BES, an inhibitor of methanogenesis, had either no effect or stimulated methylation…"	Gilmour et al. (1998b, pp. 335, 339)	++
	"In cores taken from a relatively low sulfate site, addition of sulfate stimulated methylation, and sulfide reduction, over that of unamended control treatments even though sulfide levels increased slightly (see example experiments in Fig. 4)…"	Benoit et al. (2003, p. 274)	++
	Using sediment cores from the central Everglades and an STAs: "this experiment confirms the observations from field studies that drying and rewetting of Everglades' soils produces large pulses of MeHg … Sulfate concentrations also increased dramatically in the overlying water following rewetting of the dried cores from both sites. This increase in sulfate was similar to what had been observed in field studies following the 1999 drought and burn in the northern Everglades"	Gilmour et al. (2004b, p. 2B-1-3)	+

	Using dissolved organic matter (DOM) isolates from the Florida Everglades: "The interaction between Hg, sulfide, and DOM (i.e., DOM-Hg-sulfide complex) shown in the present study provides a more likely explanation for the observed dissolution of HgS(s) in the presence of DOM"	Miller et al. (2007, p. 632)	0
	Based on incubations done in the lab using soil and surface water WCA-3A: "the low net methylation in the presence of MoO_4^{2+} indicates that Hg methylation in sediment and water from central WCA-3A is mediated primarily by SRB . . . the non-zero rate of MeHg accumulation in the presence of both MoO_4^{2+} and BES suggests that non-SRB microbial guilds may contribute to net methylation, though at a lesser rate than SRB. . ."	Dierberg et al. (2014, pp. 8 & 10)	++
Manipulation of exposure—field experiments	Based on application of different stable Hg isotopes (e.g., ^{202}Hg, ^{200}Hg, ^{199}Hg) within in situ mesocosms: "new mercury is far more available for methylation and subsequent bioaccumulation . . . sulfate dosing had less effect on old Hg than new Hg. Dosing range used in our study for sulfate (5, 10, and 20 mg/L sulfate) reproduced the observed MeHg gradient in the Everglades (i.e., positive response observed at the low dose, inhibition at the high does, and maximum MeHg observed at the mid-dosing range). . ."	Gilmour et al. (2007a, p. 2)	++
	Based on mesocosm studies conducted in 2009-early 2010 in WCA-3A, 9 km from WCA-3A15: ". . .SRB: (1) were indeed present in the native microbial consortium. . ., and (2) were sulfate-limited, at least in part, at native sulfate concentrations. . ."	Jerauld et al. (2015, p. 17)	++
Temporal gradient	"Based on the observation that THg and DOC quantity and quality were constant for burned sites but there was a tenfold increase in net methylation efficiency (defined here as the percent of THg as MeHg in sediment and porewater), it can be inferred that liberation of sulfate from sediments and secondary stimulation of SRBs was a primary driving factor of excess methylmercury production in burned/dried areas"	Krabbenhoft and Fink (2001, p. A7-8-13)	+++
	"It is evident that MeHg concentrations in Largemouth bass have recently declined at site 3A-15 (Fig. 2B-10), . . . Corresponding with this trend in MeHg concentrations are declines in sulfate (Fig. 2B-13) and DOC concentrations (Fig. 2B-14). It is now likely that sulfate concentrations at site 3A-15 are suboptimal for MeHg production by sulfate-reducing bacteria (SRB). Reductions in sulfate concentrations at site 3A-15 in recent years may account for a substantial proportion of the Largemouth bass MeHg concentration declines that are not explained by declines in atmospheric inputs of mercury"	Axelrad et al. (2005, p. 2B-23) and Krabbenhoft et al. (2001, referenced therein)	+++

(continued)

Table 3.2 (continued)

Types of evidence	Description	References	Score
	"Significant and pronounced drop in mosquitofish mercury concentration from 1995 to 1999 (wet season $p < 0.01$; with dry season $p \ll 0.001$) and again from 1999 to 2005 (wet season $p < 0.01$; dry season $p < 0.01$)." This coincided with a significant decline in the area of the Everglades marshes >1.0 mg/L sulfate; from 66.1% in 1995 down to 57.3% in 2005	Scheidt and Kalla (2007, p. 77)	+++
Stressor-response relationships from simulation models	Based on modeling using structural equations and path analysis: "The greatest decrease in water methyl and fish mercury concentrations resulted from a reduction in all three input constituents—total phosphorus, sulfate, and total mercury concentrations. . ."	Stober et al. (2001, pp. 8–10)	+
	"The model-estimated decline in HgS^0 with increasing sulfide was consistent with the observed decline in bulk sediments MeHg"	Benoit et al. (1999a, p. 951)	0
	"The model was able to accurately reproduce the observed aqueous and solid phase Hg data, and total MeHg concentrations after adjustment of the scalable variables to match sulfate and dissolved Hg profiles . . . The final application of the diagenetic model was to site 3A15 in WCA 3A, . . . Model outputs accurately predicted the concomitant declines in sulfate reduction rate and MeHg production at this site over time, providing mechanistic support for the hypothesis that sulfate declines are driving at least part of the observed decline in MeHg at this site"	Gilmour et al. (2008, p. iii) SFER Append 3B-3	++
	Based on results from structural equation modeling: "A direct, positive connection between sulfate methylation potential and *Gambusia* Hg concentrations also is indicated by the model and is conceptually similar to the nonlinear relationship observed by Gabriel et al. (2014) between sulfate and Hg concentrations in several different trophic levels of fish (including *Gambusia*) in the Everglades"	Pollman (2014, p. 67)	++
Verified predictions	"Flow rate and water depth were managed as a means to alter sulfur biogeochemistry and, thereby, reduce in situ mercury methylation"	Rumbold and Fink (2006, p. 115)	++

At the outset there was controversy over the source of the sulfate (Orem et al. 1997, 2011; Schueneman 2001; Renner 2001; Bates et al. 2002; Gabriel et al. 2008). Some of the controversy was fed by miscommunications resulting from the use of imprecise language (i.e., vagueness), in some cases perhaps intentionally. The situation became even more contentious in 2004 when sulfate contamination (and resulting porewater sulfide) was also linked to two other longstanding environmental problems in the Everglades: phosphate (P) cycling and cattail expansion (Orem 2004; Axelrad et al. 2006). The issue erupted into a firestorm in 2007 with publication of a special report assessing sulfur sources, trends and effects in the Everglades (Gilmour et al. 2007b) as an appendix to the state's annual report on the Everglades (i.e., South Florida Environmental Report or SFER; for additional information on historical use and sources of sulfate, see Chap. 2, Volume I and Chap. 2, this volume). An external, peer-review panel charged to review the report had a mixed response. Reviewer's comments on the main document (Axelrad et al. 2007) were generally favorable and included the following statement (Jordan et al. 2007, p. 23) "*Since sulfate contamination affects nearly 30% of the Everglades, it requires additional study, including the development of criteria and management goals to reach safe levels.*" However, in reviewing the appendix, the Panel concluded that "*The strength of the conclusions is not supported by the data, and the overall picture is not clear*" (Jordan et al. 2007, p. 23). They (Jordan et al. 2007, p. 25) went on to state that "*sulfate and the mercury problems in South Florida are closely related. The sulfate/mercury methylation relationship needs to be understood in order to manage the mercury problem in South Florida.*" In the end, the Panel appeared to reverse itself, however, by their recommendation (Jordan et al. 2007, p. 25) that "*indicators of the mercury methylation problem other than sulfate concentrations should be developed for management purpose.*"

With a few exceptions, including a journal article authored by a staffer after leaving the SFWMD (Gabriel et al. 2014), 2007 was a turning point for the state agencies, with research by FDEP/SFWMD and their consulting environmental laboratory increasingly focused on simply repeating previous sulfate studies or on alternative candidate causes for the spatial and temporal variability in Hg methylation and biomagnification in the Everglades. The publication by the former staff member (Gabriel et al. 2014) also sparked a new vigorous debate in the published literature with several published reinterpretations and comments followed by replies to comments (Julian 2014; Pollman and Axelrad 2014a, b; Julian et al. 2015a; Gabriel et al. 2015). The peer-reviewers of the 2014 SFER report, which was comprised of different members than the 2007 Panel, were critical of the report for the minimal discussion devoted to the sulfate issue as compared to previous reports (SFWMD 2014a). The response by agency authors (SFWMD 2014b, p. 1-3-12) was that "*based on the available data, FDEP and SFWMD have concluded that it is not currently possible to quantitatively identify the causative factors and linkages between MeHg and sulfate.*" Subsequently, FDEP/SFWMD have offered a number of alternative candidate causes for the spatial and temporal variability in Hg biomagnification, including the possibility that microbes other than sulfate-reducing bacteria (SRB) may be responsible for Hg methylation; that water quality parameters other than sulfate may have as much or more influence in the accumulation of Hg, or that food web dynamics and habitat influences the location of hotspots (Julian and Gu 2015; Julian et al. 2015b, 2016).

The objective of this paper is to re-examine the cause of the observed susceptibility of the Everglades to high rates of Hg biomagnification and its extreme geographic and temporal variability using a formal and transparent causal analysis. There is little debate that the Everglades receives high Hg deposition relative to national averages but that variation in deposition across the Everglades is insufficient to explain the observed geographic and temporal patterns in biomagnification (for review of atmospheric deposition, see chapters in Volume I). Accordingly, this paper will instead focus on candidate causes for the latter, particularly the role of sulfate.

The concept of causation plays a central role in science, particularly epidemiology and ecoepidemiology (for review, see Suter et al. 2002, 2010). It is therefore surprising that formal causal analyses using consistent criteria are conducted so infrequently, particularly since such criteria have been available since 1965 when introduced in a seminal paper by Sir Austin Bradford Hill (1965). Yet, establishing causation to the satisfaction of others is not as simple as we tend to assume. One of the best examples of this is the difficulty that Bradford Hill and Richard Doll and others had in convincing people of the causal relationship between smoking and lung cancer (stemming from their work in the early 1950s). Establishing causality between environmental stressors and effects on ecosystems can be even more difficult due to the high variability of environmental factors combined with synergistic and cumulative interactions among these factors (Adams 2003). As stated by Fox (1991) "*assessing causality is neither simple nor straightforward, particularly in environmental impacts, which are often not amenable to classical statistics*." While we often attempt to seek the real or principal cause, we must also recognize that there is not always a one-to-one relationship between a factor and its effects. Instead, a causal relationship exists when evidence indicates that the factors form part of the complex of circumstances that increases the probability of the outcome (Lilienfeld and Lilienfeld 1980). Finally, Suter (2007) cautions that it requires expert judgment applied in a consistent and transparent manner and based on the totality of evidence, of which the result of any single study is only a component. Returning and using the link between smoking and lung cancer as an illustration: despite a few, well-funded studies arguing to the contrary, we now all accept that smoking can cause cancer, but it took several decades.

3.2 Methods

This causal analysis employs the framework developed by USEPA as part of a causal analysis/diagnosis decision information system (CADDIS) for exploring causation (Suter et al. 2002, http://www.epa.gov/caddis) that adapted most of Hill's criteria to guide judgement. In essence, this paper assembled existing evidence and organized it into categories associated with Hill's criteria (e.g., spatial/temporal co-occurrence, stressor response relationship in the field, manipulation of exposure, laboratory tests using site media, temporal sequence, verified predictions, etc.). The studies were divided into two major groups based on where they were generated. Lines of evidence from the Everglades were evaluated first, because studies from within the system are often the most compelling and relevant. However, it is beneficial to also include evidence from other systems, so they were also summarized here (although the list was not meant to be exhaustive). It is noteworthy that not all of Hill's criteria were assessed for each candidate cause. As the USEPA (http://www.epa.gov/caddis) guidance explains, it is unlikely that each of Hill's criteria will have been tested for a particular causal analysis. Once assembled, each line of evidence was weighted using expert judgment based on whether it supported or weakened the candidate cause and based on its quality (e.g., study design, relevance, performance, statistical analysis, and potential for bias). By convention, weights were represented by + or − symbols with +++, −−− for convincingly supports or weakens; ++, −− for strongly supports or weakens; +, − for somewhat supports or weakens; or 0 having no effect on the argument. The scores of various lines of evidence were not summed. Instead, they were evaluated for consistency by looking at the overall pattern of scores. Examples of recent applications of this approach have examined the possible causation for: (1) depressed reproduction of clams in San Francisco Bay (Brown et al. 2003); (2) reduced overwinter survivability of commercial bees in California (Staveley et al. 2014); and (3) the decline of smallmouth bass in the Susquehanna and Juniata Rivers (Shull and Pulket 2015).

3.3 Analysis

3.3.1 Candidate Cause: Balance Between Sulfate and Sulfide

3.3.1.1 Mechanistically Plausible Cause: The Relationship Between the Causal Agent and Effect Should Be Consistent with Known Principles of Biology, Chemistry and Physics

Several laboratory experiments support the mechanistic plausibility of Hg methylation by sulfate-reducing bacteria (SRB, Table 3.3). Early studies, however, also showed that under certain environmental conditions, particularly low redox potential, sulfate addition and a buildup of sulfide could inhibit Hg methylation (Craig and Moreton 1983; Compeau and Bartha 1984; Berman and Bartha 1986; Winfrey and Rudd 1990). Although sulfide inhibition was originally thought to be due to removal of Hg from solution via enhanced precipitation of HgS(s), Benoit et al. (1999b) developed an alternative theory that under highly sulfidic environments charged Hg-S complexes form that are not taken up as readily as neutral complexes by passive diffusion across the cell membrane. Their chemical equilibrium model estimated a decline in HgS^0 with increasing sulfide that was consistent with the observed decline in bulk sediment MeHg in the Everglades (Benoit et al. 1999b).

Table 3.3 Evidence supporting or refuting a causative link between sulfur cycle and Hg methylation from studies done outside the Everglades (note, this list is by no means exhaustive)

Types of evidence	Description	References
Mechanistically plausible	Sulfate-reducing bacteria were first identified as important Hg methylators in marine and freshwater environments in the 1980s and early 1990s	Compeau and Bartha (1985), Gilmour et al. (1992) and King et al. (2000)
	Sulfide inhibition	Compeau and Bartha (1985), Benoit et al. (1999a, b) and Jay et al. (2000)
Manipulation of exposure—field experiments	Dosing experiments in the Experimental Lakes Area (ELA): "demonstrated that the in situ addition of sulphate to peat and peat pore water resulted in a significant increase in pore water MeHg concentrations"	Branfireun et al. (1999, p. 743)
	"Sulfate was added to half of an experimental wetland at the Marcell Experimental Forest located in northeastern Minnesota . . . Our results demonstrate enhanced methylation and increased MeHg concentrations within the wetland and in outflow from the wetland suggesting that decreasing sulfate deposition rates would lower MeHg export from wetlands"	Jeremiason et al. (2006, p. 3800)
	Four years after sulfate additions to Marcell Experimental Forest ceased (and sulfate compounds became more recalcitrant) MeHg declined	Coleman Wasik et al. (2012)
	Based on 16 year study Little Rock Lake experimentally acidified with H2SO4 "In hypolimnetic waters, we observed a direct correlation between the maximum MeHg concentration and the sulfate deficit for each year . . . and a direct correlation between MeHg and sulfide concentration"	Watras et al. (2006, p. 257) Also see Harmon et al. (2004), Mitchell et al. (2008) and Myrbo et al. (2017)
Spatial gradient from the field	Gradient effect: sulfate stimulation of methylation, sulfide inhibition or both	Craig and Moreton (1983), Berman and Bartha (1986), Hammerschmidt and Fitzgerald (2004), Muresan et al. (2007) and Bailey et al. (2017)
Spatial gradient-cont.	"The modeling results for lake water and yellow perch indicate the influence of ecosystem variables associated with the microbial production and abundance of methylmercury (lake pH, dissolved sulfate, and connected wetlands) on methylmercury concentrations in water and fish"	Wiener et al. (2006)

(continued)

Table 3.3 (continued)

Types of evidence	Description	References
	"In this study, we compare spatial and temporal patterns in MeHg and associated geochemistry in two wetlands receiving contrasting loads of sulfate .... large input of sulfate to a chronically sulfate-impacted system led to significantly lower potential relative methylation rates"	Johnson et al. (2016, p. 725)
Time gradient	"In the lakes of Isle Royale, U.S.A., reduced rates of sulfate deposition since the Clean Air Act of 1970 have caused mercury concentrations in fish to decline to levels that are safe for human consumption, even without a discernible decrease in mercury deposition"	Drevnick et al. (2007, p. 7266)
Stressor-response relationships from simulation models	The watershed analysis risk management framework (WARMF) model was applied to Marcell Experimental Forest: "Quadrupling the sulfate deposition would increase the MeHg output by 216%"	Chen and Herr (2010)

This theory was later further supported by additional laboratory tests (Benoit et al. 1999a; Jay et al. 2000). More recently, the model has been revised to include the influence of dissolved organic matter (DOM) on inorganic Hg bioavailability under sulfidic conditions (Miller et al. 2007).

At this juncture the reader is encouraged to review the information presented in Chaps. 1, 2 and 4 (this volume) to gain an appropriate appreciation of the complex of circumstances that increases the probability of Hg methylation by SRB. Simply stated, microbial Hg methylation first requires a source of bioavailable inorganic Hg (which will be dependent on its oxidation state and its associated ligand: DOM, S^0, S^{2-}, DOM-Hg-sulfide complex, among others, all of which are concentration and environment dependent, particularly on DOM composition and reactivity). It also requires a source of labile carbon. Many different microbes have been shown to be capable of methylation (Benoit et al. 2003; Gilmour et al. 2013; Podar et al. 2015). For SRB to be the favored pathway obviously requires a bioavailable form of oxidized sulfur to serve as an electron acceptor. However, sulfate reduction will be favored only after other electron accepters higher on the redox ladder (i.e., greater energy-yielding substrates remaining in the oxidized form) are exhausted from the environment. These electron acceptors include in particular oxygen; thus the potential methylating environment must be sub-oxic with negative redox potential (for additional information, see Chaps. 1 and 5, this volume). Net methylation, which ultimately constrains bioaccumulation of Hg by biota, is dependent on the balance between methylation and demethylation; this balance reflects which of the two microbial pools are favored and is also controlled by redox conditions (Compeau and Bartha 1984; Berman and Bartha 1986). Finally, once sulfate loading and environmental conditions, particularly redox conditions (as sulfate is reduced, redox potential can become very negative, as long as oxygen diffusion rate is minimal), allow sulfide to build up in the sediments, sulfide inhibition of methylation may occur.

3.3.1.2 Spatial Gradient: Effects Should Decline or Increase as the Stressor Declines or Increases Over Space

If sulfate is stimulating SRB to methylate Hg, the rate of methylation, as evidenced by MeHg availability and magnitude of biomagnification, would be expected to be greater at sites with higher surface water sulfate (i.e., concentration-response); however, this would also be dependent on the presence of labile carbon and suitable redox conditions in the sediment (or periphyton) where most of the methylation occurs. Furthermore, as just discussed, this is a non-linear process; as sulfide accumulates in porewater, methylation may be inhibited depending again on redox potential (i.e., the so-called "inverse relationship").

As already mentioned, early surveys of Hg in the Everglades all remarked on the strong geographical variability in aqueous Hg species and in the magnitude of biomagnification, with concentrations generally highest in the less eutrophic areas located within WCA-3A (Cleckner et al. 1998; Gilmour et al. 1998b; Hurley et al. 1998). Because they sampled over a much larger spatial area in their R-EMAP study (for more information on this study, see Chap. 6, Volume III), the USEPA (Stober et al. 1996) identified hotspots further south in Shark River Slough of ENP, as well as confirming those in WCA-3A. A high degree of correspondence between mosquitofish Hg (and feather Hg) and surface water sulfate was evident in the large R-EMAP data set from 1995 to 1996 (Figs. 8.45 and 8.54 in Stober et al. 1998). USEPA also demonstrated statistical relationships among these metrics using principal component analyses (Stober et al. 1998). This link between sulfate and MeHg was demonstrated repeatedly in subsequent R-EMAP sampling events (for references, see Table 3.2).

Gilmour et al. (1998b) reported a gradient in MeHg in surficial sediments with the lowest concentrations in the Everglades Nutrient Removal Project (ENR; the prototype Stormwater Treatment Area now known as STA-1W) and WCA-2A and increasing southward to WCA-3A. They found an inverse relationship between sulfate concentration and %MeHg in sediment with highest concentrations of sulfate in ENR and decreasing southward. Perhaps, more importantly, they found that the sulfide gradient in surficial sediments was even steeper than the sulfate gradient across the system. As discussed in more detail below, incubation of a subset of these sediment cores with radiotracer ^{203}Hg revealed that methylation rates were consistent with the pattern in %MeHg in sediments (Gilmour et al. 1998b). Based on these results, Gilmour et al. (1998b, p. 341) concluded that "*methylmercury production in the eutrophic northern Everglades appears to mimic that in an estuary, where sulfate and sulfate reduction rates are high and sulfide produced by sulfate reduction inhibits MeHg production*." Over the years, the Stormwater Treatment Areas (e.g., 1W, 1E, 2, 3&4, 5 and 6) operated by SFWMD have consistently had some of the lowest MeHg concentrations in water, sediment and biota within the Everglades (Germain 2014), likely due to sulfide inhibition of methylation (Gilmour et al. 1998b; Fink 2004; Rumbold and Fink 2006; Rumbold et al. 2007). High frequency sampling during start-up or an occasional quarterly or semi-annual sampling event have, however, fortuitously captured spikes in MeHg in water or mosquitofish of a treatment area following dry outs and reflooding (Fink 2004; Rumbold and Fink 2006; Rumbold et al. 2007; this will be discussed in greater detail below).

There are other areas of the Everglades besides the STAs where biota are low in Hg. We must, however, differentiate areas low in MeHg that are caused by too much sulfide, such as the STAs and highly eutrophic, sulfidic areas in WCA-2, from areas where too little sulfate is limiting Hg methylation. Based on their work both in the field and from spiking sediment cores taken from various areas of the Everglades, Gilmour and Krabbenhoft (2001, p. A7-4-7; also see Gilmour et al. 2007a) concluded that low sulfate concentrations found in Taylor Slough (e.g., ENP sites TS7 & TS9) and central WCA-1 limit the activity of SRB and therefore limit methylation rates. This is consistent with the patterns in USEPA R-EMAP data sets from 1995 and 1996, 2000, and 2005 that show areas of low sulfate and low mosquitofish Hg (Stober et al. 1998, 2001; Scheidt and Kalla 2007). A more recent study with high density sampling within ENP also reported that low sulfate concentrations (less than 1 mg/L) outside of Shark River Slough limited Hg methylation (Krabbenhoft et al. 2010 as cited by Gu et al. 2012). This would be consistent with the earlier report of patterns of sulfate in surface water (McPherson et al. 2003) and in sediment sulfide (Chambers and Pederson 2006). It would also be consistent with a study done along a transect into Florida Bay from 2000 to 2002 (Rumbold et al. 2011) that found low MeHg in sediments from marshes of Taylor Slough but higher MeHg in marshes of the C111 drainage system; these two systems exhibit a gradient in surface water sulfate (McPherson et al. 2003). Furthermore, sulfate limitation within the deep interior of WCA-1 is also in agreement with the consistently low Hg observed in bass from WCA-1 (Axelrad et al. 2013) and with a report of low Hg in alligators from WCA-1 (Rumbold et al. 2002).

The data sets discussed above have also been used in developing predictive, statistical models. USEPA (Scheidt and Kalla 2007), for example, found statistically significant relationships between bioaccumulation factors (i.e., BAFm = concentration of Hg in mosquitofish divided by MeHg concentration in surface water; for review of BAFs, see Chap. 7, this volume) and surface water DOC, porewater sulfide, porewater sulfate and MeHg in periphyton. Using the same USEPA data set, Pollman (2012, 2014) reported that the relationship between tissue Hg concentration in mosquitofish and surface water sulfate across the Everglades was a unimodal distribution, i.e., levels of tissue Hg increased with increasing sulfate, reaching a peak and then decreased with increasing sulfate, presumably due to a buildup of sulfide and inhibition of methylation. These results were later challenged by Julian (2013) when, using the same USEPA data, he failed to find a unimodal relationship between the BAF for mosquitofish and surface water sulfate across the Everglades. As commented upon by Pollman and Axelrad (2014a), Julian's failure to identify a unimodal relationship reflected a series of fundamental flaws in his analysis, including modeling calculated BAFs as opposed to direct tissue Hg concentrations and poor model specificity. Using a more appropriate model framework coupled with regression model diagnostics to test underlying model assumptions, Pollman and Axelrad (2014a) showed that the relationship between water-column MeHg concentrations was indeed unimodal. Later, based on data collected by SFWMD, Gabriel et al. (2014) also reported finding unimodal distributions in tissue Hg concentrations in three different species of fish (e.g., mosquitofish, sunfish and bass) and surface water sulfate similar to that of Pollman (2012). Julian et al. (2015a) later also criticized the approach taken by Gabriel et al. (2014) and dismissed their conclusions. Gabriel et al. (2015) later responded to the criticisms and stood by their conclusions.

In a re-analysis of the relationship between tissue Hg in Largemouth bass (LMB) and sulfate (based on the same SFWMD data set as used by Gabriel et al. 2014), Julian and Gu (2015, p. 202) concluded that "*sulfate has little association with Hg concentrations in LMB. It is suggested that other water quality variables including water pH, alkalinity and specific conductance may have as much or more influence in the accumulation of Hg in LMB.*" Their analysis, however, suffered from pseudoreplication when they treated each fish caught at the same site each year as the measurement unit rather than as a replicate and then compared all these pseudoreplicates (thereby inflating n) against an annual mean surface water sulfate concentration at that particular site (cf., Gabriel et al. 2014, which correctly used median tissue Hg for fish collected during a sampling event for a given site). Consequently, conclusions from Julian and Gu (2015) are likely misleading. Additionally, as discussed below, they (as well as others) also mismatched measurement time scales (i.e., mean of surface water sulfate collected quarterly and tissue Hg in a LMB that represents exposure integrated over years). A mismatch in measurement time scales is a chronic problem of many of the statistical models and is the root of still another criticism made by Julian et al. (2015a) of Gabriel et al. (2014). They (Julian et al. 2015a, p. 4) argued that "*there was high variability in the fish THg* [total Hg] *data to the extent that no predictive pattern could be seen in the data, and there were several instances of high fish THg concentrations in both high (>30 mg/L) and low (<1 mg/L) sulfate concentrations.*" The data set used by Gabriel et al. (2014) consisted of 679 Largemouth bass, 2559 sunfish (both of which were used to calculate site medians), 484 mosquitofish composite samples and 2360 surface water sulfate samples from 12 stations collected over 11 years. Consequently, having several instances of data that did not fit the model should not be surprising. In fact, Gabriel et al. (2014, p. 590) writes "*considering the large number of factors that can influence Hg methylation and bioaccumulation and sulfate concentration, it is quite surprising to observe any relationship between surface water sulfate and fish THg.*" While the authors are correct in that there are many factors involved in the complex of circumstances that increases the probability of methylation by SRB, the larger problem is that many of these processes operate over different spatial scales and temporal rates and are not being measured with the necessary spatial and temporal resolution. First and foremost, with the exception of Scheidt and Kalla (2007), all of the above analyses relied on surface water sulfate concentrations rather than some measure (or combination of variables) that would be a better proxy for the balance of sulfate-sulfide and redox conditions (i.e., must be sub-oxic) in sediment porewater or periphyton that would favor methylation. Failure by some investigators (and resource managers) to properly recognize and precisely communicate the relationship between surface water sulfate concentration, the necessary redox conditions (either in sediments or periphyton) and the potential for sulfide buildup to influence SRB methylation was and continues to be a major source of miscommunication in solving Florida's Hg problem. Gabriel et al. (2014, p. 587) writes "*we chose surface water as a means to observe sulfate level since it is less labor-intensive to sample and more cost-effective to monitor and control compared to sediment porewater.*" As previously stated, this has been a chronic problem, particularly of SFWMD monitoring programs.

Although surface-water sulfate concentration has been shown to be statistically correlated with porewater sulfide the relationship is less than exact (even if measured at the same time), in part because of measurement error and, in part because sulfide concentrations undergo other reactions within the porewater. Further, as mentioned above, the balance between sulfate and sulfide and SRB methylation potential will depend on the rate of O_2 diffusion into sediments and redox conditions that are extremely dynamic. Redox conditions in Everglades sediments have been shown to be highly variable even on a diel cycle and this variability was found to be greatest for the top 5 cm of sediments under low water conditions (Thomas et al. 2009); this also happens to be the principal site of methylation. Clearly, water depths and thermal stratification will change seasonally in the Everglades and this will have an effect on O_2 diffusion rates and redox conditions (Qualls et al. 2001). Shifts in the balance between sulfate and sulfide have been observed in sediments following marsh dry-down and reflooding with a subsequent pulse of strong methylation both in the field (Krabbenhoft and Fink 2001; Rumbold et al. 2001a) and as simulated in the lab (Gilmour et al. 2004b). Water depth, however, can vary over shorter time scales (e.g., pump operation, tides) influencing O_2 and redox and have resulted in similar pulses of methylation (e.g., Rumbold and Fink 2006; Rumbold et al. 2011). Moreover, even without changes in water depth, redox conditions can vary over short time scales due to many factors including diurnal cycles in photosynthesis, water stratification (and will be dependent on wind and/or temperature) and biological oxygen demand (Qualls et al. 2001; Thomas et al. 2009; Nimick et al. 2011; for further information, see Chap. 5, this volume). Not surprisingly, select Hg species, including MeHg, have been found to show a strong diurnal variation in the water column with increasing MeHg concentrations during non-daylight periods in the Everglades and elsewhere, possibly due to the diel variation in photosynthesis and production of oxygen, daytime photodemethylation in the water column, photolytic sorption/desorption, bioadvection, or some combination (Krabbenhoft et al. 1998a, b; Naftz et al. 2011). At night, when there is no oxygen production by the periphyton, the sulfate-reducing bacteria in the thick periphyton mats may also methylate

Hg (Cleckner et al. 1999). One must also recognize that environmental conditions can lead to extremely dynamic micro-environments, in terms of oxygen diffusion and redox conditions, that would allow for pulses in sulfate reduction and Hg methylation even in what appears to be highly sulfidic conditions (i.e., which would be expected to inhibit methylation). Microenvironments might be created, for example, by (1) bioturbation (Hammerschmidt et al. 2004); (2) oxygen surrounding plant roots (Marvin-DiPasquale et al. 2003); or, (3) vertical migration of the oxycline and redox transition zone due to tidal pumping in estuarine environments or due to daytime peaks in photosynthesis (Langer et al. 2001). Microenvironments were invoked to explain how methylation occurs in sulfidic sediments of Florida Bay (Rumbold et al. 2011).

Therefore, attempts to assess the relation between point-in-time snapshots of water-column chemistry (as a proxy for porewater chemistry), that were likely collected once during the past 3 months (i.e., quarterly grab samples) with measures of MeHg production suffer from spatial and temporal mismatch. This is not a trivial problem and the mismatch in the measurement scales will generate considerable variability. This has also been observed in other regions.

In their study of wetlands in Minnesota that receive sulfate discharges from iron mining activities, Johnson et al. (2016) found short-term measurements of methylation and demethylation potentials only weakly related to longer-term measurements of MeHg accumulation in peat. They concluded that methylation and demethylation processes were highly transient through time. Yet, they did find a log-linear relationship between %MeHg in peat and porewater sulfide at sulfide concentrations $<700\ \mu g\ L^{-1}$; however, %MeHg decreased as sulfide increased above that concentration. Numerous other studies outside the Everglades have also found a spatial gradient in MeHg and sulfate concentration and, where sulfide accumulated in the sediments, inhibition of Hg methylation (Table 3.3, for review, see Munthe et al. 2007).

3.3.1.3 Laboratory Tests of Site Media: Controlled Laboratory Manipulation of Site Media Should Induce the Same Effects as Seen in the Field

The role of sulfate in Hg methylation in the Everglades was demonstrated in the laboratory using intact sediment cores by finding: (1) sulfate addition stimulated methylation in the core; and (2) when sulfate reduction was chemically inhibited, methylation was inhibited (Gilmour et al. 1998b; Benoit et al. 2003, Table 3.2). These studies also provided evidence that: (1) methanogenesis, which should be important in fresh water environments, was not important in Hg methylation processes in these cores; and (2) that sulfate was limiting in cores taken from certain areas of the Everglades. Later, using sediment cores from the central Everglades and STA-2, Gilmour et al. (2004b) found sulfate concentrations increased in overlying water if cores were dried and then re-wet and that there was a concurrent pulse in MeHg. This laboratory study, thus, confirmed what had been observed in the field following the 1999 drought (discussed in greater detail below) and during start-up of STA-2.

Based on incubations of sediment and surface water (collected from a site 10 km northwest of site 3A-15[1]) in the laboratory, Dierberg et al. (2014) also reported that methylation was mediated primarily by SRB. This line of evidence therefore supports sulfate and SRB as the cause. However, unlike Gilmour et al. (1998b), Dierberg et al. (2014) reported methylation continued to occur, at a lesser rate, in the presence of the inhibitors of both sulfate reduction and methanogenesis and concluded other microbial guilds may contribute to net methylation. They also report methylation occurred with only very small additions of sulfate. Both these findings will be discussed in greater detail in ensuing sections.

3.3.1.4 Manipulation of Exposure: Field Experiment or Management Actions That Decrease or Increase the Intensity of the Causal Agent Should Decrease or Increase the Effect

Much of the information on Hg biogeochemistry in the Everglades was gathered using an experimental approach at the mesocosm-scale (for review, see Gilmour et al. 2004a, 2007a). From dosing 18 mesocosms (located at 3A-15) with a range of sulfate concentrations (5, 10, and 20 mg/L sulfate), Gilmour et al. (2007a) reported they were able to reproduce the MeHg gradient observed in the Everglades (i.e., positive response observed at the low dose, inhibition at the high dose, and maximum MeHg observed at the mid-dosing range). Clearly, this strongly supports this candidate cause (Table 3.2).

Based on results of another mesocosm study conducted from 2009 through early 2010 in WCA-3A (a few km from 3A-15), Jerauld et al. (2015) reported sulfate amendments increased the rate of sulfate reduction and concluded that SRB were present and were sulfate-limited (the significance of this will be discussed further in Sect. 3.4). However, the authors also reported that they did not observe the expected unimodal (bell-shaped) MeHg response with respect to sulfate concentration. While some might view this as refuting the relationship, the authors (Jerauld et al. 2015) cautioned that previous studies (Gilmour et al. 2004a) had found sulfate dosing had less effect on old Hg than new Hg and they had not dosed their mesocosms with new Hg when dosing with sulfate. The authors (Jerauld et al. 2015) cautioned that Hg methylation could also have been limited due to a lack of labile carbon. What was surprising, even to the investigators, was that surface water MeHg concentration decreased following a one-time addition of inorganic Hg. They speculated that the timing of follow-up sampling (6 days after the spike) was either too late or too soon

[1]Site 3A-15 had been previously well established as a Hg bioaccumulation "hot spot" and was a site of intense interest with respect to temporal trends in fish tissue Hg concentrations and causative factors, including the role of sulfate. Dynamic modeling using the Everglades Mercury Cycling Model (E-MCM) to test various causative factors is the subject of Chap. 4, Volume III.

to capture the MeHg spike (Jerauld et al. 2015). It is noteworthy that Gilmour et al. (2004a) had reported that recently added Hg (spikes of labeled inorganic Hg) was methylated at the maximum extent within about 24–48 h after addition. Thus, Jerauld et al. (2015) likely sampled too late and missed the pulse of methylation. This further illustrates the dynamic nature of the methylation process as previously discussed and the difficulty of relying on environmental conditions measured with insufficient temporal resolution. For these reasons, the results from Jerauld et al. (2015) study were categorized as only somewhat weakening the role of sulfate in methylation (Table 3.2).

Since the pioneering Hg work in the Everglades, there have been several manipulative studies done elsewhere that convincingly demonstrate the role of the sulfate-sulfide balance as the key driver of Hg methylation (Table 3.3). For example, using mesocosms containing sediment from Rice Portage Lake in Minnesota dosed at five different sulfate treatment levels, Myrbo et al. (2017) found MeHg peaked at intermediate pore water sulfide concentration.

3.3.1.5 Temporal Gradient: Effects Should Increase or Decrease as the Stressor Increases or Decreases Over Time

If sulfate is feeding SRB and making the Everglades more susceptible to the global Hg problem, then the rate of methylation and magnitude of biomagnification should increase or decrease as sulfate loading and concentration increases or declines over time (i.e., at least within the linear portion of the concentration-response relationship).

Due to a lack of historical data on both Hg and sulfate concentrations in the Everglades, it is difficult to assess whether the increased magnitude of biomagnification in Florida's biota coincided with increased Hg deposition or as the system became contaminated with sulfate. Atmospheric deposition monitoring did not begin in south Florida until 1992, with the establishment the Florida Atmospheric Mercury Study (Guentzel et al. 1995). Prior to that we must rely on sediment records and emission estimates (see Chap. 3, Volume III for a more inclusive analysis of trends in atmospheric depositional fluxes of Hg inferred from sediment cores, including trends in the Everglades). Based on sediment cores taken from the Great Lakes, Pirrone et al. (1998) reported Hg deposition to North America increased dramatically beginning in the 1880s with subsequent peaks during the late 1940s, 1970 and again in 1989. Similarly, based on sediment cores from Florida, Rood et al. (1995) reported Hg accumulation rates increased gradually between 1900 and 1940, increased mid-century and then increased again during the 1970s and 1980s. Using a different approach, Husar and Husar (2002) used a material flows analysis based on the production, consumption, recycling and combustion of different materials using Hg stocks to estimate emission rates. Their results suggested Hg emissions in the US spiked in the mid-1940s, again in the early 1950s, peaked in the late 1960s and then fell sharply after 1990 (Husar and Husar 2002; see also Chap. 5, Volume I).

Historical analyses of Hg biomagnification suggest that Hg levels did not increase in Everglades' biota during early peaks in Hg deposition but did increase in the late 1970s and 1980s. The earliest report of elevated Hg levels in the Everglades was in a small number of biological samples collected from the WCAs and ENP between 1971 and 1973 by Ogden et al. (1974; relatively high levels were observed in only 5 out of 89 samples; comprised of an alligator egg and four bluegill from ENP). It is likely that surveys done before that time or in other areas of the Everglades were not looking for Hg. To look further back in time, we must rely on two studies done on museum specimens collected in Florida (Porcella et al. 2004; Frederick et al. 2004). Although hampered by poor temporal resolution and small sample size, a survey of raccoon museum specimens by Porcella et al. (2004) reported relatively low Hg levels in hair from pelts of raccoons from Shark River Slough from 1947 to 1948 (mean 11.2 mg/kg). For comparison, Roelke et al. (1991) reported raccoons sampled in 1990 had very high Hg levels in their fur (mean of 72 mg/kg). On the basis of surveys of Hg in bird feathers from museum specimens, which had better temporal resolution (including during the late 1940s and 1950s), Frederick et al. (2004, p. 1477; for more information on trends in bird feather, see Chap. 2, Volume III) concluded that "*mercury concentrations in feathers of aquatic birds from south Florida were consistently low across four species until rapid increases during some point after the late 1970s*." The timing of this rapid rise in Hg biomagnification in the birds would have followed completion of the construction of the EAA, nutrient and sulfur infiltration into the WCAs (as evidenced by sediment cores and observed changes in water quality), and changes in vegetation, O_2 dynamics and sediment redox that would have increased Hg methylation efficiency (for review of timing of these changes, see Chap. 2, Volume I). It should also be noted, however, that the increase in bird feather Hg concentrations in the late 1970s also corresponded to the end of an approximately 35 year long period of generally sustained increases in inferred local emissions coupled with an abrupt increase in local emissions beginning in the mid-1970s (see Fig. 2 in Chap. 3, Volume I).

An even stronger case for temporal gradient can be made based on recent observations indicating that the magnitude of biomagnification has declined in certain areas as surface water sulfate concentrations have declined (note, the reason for the sulfate declines are, as yet, uncertain but may simply represent a fortuitous event related to water redistribution). Several investigators have reported declines in Hg in biota, e.g., fish, birds and alligators, from certain areas in the 2000s as compared to peaks in late 1990s (Lange et al. 2000, 2005; Rumbold et al. 2001b, Rumbold 2005b; Frederick et al. 2002). While initially these declines were thought to be the result of decreases in local Hg emissions (Atkeson and Axelrad 2004) there was no corresponding obvious change in local deposition (for review, see Rumbold et al. 2006); however, more recent analysis may suggest at least some variation in deposition during this time period (see both Chaps. 1 and 3, Volume III). Krabbenhoft et al. (2004, as cited by Axelrad et al. 2005) was the first to report that Hg declines in Largemouth bass at 3A-15 coincided with declines in surface water sulfate and DOC (as mentioned previously, a mechanistic model analysis of the trends observed at 3A-15 is presented in Chap. 4, Volume III). Axelrad et al.

(2005, p. 2B-23) concluded that "*It is now likely that sulfate concentrations at site 3A-15 are suboptimal for MeHg production by sulfate-reducing bacteria (SRB).*" Similarly, USEPA (Scheidt and Kalla 2007) also reported coincident declines in tissue Hg in mosquitofish and declines in surface water sulfate across the Everglades. They attempted to identify constituents relevant to the pronounced drop in mosquitofish Hg and reportedly "*found MeHg in surface water was most correlated with MeHg in floating periphyton ($r = 0.87$, $p < 0.001$) and surface water sulfate ($r = 0.65$, $p < 0.001$). MeHg in benthic periphyton was highly correlated with surface water sulfate during the wet season ($r = 0.87$, $p < 0.001$).*"

Time order can also be assessed by examining pulses of methylation that followed increases in surface water sulfate as a result of drydown and reflooding and, in some cases, burning of peat at several Everglades sites during a severe drought in 1999. Based on the results of a temporally intensive sampling campaign at 13 sites over several months, Krabbenhoft and Fink (2001, p. A7-8-12; for more discussion of pulse following burn, see Chap. 2, this volume) concluded "*...THg and DOC quantity and quality were constant for burned sites but there was a ten-fold increase in net methylation efficiency (defined here as the percent of THg as MeHg in sediment and porewater), it can be inferred that liberation of sulfate from sediments and secondary stimulation of SRBs was a primary driving factor of excess methylmercury production in burned/dried areas.*"

In addition, similar declines in biomagnification have been observed elsewhere following halting sulfate addition to large mesocosms (Coleman Wasik et al. 2012) and following declines in sulfate loading or surface water concentration in the field (Drevnick et al. 2007; Meyer et al. 2011). For example, in Little Rock Lake, WI (where sulfate concentrations were considered limiting for Hg methylation mediated by sulfate-reducing bacteria), experimental additions of sulfate to the lake led to increases in both SRB abundance and hypolimnetic MeHg concentrations; cessation of sulfate additions led to concomitant declines in both sulfate and MeHg concentrations, with the greatest reductions in MeHg corresponding to the sub-basin in the lake with the greatest decline in sulfate (Hrabik and Watras 2002).

3.3.1.6 Stressor-Response Relationships from Ecological Simulation Models: The Causal Agent in the Case Is at Levels That Are Associated with Similar Effects in Mathematical Models That Simulate Ecological Processes

Information gathered from these studies have been incorporated and used in the development of process-based models (Harris et al. 2001; Gilmour et al. 2008). The most recent effort attempted to construct a diagenetic model and couple it with the Everglades Mercury Cycling Model (E-MCM) to predict responses to changing sulfate concentrations in the Everglades. As reported by Gilmour et al. (2008), the model was able to accurately reproduce the observed aqueous and solid phase Hg data, and total MeHg concentrations after adjustment of the scalable variables to match sulfate and dissolved Hg profiles. Moreover, the model was also able to predict the concomitant declines in MeHg production and sulfate reduction rate as sulfate concentrations declined at WCA-15A, providing mechanistic support for the conclusions of Krabbenhoft et al. (2004, as cited by Axelrad et al. 2005) and Scheidt and Kalla (2007) as discussed above.

Process models developed based on data collected outside the Everglades, which also include sulfate reduction as a major driver, have also been found to make accurate predictions (Chen and Herr 2010, Table 3.3).

3.3.1.7 Verified Predictions: Knowledge of the Causal Agent's Mode of Action Permits Prediction of Unobserved Effects That Can Be Subsequently Confirmed

As mentioned previously, the conceptual model that methylation rate was controlled by the balance between sulfate and sulfide was incorporated into a mitigation strategy developed by the SFWMD (Rumbold and Fink 2006). During the start-up phase of STA-2 in 2000, two of the three flow cells quickly satisfied the start-up criteria (i.e., outflow concentrations < inflow). In contrast, the third cell exhibited anomalously high MeHg concentrations in the water column and in fish despite receiving the same inflows as the other two cells (note, this remains one of the best examples of variability in methylation potential in the Everglades). During start-up phase water-column concentrations in this cell reached 32 ng THg/L and an unprecedented 20 ng MeHg/L. Tissue Hg in resident fishes reached levels as high as 430 ng/g in mosquitofish, *Gambusia holbrooki*, 930 ng/g in sunfish, *Lepomis* spp., and 2000 ng/g in Largemouth bass, *Micropterus salmoides*. The anomalous MeHg production was thought to be the result of transient pump operation that temporarily lowered surface water stage allowing sediments to partially oxidize (and release sulfate) that in turn caused MeHg production to spike. Based on the conceptual model, a recommendation was made to simply increase the control elevation of the weir boxes in front of the outflow culverts to reduce drawdown, thereby stabilizing redox conditions and allow sulfide to build up to inhibitory levels. This strategy was successful as evidenced by subsequent declines in MeHg in both the water column and fish (Rumbold and Fink 2006).

3.3.2 Candidate Cause: Microbes Other Than SRB as Hg Methylators

FDEP/SFWMD have recently raised the possibility that other non-SRB bacteria (i.e., that would not require sulfate) are playing an important role in methylating Hg in the Everglades (Julian et al. 2015a, b, 2016). For example, Julian et al. (2015a, p. 3) recently stated that "*current research also show other microbes capable of producing MeHg not related to sulfate including but not limited to iron (Fe) reducing bacteria, methanogens, and others* (Bae et al. 2014)." Bae et al. (2014) had reported finding hgcA, a gene encoding a protein essential for Hg methylation, in diverse phyla from sediments and water taken from WCA-3A. They further reported that the phylogenetic tree for their gene sequence was in good agreement with the sequence, hgcAB, that had only recently been shown to be good predictor of Hg methylation capability by Gilmour et al. (2013). However, it must be noted that Gilmour et al. (2013) found the rates of methylation and MeHg production to vary substantially among the five clades of bacteria containing the gene. Thus, while other microbes capable of Hg methylation are likely present in the Everglades (i.e., mechanistic probability), they may methylate at very different rates than SRB (for further discussion of syntrophs as possible methylators, see Chap. 2, this volume). Bacteria other than SRB have been known to methylate Hg for some time. Iron-reducing bacteria, for example, were shown to methylate Hg in the field (Warner et al. 2003) and pure culture (Kerin et al. 2006). However, Warner et al. (2003) found methylation rates under Fe-reducing conditions to be lower than those observed in sulfate-reducing sediments. As early as 2005, the USGS ACME team had conducted a mesocosm study in the Everglades to assess the role of iron-reducing bacteria on net methylation by additions of Fe oxyhydroxide and stable Hg isotope (Gilmour et al. 2007a). They found the response to Fe(III) additions to be mixed. Iron enhanced MeHg concentrations in surface waters at low doses but inhibited MeHg production in soils at the highest dose (Gilmour et al. 2007a). Iron-reducing bacteria could explain the observation by Dierberg et al. (2014) that "*non-zero methylation*" was occurring in sediment cores from WCA-3A in the presence of chemical inhibitors that should have inhibited both the SRBs and the methanogens. However, it should be noted that both Gilmour et al. (2007a) and Dierberg et al. (2014) reported SRB were the dominant methylators. More importantly, several investigators report that iron concentrations are very low in most Everglades sediments (Gilmour et al. 1998b; Snyder 1992 as cited by Fink et al. 1999; Qualls et al. 2001; Chambers and Pederson 2006; Julian et al. 2016). The scarcity of iron in Everglades' sediments would appear to be a break in the causal pathway that would lessen the likelihood that iron-reducing bacteria are the cause of the susceptibility and geographical variability of the Everglades Hg problem.

3.3.3 Candidate Cause: Differences in Food Webs Are Responsible for Extreme Geographical and Temporal Variability in Biomagnified Hg

FDEP/SFWMD have also recently raised the possibility that Hg hotspots in the Everglades are due to differences in diet and food web dynamics (Julian and Gu 2015; Julian et al. 2016). For example, Julian and Gu (2015, p. 212) recently stated that "*the architecture of a particular species of aquatic plant may contribute to differences in diets of adult and juvenile LMB, which in turn change the THg level of*

LMB." Similarly, Julian et al. (2016) states "*initial data suggests that food web dynamics and habitat influences mosquitofish THg concentrations.*"

Clearly, diet and trophic position will affect Hg biomagnification within (ontogenetic changes) and among species (for review, see Chaps. 8 and 12, this volume). There are examples in the Everglades where spatial differences in diet for a given species have affected their Hg burden (Cleckner et al. 1998; Loftus 2000, for review, see Chaps. 8 and 12, this volume). The question, however, is whether geographic variation in food webs across the Everglades are responsible for the extreme temporal and geographic variability in the magnitude of biomagnification.

Loftus (2000) found Hg concentrations in various taxa in ENP to be related to trophic position as determined by literature sources, stomach contents and δ15N. He also found that hydroperiod played a role in Hg uptake in ENP; however, its effect varied with season and specific location. USEPA (Stober et al. 2001) directly addressed the idea that diet was responsible for geographical variation in mosquitofish Hg by examining the gut contents of 2784 mosquitofish collected from 259 sites across the Everglades. They found that (pp. 6–23) "*though mosquitofish consume a variety of animal prey, all the animal types had similar trophic scores.*" Furthermore, statistical modeling by USEPA (Stober et al. 2001) revealed regional and temporal variation explained only 13.5% of the total variance in trophic position. Interestingly, mosquitofish from Taylor Slough (which contained relatively low Hg) had a higher trophic score than those from Shark Slough (which had high Hg). USEPA (Stober et al. 2001, pp. 6–24) concluded that "*the hypothesis that trophic position could be used to explain Hg concentration in mosquitofish was not supported.*" Later a study by Kendall et al. (2003) using stable isotopes of nitrogen (δ15N) in mosquitofish as an indicator of food web position (rather than stomach contents) reportedly found no statistically significant difference in δ15N values for fish from high and low nutrient sites. They concluded that there was no evidence for shorter food chains at high nutrient sites. Likewise, Bemis et al. (2003) found that while δ15N was a good local predictor of THg in LMB, when data were pooled across sites, δ15N only explained a small amount of the variance (~5%) in THg. Instead, location (i.e., latitude, ~38%) and fish length (~15%) explained the majority of the variance in Hg (Bemis et al. 2003). Yet Julian and Gu (2015, p. 2011) continue to speculate that "*LMB could potentially reside at a higher trophic position within ENP than other regions in the EvPA as indicated by higher tissue Hg concentrations*". They go on to say that "*unfortunately very little isotope data is available to confirm this hypothesis.*" Stable isotope data are, however, available for biota in ENP (Loftus 2000; Rumbold et al. 2002). Loftus (2000, p. 124) reported that "*δ15N for these Everglades fishes were similar to the ranges for fishes from nearby Lake Okeechobee.*" Rumbold et al. (2002) conducted a survey of THg and stable isotopes in alligators across the Everglades including ENP. Based on their analysis, they concluded that differences in δ15N were not sufficient to account for the elevated THg concentrations in alligators from ENP. More recently, Rumbold et al. (2018) measured THg and δ15N across the entire food web at a hotspot in ENP along the Shark River Estuary that has some of the highest Hg levels in Florida. They used this information to calculate a trophic magnification slope (TMS), from the regression of log Hg concentration versus δ15N (as a measure of the efficiency with which a local food web can biomagnify Hg). The TMS did not differ among reaches of the estuary and, thus, basal Hg entering the food web was thought to more important in determining THg concentration in top predator fish (Rumbold et al. 2018).

The results of Rumbold et al. (2018) are also consistent with earlier TMS studies conducted elsewhere in which spatial variations observed in bioaccumulated Hg may largely be a result of variability in the amount of MeHg entering the base of the food web rather than differences in food web structure (Kidd et al. 2003; for review, see Rigct et al. 2007). While more recent studies have had the power to discern statistical differences in slopes between habitats (Kidd et al. 2012), basal Hg entering the food web often shows greater variability and is a better predictor for Hg concentration in top predatory fish (for review, see Chaps. 7 and 12, this volume).

3.3.4 Candidate Cause: Dissolved Organic Matter (DOM) May Have as Much or More Influence in the Accumulation of Hg Than Sulfate

Several studies done in the Everglades have found that the biogeochemistry of Hg is influenced by DOM in a number of ways, including: (1) complexation and transport; (2) oxidation-reduction reactions; (3) precipitation, aggregation, and dissolution of mineral phases (e.g., metacinnabar); (4) microbial Hg methylation; (5) photodegradation of MeHg and photoreduction of Hg(II); and (6) biouptake of Hg and MeHg and tropic transfer of Hg/MeHg (Table 3.2; for review, see Chap. 4, this volume). With regard to microbial Hg methylation, DOM can serve as a carbon source for methylators and so changes in its concentration or quality (i.e., lability or susceptibility to microbial degradation) can impact the microbial process. As previously discussed, DOM can also serve as a ligand for inorganic Hg affecting its bioavailability (either increasing or decreasing it depending on DOM's molecular weight, aromaticity, or sulfur content). For instance, higher molecular weight DOM with increased aromaticity and which is highly sulfidized (found in sulfidic environments) is thought to enhance Hg bioavailability (for review, see Chap. 4, this volume).

Everglades surface water is naturally high in DOM. Hence, it will always be a potential source of reduced carbon for methylators and other microbes. Its sources and quality in the Everglades have, however, been altered (for review, see Chap. 2, Volume I and Chap. 4, this volume) and, thus, there may be opportunities for management in that regard (for review, see Chap. 4, this volume). Alternatively, in addition to the benefit of starving the SRBs, reducing sulfate contamination could lower the sulfur content of Everglades' DOM thereby reducing its Hg-bioavailability enhancement capabilities.

3.4 Management Relevant Conclusions

The guidance provided by USEPA on assessing causality (http://www.epa.gov/caddis) states that confidence in the argument for or against a candidate cause is increased when many types of evidence consistently support or weaken it. It further suggests that increasing the number of lines of evidence will decrease the likelihood of being misled by any one faulty study or data set. This review presents many types of evidence (e.g., laboratory tests, mesocosms, spatial gradients, temporal gradients, simulation models, verified predictions) each with multiple studies or lines of evidence (Table 3.2) that suggest that it is the balance between sulfate and sulfide that is the key control of net Hg methylation rate and the dominating cause for the extreme temporal and geographic differences in the magnitude of biomagnification in the Everglades. This is consistent with previous reviews of sulfate's role in Hg methylation in many ecosystems (Munthe et al. 2007). Although a few studies presented here purport to refute this cause, as discussed above, these lines of evidence were weakened for several reasons. Furthermore, as previously stated, assigning causality must be based on the totality of evidence not on any one study or line of evidence (Suter 2007). In this case, the consistency of the evidence from the system (Table 3.2) in combination with the consistency of association shown in studies done outside of the Everglades (Table 3.3) convincingly supports the case for sulfate. Although other candidate causes were not refuted, the evidence suggests sulfate was the leading cause. Thus, while high Hg deposition was judged to be the likely principal cause, it was not the sole cause of the Everglades susceptibility to high rates of Hg biomagnification nor its extreme geographical and temporal variability. Instead, sulfate contamination and the balance between the resulting sulfate stimulation and sulfide inhibition of Hg methylation was the probable cause. Notice that there is no assertion of definitive proof of causation. As stated by Suter et al. (2010, p. 28) "*as pragmatists, we sidestep the impossible goal of proving causation in real cases and instead attempt to develop an adequate body of evidence to justify taking action to remediate the most likely cause*."

Arguments by FDEP/SFWMD Against Controlling Sulfate to Mitigate Florida's Hg Problem

FDEP/SFWMD have argued against attempting to control sulfate (to starve the SRB) as a means to remediate the Hg problem in the Everglades. They argue that remaining uncertainties must be resolved first and identify the following interrelated questions that must be answered (Julian et al. 2014):

1. Does water column sulfate influence fish tissue Hg concentrations in a predictable manner?
2. Do other water quality parameters affect fish tissue Hg concentrations?
3. Are sulfate-reducing bacteria the sole Hg methylators or are there other bacteria groups involved?
4. Can background inputs of sulfate maintain the production of MeHg?

Drawing on the results of this formal causal analysis, these questions will each be addressed, but in reverse order.

Can Background Inputs of Sulfate Maintain the Production of MeHg?
Based on their laboratory studies (in 100 ml incubation vessels), Dierberg et al. (2014) found increased MeHg concentrations following relatively low sulfate additions (0.5–1.0 mg/L) and concluded that "*non-abatable sources of sulfate to the Everglades could support meaningful MeHg production.*" This conclusion was accepted and re-stated by FDEP/SFWMD in the state's annual report (Julian et al. 2015b). However, their acceptance of this conclusion based on bench-scale laboratory study is difficult to reconcile with previous cautions by FDEP/SFWMD regarding scaling up conclusions based on small scale tests to the entire Everglades. For example, when dismissing conclusions of Gilmour (2011) based on results from mesocosms, Julian et al. (2014, p. 3B-47) wrote "*like most mesocosm studies, extrapolating this information into actual ecosystem dynamics is laden with assumptions and often times not scalable.*" That inconsistency aside, the larger issue is what Dierberg et al. (2014) meant by "meaningful MeHg production". There is a huge distinction between measurable Hg methylation (which is almost ubiquitous in the environment) to the magnitude of net methylation necessary to sustain a biological hotspot. In further support of their conclusion that background sulfate levels could stimulate significant Hg methylation, FDEP/SFWMD (Julian et al. 2015b, p. 3B-62) point to results of statistical models that show "*several instances of relatively high fish THg levels in both high (>30 mg/L) and low (<1 mg/L) sulfate conditions.*" As previously discussed, these "several instances" among 100s of paired data may simply be the result of a mismatch in measurement time scales. Care must also be taken as to which data sets are used to identify areas of sulfate limitation. As discussed in greater detail in Chap. 2, this volume, sulfate contamination of the Everglades is widespread and FDEP/SFWMD focus has been on monitoring the nutrient impacted areas rather than attempting to identify areas where sulfate was limiting. The last major study of Hg biogeochemistry undertaken by SFWMD (Julian and Gu 2016) focused on repeating studies done by USGS. Julian and Gu (2016, p. 3B-1-2) wrote that the "*sampling locations were parsed into hotspot and non-hotspot locations.*" However, the selected sites were in STA-1w (ENR302), WCA-2A (F1 and U3) and WCA-3 (WCA315). The previously studied non-hotspots were known to have high sulfides (i.e., low MeHg due to inhibition).

Based on the evidence presented here on spatial and temporal gradients (Table 3.2), it is clear that cold spots (i.e., where local biota have much lower Hg) have and continue to be documented in the Everglades where low sulfate is limiting Hg methylation.

Are Sulfate-Reducing Bacteria the Sole Hg Methylators or Are There Other Bacteria Groups Involved?
As discussed above, FDEP/SFWMD have recently repeatedly raised the possibility that other non-SRB bacteria are playing an important role in methylating Hg in the Everglades (Julian et al. 2015a, b, 2016). The evidence presented here from laboratory and mesocosm studies clearly show that other microbes are capable of Hg

methylation (Gilmour et al. 2007a, 2013); however, most investigators agree that SRB are the dominant methylators of Hg in the Everglades (Gilmour et al. 2007a; Dierberg et al. 2014). Furthermore, the presence of cold spots or low Hg areas within the spatial and temporal sulfate gradients discussed above would also suggest that other microbes capable of methylation (in particular the methanogens) do not ascend as important methylators when SRB are limited.

Do Other Water Quality Parameters Affect Fish Tissue Hg Concentrations? Julian et al. (2015b, p. 3B-62) state that "*too many factors influence the biogeochemistry of* SO_4^{2-} *and Hg in natural ecosystems to allow any consistent and identifiable areas of high and low fish THg across the spectrum of surface water* SO_4^{2-} *concentrations. Therefore, without the ability to quantitatively link surface water* SO_4^{2-} *and ambient MeHg and THg in fish tissue, an ecosystem-wide sulfur strategy as a management approach to reduce MeHg risk in the EPA is not warranted.*" There is no disputing that Hg methylation and bioaccumulation is a function of numerous interdependent processes that form the complex of circumstances that increases the probability of Hg methylation by SRB. Therefore, simple models based on a single variable may not be sufficiently rigorous to have predictive power. Because MeHg cycling and trophic transfer are processes that involve multiple variables, simple models can suffer from model misspecification. Model misspecification can occur when the actual form of the relationship between the independent and dependent variable violates the underlying assumptions of the statistical model; it can also arise when important independent variables are excluded from the model. Moreover, the relationship between, for example, sulfate and Hg accumulation in fish is not wholly direct but is mediated through the effect of sulfate on other variables. Recognition of these led Pollman (2014) to use structural equation modeling (SEM) to model both surface water MeHg concentrations and mosquitofish Hg concentrations in the Everglades as a function of surface water sulfate and dissolved organic carbon, porewater sulfide, sediment organic matter content, and sediment THg concentrations as the driving external variables. Underlying model assumptions were formally tested by Pollman (2014) to help minimize model misspecification problems and the analysis demonstrated that the effects of sulfate and sulfide dynamics on mosquitofish Hg concentrations were more important than DOC or sediment THg and organic matter content.

Multivariate techniques have been employed to assess the relative importance of various divers in Hg accumulation in other regions, such as New Hampshire and Vermont lakes (Kamman et al. 2004; Shanley et al. 2012), in San Francisco Bay (Luengen and Flegal 2009), and in North Carolina (Sackett et al. 2009) and South Carolina (Glover et al. 2010). Moreover, while many of these relationships may be intrinsically linear and may be assessed using linear regression models after appropriate transformations, segmented or broken-line regression should be considered to better assess the non-linearity in several of the processes.

Does Water Column Sulfate Influence Fish Tissue Hg Concentrations in a Predictable Manner?

FDEP/SFWMD (Julian et al. 2014, p. 3B-49) state that "*any water quality criterion must be supported by strong and consistent field relationships; the sulfur and MeHg linkage is not directly quantifiable or reasonably predictable.*" They often highlight the variability in statistical models attempting to link surface water sulfate and Hg in fish. Commenting on an analysis by Corrales et al. (2011, which was done in support of a sulfate water quality target), Julian et al. (2014, p. 3B-47) wrote "*Figure 3B-20 is a quartic regression line; while significant from a statistical viewpoint, it has little utility for environmental decision-making due to the very low R2 values.*" When they again critiqued the results of that analysis, Julian et al. (2015b, p. 3B-64) wrote "*the low R2 values also imply the complexity of the Everglades ecosystem, underscoring that no one-variable empirical model can capture the complexities of the Everglades Hg cycle. Overall, these data provide no evidence that reducing* SO_4^{2-} *concentrations can be expected to lower Hg bioaccumulation in fish.*"

As previously stated, our predictive modeling efforts should not be limited to a "*one-variable empirical model*"; however, caution must be used in developing more complex models. There are several problems with the model offered by Julian et al. (2014) that was shown not to have predictive power. The most immediate are related to model misspecification: (1) no underlying basis for their use of a fourth-order polynomial; and (2) their choice to model the ratio of surface water MeHg to THg rather than modeling MeHg concentrations directly. This latter fact unnecessarily complicates the interpretation of the dependent variable and makes the application of their results to critique other analyses problematic (cf., Pollman and Axelrad 2014a, b). In addition, fitting the fourth-order polynomial as implemented by Julian et al. (2014) is unduly influenced by leverage exerted by small numbers of observations at the higher end of the sulfate distribution, resulting in nonsensical, spurious functional aspects to the sulfate relationship. Further, it also suffers from the failure to include other variables of importance, which by their inclusion, would change the fitted relationship between sulfate and the response variable (and can strengthen its significance as well).

Most importantly, as discussed previously, sulfate reduction and Hg methylation are dynamic processes that can change rapidly over space and time. Where spatial and temporal resolution of measurements are poor (e.g., using surface water concentrations collected on a quarterly basis as a proxy for porewater sulfide at any given moment) the mismatch in the measurement scales will generate considerable variability. If we are truly committed to improving the predictive power of our models to allow us to develop and implement strategies to control the Everglades' Hg problem, then we must apply new and improved technologies for measuring these processes at appropriate spatial scales and temporal rates (e.g., Vorenhout et al. 2011; Superville et al. 2013). Ultimately, the question boils down to the resources that managers are willing to devote to reduce uncertainty. However, we must also be cautious of falling into an endless loop of attempting to satisfy managers that cry "*we need more science*" as an excuse for not taking any action.

3.5 Final Thoughts

In his seminal address, Sir Austin Bradford Hill (1965, p. 300) wrote the following, likely due to the frustration of witnessing the counter arguments made over decades against the link between smoking and cancer: "*All scientific work is incomplete—whether it be observational or experimental. All scientific work is liable to be upset or modified by advancing knowledge. That does not confer upon us a freedom to ignore the knowledge we already have, or to postpone the action that it appears to demand at a given time.*"

References

Adams SM (2003) Establishing causality between environmental stressors and effects on aquatic ecosystems. Hum Ecol Risk Assess 9:17–35

Aiken G, Haitzer M, Ryan JN, Nagy K (2003) Interactions between dissolved organic matter and mercury in the Florida Everglades. Journal du Physique, IV 107:29–32

Atkeson TD, Axelrad DM (2004) Chapter 2B: Mercury monitoring, research and environmental assessment. In: 2004 Everglades consolidated report. South Florida Water Management District, West Palm Beach, FL, pp 2B:1–28

Atkeson TD, Parks P (2001) Chapter 7: Everglades mercury problem. In: 2001 Everglades consolidated report. South Florida Water Management District, West Palm Beach, FL

Axelrad DM et al (2005) Chapter 2B: Mercury monitoring, research and environmental assessment in South Florida. In: 2005 South Florida environmental report—volume I. South Florida Water Management District, West Palm Beach, FL

Axelrad DM et al (2006) Chapter 2B: Mercury monitoring, research and environmental assessment in South Florida. In: 2006 South Florida environmental report—volume I. South Florida Water Management District, West Palm Beach, FL

Axelrad DM et al (2007) Chapter 3B: Mercury monitoring, research and environmental assessment. In: 2007 South Florida environmental report—volume I. South Florida Water Management District, West Palm Beach, FL

Axelrad DM et al (2013) Chapter 3B: Mercury and sulfur environmental assessment for the Everglades. In: 2013 South Florida environmental report. South Florida Water Management District, West Palm Beach, FL

Bae H, Dierberg FE, Ogram A (2014) Syntrophs dominate sequences associated with the mercury-methylating gene hgcA in the water conservation areas of the Florida Everglades. Appl Environ Microbiol 80:6517–6526

Bailey LT, Mitchell CP, Engstrom DR, Berndt ME, Wasik JKC, Johnson NW (2017) Influence of porewater sulfide on methylmercury production and partitioning in sulfate-impacted lake sediments. Sci Total Environ 580:1197–1204

Bates AL, Spiker EC, Holmes CW (1998) Speciation and isotopic composition of sedimentary sulfur in the Everglades, Florida, USA. Chem Geol 146:155–170

Bates AL et al (2002) Tracing sources of sulfur in the Florida Everglades. J Environ Qual 31:287–299

Bemis BE, Kendall C, Lange T, Campbell L (2003) Using nitrogen and carbon isotopes to explain mercury variability in Largemouth bass. Greater Everglades Ecosystem Restoration (GEER) Meeting, April 2003, Palm Harbor, FL. Program and Abstracts

Benoit JM, Gilmour CC, Mason RP, Heyes A (1999a) Sulfide controls on mercury speciation and bioavailability in sediment pore waters. Environ Sci Technol 33:951–957

Benoit JM, Mason RP, Gilmour CC (1999b) Estimation of mercury-sulfide speciation in sediment pore waters using octanol-water partitioning and implications for availability to methylating bacteria. Environ Toxicol Chem 18:2138–2141

Benoit J, Gilmour C Heyes A et al (2003) Geochemical and biological controls over methylmercury production and degradation in aquatic ecosystems. In: Chai Y, Braids OC (eds) Biogeochemistry of environmentally important trace elements, ACS symposium series #835, American Chemical Society, Washington, DC, pp 262–297

Berman M, Bartha R (1986) Control of the methylation process in a mercury-polluted aquatic sediment. Environ Pollut B 11:41–53

Branfireun BA, Roulet NT, Kelly C, Rudd JW (1999) In situ sulphate stimulation of mercury methylation in a boreal peatland: toward a link between acid rain and methylmercury contamination in remote environments. Glob Biogeochem Cycles 13:743–750

Brown CL, Parchaso F, Thompson JK, Luoma SN (2003) Assessing toxicant effects in a complex estuary: a case study of effects of silver on reproduction in the bivalve, *Potamocorbula amurensis*, in San Francisco Bay. Hum Ecol Risk Assess 9:95–119

Chambers RM, Pederson KA (2006) Variation in soil phosphorus, sulfur, and iron pools among South Florida wetlands. Hydrobiologia 569:63–70

Chen CW, Herr JW (2010) Simulating the effect of sulfate addition on methylmercury output from a wetland. J Environ Eng 136:354–362

Cleckner LB, Garrison PJ, Hurley JP, Olson ML, Krabbenhoft DP (1998) Trophic transfer of methyl mercury in the Northern Florida Everglades. Biogeochemistry 40:347–361

Cleckner LB, Gilmour CC, Hurley JP, Krabbenhoft DP (1999) Mercury methylation in periphyton of the Florida Everglades. Limnol Oceanogr 44:1815–1825

Coleman Wasik JK et al (2012) Methylmercury declines in a boreal peatland when experimental sulfate deposition decreases. Environ Sci Technol 46:6663–6671. https://doi.org/10.1021/es300865f

Compeau G, Bartha R (1984) Methylation and demethylation of mercury under controlled redox, pH and salinity conditions. Appl Environ Microbiol 48:1203–1207

Compeau GC, Bartha R (1985) Sulfate-reducing bacteria: principal methylators of mercury in anoxic estuarine sediment. Appl Environ Microbiol 50:498–502

Corrales J, Naja GM, Dziuba C, Rivero RG, Orem W (2011) Sulfate threshold target to control methylmercury levels in wetland ecosystems. Sci Total Environ 409:2156–2162

Craig P, Moreton P (1983) Total mercury, methyl mercury and sulphide in River Carron sediments. Mar Pollut Bull 14:408–411

Dierberg FE, DeBusk TA, Jerauld M, Gu B (2014) Appendix 3B-1: evaluation of factors influencing methylmercury accumulation in South Florida marshes. In: 2014 South Florida environmental report. South Florida Water Management District, West Palm Beach, FL

Drevnick PE et al (2007) Deposition and cycling of sulfur controls mercury accumulation in Isle Royale Fish. Environ Sci Technol 41:7266–7272

Dvonch JT et al (1998) An investigation of source-receptor relationships for mercury in South Florida using event precipitation data. Sci Total Environ 213:95–108

Exponent (formerly PTI) (1998) Ecological risks to wading birds of the Everglades in relation to phosphorus reductions in water and mercury bioaccumulation in fishes. Prepared for the Sugar Cane Growers Cooperative of Florida

Fink L (2004) Appendix 2B-6: STA-6 mercury special studies interim report. In: 2004 Everglades consolidated report. SFWMD, West Palm Beach, FL

Fink L, Rawlik P (2000) Chapter 7: The Everglades mercury problem. In: 2000 Everglades consolidated report. South Florida Water Management District, West Palm Beach, FL, January

Fink L, Rumbold DG, Rawlik P (1999) Chapter 7: The Everglades mercury problem. Everglades interim report. South Florida Water Management District, West Palm Beach, FL

Fox GA (1991) Practical causal inference for ecoepidemiologists. J Toxicol Environ Health 33:359–373

Frederick PC, Spalding MG, Dusek R (2002) Wading birds as bioindicators of mercury contamination in Florida: annual and geographic variation. Environ Toxicol Chem 21:262–264

Frederick PC, Hylton B, Heath JA, Spalding MG (2004) A historical record of mercury contamination in Southern Florida (USA) as inferred from avian feather tissue. Environ Toxicol Chem 23:1474–1478

Gabriel M, Redfield G, Rumbold D (2008) Appendix 3B-2: sulfur as a regional water quality concern. In: 2008 South Florida environmental report volume 1. South Florida Water Management District, West Palm Beach, FL

Gabriel MC, Howard N, Osborne TZ (2014) Fish mercury and surface water sulfate relationships in the Everglades protection area. Environ Manag 53:583–593

Gabriel MC, Axelrad D, Orem W, Osborne TZ (2015) Response to Julian et al. (2015) "Comment on and reinterpretation of Gabriel et al. (2014) 'Fish mercury and surface water sulfate relationships in the Everglades protection area'". Environ Manag 55:1227–1231

Germain G (2014) Appendix 3-1: annual permit report for Everglades stormwater treatment areas. In: 2014 South Florida environmental report—volume 1. South Florida Water Management District, West Palm Beach, FL

Gilmour CC (2011) A review of the literature on the impact of sulfate on methylmercury in sediments and soils. Prepared for Florida Department of Environmental Protection, Tallahassee, FL

Gilmour CC, Henry EA (1991) Mercury methylation in aquatic systems affected by acid deposition. Environ Pollut 71(2–4):131–169

Gilmour CC, Krabbenhoft DP (2001) Appendix 7-4: status of methylmercury production studies. In: 2001 Everglades consolidated report. South Florida Water Management District, West Palm Beach, FL

Gilmour CC, Henry EA, Mitchell R (1992) Sulfate stimulation of mercury methylation in freshwater sediments. Environ Sci Technol 26:2281–2287

Gilmour CC, Heyes A, Benoit JM, Riedel GS, Bell JT (1998a) Distribution and biogeochemical control of mercury methylation in the Florida Everglades. Report to the SFWMD. Contract #C-7690, West Palm Beach, FL, 38 p

Gilmour CC, Riedel GS et al (1998b) Methylmercury concentrations and production rates across a trophic gradient in the Northern Everglades. Biogeochemistry 40:327–345

Gilmour CC, Krabbenhoft DP, Orem WO (2004a) Appendix 2B-3: mesocosm studies to quantify how methylmercury in the Everglades responds to changes in mercury, sulfur, and nutrient loading. In: 2004 Everglades consolidated report. South Florida Water Management District, West Palm Beach, FL

Gilmour CC, Krabbenhoft DP, Orem W, Aiken G (2004b) Appendix 2B-1: influence of drying and rewetting on mercury and sulfur cycling in Everglades and STA soils. In: 2004 Everglades consolidated report. South Florida Water Management District, West Palm Beach, FL, 19 p

Gilmour CC, Krabbenhoft D, Orem W, Aiken G, Roden E (2007a) Appendix 3B-2: status report on ACME studies on the control of mercury methylation and bioaccumulation in the Everglades. In: 2007 South Florida environmental report—volume I. South Florida Water Management District

Gilmour CC, Orem W, Krabbenhoft D, Mendelssohn IA (2007b) Appendix 3B-3. preliminary assessment of sulfur sources, trends and effects in the Everglades. In: 2007 South Florida environmental report—volume I. South Florida Water Management District, West Palm Beach, FL

Gilmour CC, Roden E, Harris R (2008) Appendix 3B-3: approaches to modeling sulfate reduction and methylmercury production in the Everglades. In: 2008 South Florida environmental report—volume I. South Florida Water Management District

Gilmour CC, Podar M et al (2013) Mercury methylation by novel microorganisms from new environments. Environ Sci Technol 47:11810–11820

Glover JB et al (2010) Mercury in South Carolina fishes, USA. Ecotoxicology 19:781–795

Gu B, Axelrad DM, Lange T (2012) Chapter 3B: Regional mercury and sulfur monitoring and environmental assessment. In: 2012 South Florida environmental report—volume I. South Florida Water Management District, West Palm Beach, FL

Guentzel J, Landing W, Gill G, Pollman C (1995) Atmospheric deposition of mercury in Florida: the FAMS project (1992–1994). Water Air Soil Pollut 80:393–402

Hammerschmidt CR, Fitzgerald WF (2004) Geochemical controls on the production and distribution of methylmercury in near-shore marine sediments. Environ Sci Technol 38:1487–1495

Hammerschmidt C, Fitzgerald W, Lamborg C, Balcom P, Visscher P (2004) Biogeochemistry of methylmercury in sediments of Long Island Sound. Mar Chem 90:31–52

Harmon S, King J, Gladden J, Chandler GT, Newman L (2004) Methylmercury formation in a wetland mesocosm amended with sulfate. Environ Sci Technol 38:650–656

Harris R, Pollman CD, Hutchinson D, Beal D (2001) Appendix 7-3: status of Everglades mercury cycling model. In: 2001 Everglades consolidated report. South Florida Water Management District, West Palm Beach, FL

Hill AB (1965) The environment and disease: association or causation? Proc R Soc Med 58:295–300

Hrabik TR, Watras CJ (2002) Recent declines in mercury concentration in a freshwater fishery: isolating the effects of de-acidification and decreased atmospheric mercury deposition in Little Rock Lake. Sci Total Environ 297:229–237

Hurley JP, Krabbenhoft DP, Cleckner LB, Olson ML, Aiken GR, Rawlik PS (1998) System controls on the aqueous distribution of mercury in the Northern Florida Everglades. Biogeochemistry 40:293–311

Husar JD, Husar RB (2002) Trends of anthropogenic mercury mass flows and emissions in Florida. FDEP final report, PO# S3700 303975, pp 1–74

Jay JA, Morel FM, Hemond HF (2000) Mercury speciation in the presence of polysulfides. Environ Sci Technol 34:2196–2200

Jerauld M, Dierberg FE, DeBusk WF, DeBusk TA (2015) Appendix 3B-1: evaluation of factors influencing methylmercury accumulation in South Florida Marshes. In: 2015 South Florida environmental report. South Florida water Management District, West Palm Beach, FL

Jeremiason JD et al (2006) Sulfate addition increases methylmercury production in an experimental wetland. Environ Sci Technol 40:3800–3806

Johnson NW, Mitchell CP, Engstrom DR, Bailey LT, Wasik JKC, Berndt ME (2016) Methylmercury production in a chronically sulfate-impacted sub-boreal wetland. Environ Sci Process Impacts 18:725–734

Jordan JL, Armstrong NE, Burger J, Burkholder J, Hsieh YP, Meganck R, Donk EV, Ward R (2007) Appendix 1A-5: final report of the peer review panel concerning the 2007 South Florida environmental report. In: 2007 South Florida environmental report. South Florida Water Management District, West Palm Beach, FL

Julian P (2013) Mercury bio-concentration factor in mosquito fish (*Gambusia* spp.) in the Florida Everglades. Bull Environ Contam Toxicol 90:329–332

Julian P (2014) Reply to "Mercury bioaccumulation and bioaccumulation factors for everglades mosquitofish as related to sulfate: a re-analysis of Julian II (2013)". Bull Environ Contam Toxicol 93:517

Julian P, Gu B (2015) Mercury accumulation in largemouth bass (*Micropterus salmoides* Lacépède) within marsh ecosystems of the Florida Everglades, USA. Ecotoxicology 24:202–214

Julian P, Gu B (2016) Appendix 3B-1: Everglades mercury hotspot study: preliminary analysis. In: 2016 South Florida environmental report. South Florida Water Management District, West Palm Beach, FL

Julian P, Gu B et al (2014) Chapter 3B: Mercury and sulfur environmental assessment for the Everglades. In: 2014 South Florida environmental report. South Florida Water Management District, West Palm Beach, FL

Julian P, Gu B, Redfield G (2015a) Comment on and reinterpretation of Gabriel et al. 2014a. 'Fish mercury and surface water sulfate relationships in the Everglades Protection Area'. Environ Manag 55:1–5

Julian P, Gu B, Redfield G, Weaver K et al (2015b) Chapter 3B: Mercury and sulfur environmental assessment for the Everglades. In: 2015 South Florida environmental report. South Florida Water Management District, West Palm Beach, FL

Julian P, Gu B, Redfield G, Weaver K (2016) Chapter 3B: Mercury and sulfur environmental assessment for the Everglades. In: 2016 South Florida environmental report. South Florida Water Management District, West Palm Beach, FL

Kamman NC, Lorey PM, Driscoll CT, Estabrook R, Major A, Pientka B, Glassford E (2004) Assessment of mercury in waters, sediments, and biota of New Hampshire and Vermont Lakes, USA, sampled using a geographically randomized design. Environ Toxicol Chem 23:1172–1186

Kendall C, Bemis BE, Trexler J, Lange T, Stober JQ (2003) Is food web structure a main control on mercury concentrations in fish in the Everglades? Greater Everglades ecosystem restoration (GEER) meeting, April 2003, Palm Harbor, FL. Program and Abstracts

Kerin E et al (2006) Mercury methylation among the dissimilatory iron-reducing bacteria. Appl Environ Microbiol 72:7912–7921

Kidd KA, Bootsma HA, Hesslein RH, Lyle Lockhart W, Hecky RE (2003) Mercury concentrations in the food web of Lake Malawi, East Africa. J Great Lakes Res 29:258–266

Kidd KA, Muir DC, Evans MS, Wang X, Whittle M, Swanson HK, Johnston T, Guildford S (2012) Biomagnification of mercury through lake trout (*Salvelinus namaycush*) food webs of lakes with different physical, chemical and biological characteristics. Sci Total Environ 438:135–143

King JK, Kostka JE, Frischer ME, Saunders FM (2000) Sulfate-reducing bacteria methylate mercury at variable rates in pure culture and in marine sediments. Appl Environ Microbiol 66:2430–2437

Krabbenhoft DP, Fink L (2001) Appendix 7-8: the effect of dry down and natural fires on mercury methylation in the Florida Everglades. In: 2001 Everglades consolidated report. South Florida Water Management District, West Palm Beach, FL, 14 p

Krabbenhoft DP, Hurley JP, Olson ML, Cleckner LB (1998a) Diel variability of mercury phase and species distributions in the Florida Everglades. Biogeochemistry 40:311–325

Krabbenhoft DP et al (1998b) Methylmercury dynamics in littoral sediments of a temperate seepage lake. Can J Fish Aquat Sci 55:835–844

Krabbenhoft D et al (2001) Mercury cycling in the Florida Everglades: a mechanistic study. Verh Internat Verein Limnol 27:1657–1660

Krabbenhoft D, Orem W, Aiken G, Gilmour C (2004) Unraveling the complexities of mercury methylation in the Everglades: the use of mesocosms to test the effects of "new" mercury, sulfate, and organic carbon. In: 7th international conference on mercury as a global pollutant, June 27–July 2, 2004, Ljubljana, Slovenia

Krabbenhoft DP et al (2010) The influence of canal water releases on the distribution of mercury, methylmercury, sulfate and dissolved organic carbon in Everglades National Park: implications for ecosystem restoration. Greater Everglades Ecosystem Restoration (GEER) Meeting, July 2010, Naples, FL. Program and Abstracts

Landing WW et al (1995) Relationships between the atmospheric deposition of trace elements, major ions, and mercury in Florida: the FAMS project (1992–1993). Water Air Soil Pollut 80:343–352

Lange TR, Richard DA, Royals HE (1999) Trophic relationships of mercury bioaccumulation in fish from the Florida Everglades. Final annual report. Florida game and fresh water fish commission, fisheries research laboratory, Eustis, FL. Prepared for the Florida Department of Environmental Protection, Tallahassee, FL

Lange TR, Richard DA, Royals HE (2000) Long-term trends of mercury bioaccumulation in Florida's largemouth bass. In: Abstracts of the annual meeting of the South Florida mercury science program, Tarpon Springs, FL, 8–11 May 2000

Lange TR, Richard DA, Sargent B (2005) Annual fish mercury monitoring report, august 2005. Long-term monitoring of mercury in largemouth bass from the Everglades and peninsular Florida. Florida Fish and Wildlife Conservation Commission, Eustis, FL
Langer CS, Fitzgerald WF, Visscher PT, Vandal GM (2001) Biogeochemical cycling at Barn Island salt marsh, Stonington, CT, USA. Wetl Ecol Manag 9:295–310
Lilienfeld AM, Lilienfeld AB (1980) Foundations of epidemiology, 2nd edn. Oxford University Press, New York
Loftus WF (2000) Accumulation and fate of mercury in an Everglade aquatic food web. PhD dissertation, Florida International University, Miami, FL
Luengen AC, Flegal AR (2009) Role of phytoplankton in mercury cycling in the San Francisco Bay estuary. Limnol Oceanogr 54:23–40
Marvin-DiPasquale M, Agee J, Bouse R, Jaffe B (2003) Microbial cycling of mercury in contaminated pelagic and wetland sediments of San Pablo Bay, California. Environ Geol 43:260–267
McPherson BF, Miller RL, Sobczak R, Clark C (2003) Water quality in Big Cypress National Preserve and Everglades National Park, 1960–2000. U.S. Geological Survey Fact Sheet FS-097-03
Meyer MW, Rasmussen PW, Watras CJ, Fevold BM, Kenow KP (2011) Bi-phasic trends in mercury concentrations in blood of Wisconsin common loons during 1992–2010. Ecotoxicology 20:1659–1668
Miller CL, Mason RP, Gilmour CC, Heyes A (2007) Influence of dissolved organic matter on the complexation of mercury under sulfidic conditions. Environ Toxicol Chem 26:624–633
Mitchell CP, Branfireun BA, Kolka RK (2008) Assessing sulfate and carbon controls on net methylmercury production in peatlands: an in situ mesocosm approach. Appl Geochem 23:503–518
Munthe J et al (2007) Recovery of mercury-contaminated fisheries. AMBIO J Hum Environ 36:33–44
Muresan B, Cossa D, Jézéquel D, Prévot F, Kerbellec S (2007) The biogeochemistry of mercury at the sediment–water interface in the Thau lagoon. 1. Partition and speciation. Estuar Coast Shelf Sci 72:472–484
Myrbo A, Swain E, Johnson N, Engstrom D, Pastor J, Dewey B, Monson P, Brenner J, Dykhuizen Shore M, Peters E (2017) Increase in nutrients, mercury, and methylmercury as a consequence of elevated sulfate reduction to sulfide in experimental wetland mesocosms. J Geophys Res Biogeo 122:2769–2785
Naftz DL, Cederberg JR, Krabbenhoft DP, Beisner KR, Whitehead J, Gardberg J (2011) Diurnal trends in methylmercury concentration in a wetland adjacent to Great Salt Lake, Utah, USA. Chem Geol 283:78–86
Nimick DA, Gammons CH, Parker SR (2011) Diel biogeochemical processes and their effect on the aqueous chemistry of streams: a review. Chem Geol 283:3–17
Ogden J, Robertson W Jr, Davis G, Schmidt T (1974) Pesticides, polychlorinated biphenyls and heavy metals in upper food chain levels, Everglades National Park and vicinity. South Florida environmental project: ecological report no. DI-SFEP-74-16, National Technical Information Service, US Department of Commerce
Orem WH (2004) Impacts of sulfate contamination on the Florida Everglades ecosystem. USGS Fact Sheet FS 109-03
Orem WH, Lerch HE, Rawlik P (1997) Geochemistry of surface and pore water at USGS coring sites in wetlands of South Florida: 1994 and 1995. U.S. Geological Survey Open-File Report 97–454, pp 36–39
Orem W, Gilmour C et al (2011) Sulfur in the South Florida ecosystem: distribution, sources, biogeochemistry, impacts and management for restoration. Crit Rev Environ Sci Technol 41:249–288
Patrick Jr W, Gambrell R, Parkpian P, Tan F (1994) Mercury in soils and plants in the Florida Everglades sugarcane area. Mercury pollution: integration and synthesis. Lewis Publishers, Boca Raton, FL, p 609

Pirrone N, Allegrini I, Keeler GJ, Nriagu JO, Rossmann R, Robbins JA (1998) Historical atmospheric mercury emissions and depositions in North America compared to mercury accumulations in sedimentary records. Atmos Environ 32:929–940

Podar M et al (2015) Global prevalence and distribution of genes and microorganisms involved in mercury methylation. Sci Adv 1:e1500675

Pollman CD (2012) Modeling sulfate and gambusia mercury relationships in the Everglades—final report. Florida Department of Environmental Protection, Tallahassee, FL. Aqua Lux Lucis, Gainesville, FL

Pollman CD (2014) Mercury cycling in aquatic ecosystems and trophic state-related variables—implications from structural equation modeling. Sci Total Environ 499:62–73

Pollman CD, Axelrad DM (2014a) Mercury bioaccumulation and bioaccumulation factors for everglades mosquitofish as related to sulfate: a re-analysis of Julian II (2013). Bull Environ Contam Toxicol 93:509–516

Pollman CD, Axelrad DM (2014b) Mercury bioaccumulation factors and spurious correlations. Sci Total Environ 496:vi–xii

Pollman CD et al (1995) Overview of the Florida Atmospheric Mercury Study (FAMS). Water Air Soil Pollut 80:285–290

Porcella D, Zillioux E, Grieb T, Newman J, West G (2004) Retrospective study of mercury in raccoons (*Procyon lotor*) in South Florida. Ecotoxicology 13:207–221

Qualls RG, Richardson CJ, Sherwood LJ (2001) Soil reduction-oxidation potential along a nutrient-enrichment gradient in the Everglades. Wetlands 21:403–411

Ravichandran M, Aiken GR, Reddy MM, Ryan JN (1998) Enhanced dissolution of cinnabar (mercuric sulfide) by dissolved organic matter isolated from the Florida Everglades. Environ Sci Technol 32:3305–3311

Renner R (2001) Everglades mercury debate. Environ Sci Technol 35:59A–60A

Riget F et al (2007) Transfer of mercury in the marine food web of West Greenland. J Environ Monit 9:877–883

Roelke ME, Schultz DP, Facemire CF, Sundlof SF, Royals HE (1991) Mercury contamination in Florida panthers. Report of the Florida Panther Technical Subcommittee to the Florida Panther Interagency Committee

Rood B, Gottgens J, Delfino J, Earle C, Crisman T (1995) Mercury accumulation trends in Florida Everglades and savannas marsh flooded soils. In: Mercury as a global pollutant. Springer, Berlin, pp 981–990

Rumbold DG (2005a) Optimization of the District's mercury monitoring networks: a proposed strategy. South Florida Water Management District, West Palm Beach, FL. Dated 17 Feb 2005

Rumbold DG (2005b) A probabilistic risk assessment of the effects of methylmercury on great egrets and bald eagles foraging at a constructed wetland in South Florida relative to the Everglades. Hum Ecol Risk Assess 11:365–388

Rumbold D, Fink L (2003a) Report on project HGOS: program to monitor concentrations of total mercury and methylmercury in surface water at various water control structures and marshes in South Florida. South Florida Water Management District, West Palm Beach, FL, 12 p

Rumbold DG, Fink L (2003b) Annual permit compliance monitoring report for mercury in downstream receiving waters of the Everglades Protection Area. Appendix 2B-3 in 2003 Everglades consolidated report. South Florida Water Management District, West Palm Beach, FL

Rumbold DG, Fink LE (2006) Extreme spatial variability and unprecedented methylmercury concentrations within a constructed wetland. Environ Monit Assess 112:115–135

Rumbold DG, Fink L, Laine K, Matson F, Niemczyk S, Rawlink P (2001a) Appendix 7-9: annual permit compliance monitoring report for mercury in stormwater treatment areas and downstream receiving waters of the Everglades Protection Area. In: 2001 Everglades consolidated report. South Florida Water Management District, West Palm Beach, FL

Rumbold DG, Niemczyk SL, Fink LE, Chandrasekhar T, Harkanson B, Laine KA (2001b) Mercury in eggs and feathers of great egrets (*Ardea albus*) from the Florida Everglades. Arch Environ Contam Toxicol 41:501–507

Rumbold DG et al (2002) Levels of mercury in alligators (Alligator mississippiensis) collected along a transect through the Florida Everglades. Sci Total Environ 297:239–252

Rumbold DG et al (2006) Appendix 4-4: annual permit compliance monitoring report for mercury in stormwater treatment areas. In: 2006 South Florida environmental report, volume 1. South Florida Water Management District, West Palm Beach, FL

Rumbold DG et al (2007) Appendix 5-5: annual permit compliance monitoring report for mercury in stormwater treatment areas. In: 2007 South Florida environmental report—volume 1. South Florida Water Management District, West Palm Beach, FL

Rumbold DG et al (2011) Source identification of Florida Bay's methylmercury problem: mainland runoff versus atmospheric deposition and in situ production. Estuar Coasts 34:494–513

Rumbold DG, Lange TR, Richard D, DelPizzo G, Hass N (2018) Mercury biomagnification through food webs along a salinity gradient down-estuary from a biological hotspot. Estuar Coast Shelf Sci 200:116–125

Sackett DK, Aday DD, Rice JA, Cope WG (2009) A statewide assessment of mercury dynamics in North Carolina water bodies and fish. Trans Am Fish Soc 138:1328–1341

Scheidt DJ, Kalla PI (2007) Everglades ecosystem assessment: water management and quality, eutrophication, mercury contamination, soils and habitat: monitoring for adaptive management: a R-EMAP status report. USEPA region 4, Athens, GA. EPA 904-R-07-001, 98 p

Schueneman TJ (2001) Characterization of sulfur sources in the EAA. Soil Crop Sci Soc Fla Proc 60:49–52

SFWMD (2014a) Appendix 1-2: peer-review and public comments on draft volume I. In: 2014 South Florida environmental report. South Florida Water Management District, West Palm Beach, FL

SFWMD (2014b) Appendix 1-3: authors' responses to peer-review panel and public comments 2014 South Florida environmental report. South Florida Water Management District, West Palm Beach, FL

Shanley JB et al (2012) MERGANSER: an empirical model to predict fish and loon mercury in New England Lakes. Environ Sci Technol 46:4641–4648

Shull D, Pulket M (2015) Causal analysis of the smallmouth bass decline in the Susquehanna and Juniata Rivers. Pennsylvania Department of Environmental Protection, Harrisburg, PA

Staveley JP, Law SA, Fairbrother A, Menzie CA (2014) A causal analysis of observed declines in managed honey bees (*Apis mellifera*). Hum Ecol Risk Assess Int J 20:566–591

Stober QJ et al (1996) South Florida ecosystem assessment interim report. United States Environmental Protection Agency report # 904-R-96-008, 27 p

Stober QJ et al (1998) South Florida ecosystem assessment: monitoring for ecosystem restoration. Final technical report—phase I. EPA 904-R-98-002. USEPA Region 4 Science and Ecosystem Support Division and Office of Research and Development. Athens, GA, 285 p. Plus appendices

Stober QJ et al (2001) South Florida ecosystem assessment: phase I/II—Everglades stressor interactions: hydropatterns, eutrophication, habitat alteration, and mercury contamination. U.S. Environmental Protection Agency report # 904-R-01-002, 63 p

Superville P-J, Pižeta I, Omanović D, Billon G (2013) Identification and on-line monitoring of reduced Sulphur species (RSS) by voltammetry in oxic waters. Talanta 112:55–62

Suter GW (2007) Ecological risk assessment, 2nd edn. CRC, Boca Raton, FL, p 643

Suter GW, Norton SB, Cormier SM (2002) A methodology for inferring the causes of observed impairments in aquatic ecosystems. Environ Toxicol Chem 21:1101–1111

Suter GW, Norton SB, Cormier SM (2010) The science and philosophy of a method for assessing environmental causes. Hum Ecol Risk Assess 16:19–34

Thomas CR, Miao S, Sindhoj E (2009) Environmental factors affecting temporal and spatial patterns of soil redox potential in Florida Everglades wetlands. Wetlands 29:1133–1145

Vorenhout M, van der Geest HG, Hunting ER (2011) An improved datalogger and novel probes for continuous redox measurements in wetlands. Int J Environ Anal Chem 91:801–810

Ware FJ, Royals H, Lange T (1990) Mercury contamination in Florida largemouth bass. Proc Ann Conf Southeast Assoc Fish Wildlife Agencies 44:5–12

Warner KA, Roden EE, Bonzongo J-C (2003) Microbial mercury transformation in anoxic freshwater sediments under iron-reducing and other electron-accepting conditions. Environ Toxicol Chem 37:2159–2165

Watras C, Morrison K, Regnell O, Kratz T (2006) The methylmercury cycle in Little Rock Lake during experimental acidification and recovery. Limnol Oceanogr 51:257–270

Wiener JG et al (2006) Mercury in soils, lakes, and fish in Voyageurs National Park (Minnesota): importance of atmospheric deposition and ecosystem factors. Environ Toxicol Chem 40 (20):6261–6268

Winfrey MR, Rudd JW (1990) Environmental factors affecting the formation of methylmercury in low pH lakes. Environ Toxicol Chem 9:853–869

Chapter 4
Dissolved Organic Matter Interactions with Mercury in the Florida Everglades

Andrew M. Graham

Abstract The purpose of this chapter is to review the interactions of mercury with dissolved organic matter (DOM) in the Florida Everglades. Attention is given to the role of DOM to the complexation of inorganic Hg and methylmercury (MeHg), microbial Hg methylation, Hg and MeHg photochemistry, and MeHg bioaccumulation. This review shows that substantive changes in both DOM concentration and quality can be traced to land and water management practices in the northern Everglades, and that these perturbations to carbon cycling have likely impacted Hg biogeochemical cycling in the Everglades. The impact of sulfur enrichment on DOM quality, and its corresponding impact on microbial Hg methylation, is a special emphasis of the chapter.

Keywords Everglades · Mercury · Methylmercury · Dissolved organic matter · Sulfur · Speciation

4.1 Introduction

Strong coupling of the biogeochemical cycling of mercury (Hg) to that of carbon necessitates an understanding of the role of dissolved organic matter (DOM) in the physical, chemical, and biological processes that control Hg fate in aquatic environments. Aspects of the biogeochemical cycling of Hg impacted by DOM include: (1) aqueous complexation and transport; (2) oxidation-reduction reactions of Hg; (3) precipitation, aggregation, and dissolution of Hg-solubility controlling mineral phases (e.g., metacinnabar (β-HgS)); (4) microbial Hg methylation; (5) photodegradation of MeHg and photoreduction of Hg(II); and (6) biouptake of Hg and MeHg and trophic transfer of Hg/MeHg. Understanding of Hg-sulfide-DOM interactions and the role of DOM in photodegradation of MeHg has advanced

A. M. Graham (✉)
Grinnell College Department of Chemistry, Grinnell, IA, USA
e-mail: grahaman@grinnell.edu

© Springer Nature Switzerland AG 2019

D. G. Rumbold et al. (eds.), *Mercury and the Everglades. A Synthesis and Model for Complex Ecosystem Restoration*, https://doi.org/10.1007/978-3-030-32057-7_4

significantly since the last major review of Hg-DOM interactions (Ravichandran 2004) and is a special emphasis of this chapter.

The Florida Everglades are one of the largest freshwater wetlands in the world, receive substantial inputs of Hg from atmospheric deposition (Holmes et al. 2016), and fish, wading birds, and reptiles are known to have elevated MeHg concentrations (Cleckner et al. 1998; Rumbold et al. 2008). Significant changes in DOM flux and quality have occurred in response to hydrologic modifications and nutrient and sulfate enrichment caused by conversion of wetlands to agricultural land in the northern Everglades (Aiken et al. 2011). The response of microbial MeHg production and, ultimately, fish MeHg concentrations to changes in DOM loading and quality remain incompletely understood and are the focus of this chapter. Ultimately, we desire to understand how environmental management may impact DOM dynamics and Hg fate, transport, and bioaccumulation in the Everglades.

4.2 DOM Concentrations and Quality in the FL Everglades

4.2.1 Sources of DOM

Sources of DOM in the Florida Everglades include: (1) leaching and/or partial oxidation of accumulated peat deposits (often referred to as soil-derived DOM); (2) OM derived from sugarcane (the chief crop grown in the Everglades Agricultural Area (EAA)); (3) OM derived from senescent wetland plant biomass (mainly sawgrass, but increasing dominance of cattails in the northern Everglades); and (4) OM derived from algal biomass within wetlands (Wang et al. 2002). A strong north-south gradient, coincident with gradients in phosphorous and sulfate (Scheidt and Kalla 2007) for surface water dissolved organic carbon (DOC) and two spectroscopic measures of DOM character, fluorescence index (FI) and slope ratio (S_R) is shown in Fig. 4.1.

Surface water DOC concentrations are highest in the northern Everglades, especially in WCA-2 (Water Catchment Area 2) that receives substantial water inputs from the EAA. Average surface water DOC concentrations in the eutrophic WCA-2 generally exceed 25 mg C/L and can reach >40 mg C/L (Wang et al. 2002; Yamashita et al. 2010; Aiken et al. 2011). In contrast, surface water DOC concentrations in the more pristine and oligotrophic Everglades National Park (ENP) are substantially lower, typically between 10 and 20 mg C/L (Maie et al. 2005, 2006; Yamashita et al. 2010; Aiken et al. 2011; Chen et al. 2013). The trend of increasing fluorescent index (FI) indicates increasing contributions of microbially- or algal-derived DOM (Cory and McKnight 2005) and the increase in slope rate (S_R) reflects decreasing average molecular weight and aromaticity (Helms et al. 2008). The prevailing explanation for these observations is that elevated DOC concentrations in the northern Everglades are related to accelerated oxidation of peat in the EAA following drainage for sugarcane production (Wang et al. 2002; Qualls and

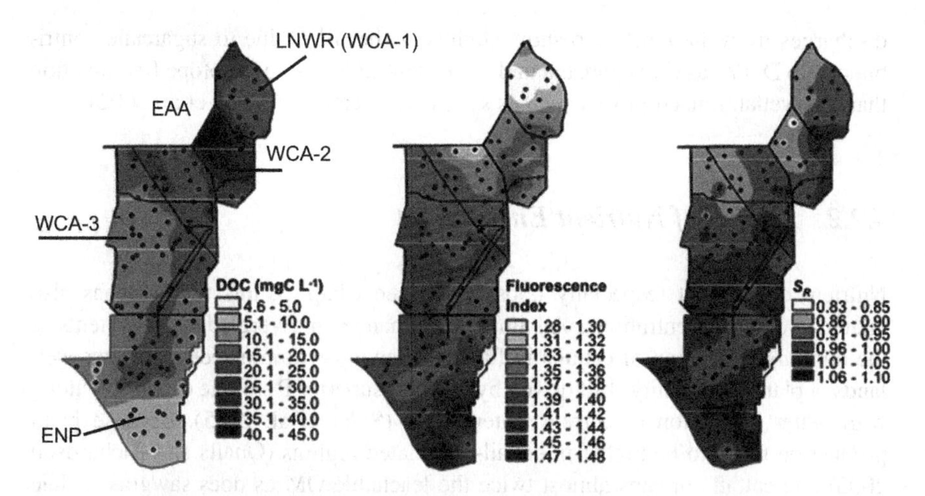

Fig. 4.1 Spatial distribution of dissolved organic carbon (DOC) and measures of DOM quality including fluorescence index (FI; the ratio of emission intensity at 470–520 nm with excitation at 370 nm) and slope ratio (S_R; the ratio of spectral slopes from 275–295 nm to 350–400 nm). FI is correlated with the proportion of DOM of microbial and/or algal origin (Cory and McKnight 2005), and S_R is inversely correlated with DOM aromaticity (Helms et al. 2008). Data were collected November 2005, corresponding to the wet season. Figure slightly modified from Yamashita et al. (2010) and used with permission. *EAA* Everglades Agricultural Area; *LNWR* Loxahatchee Conservation Area; *WCA* Water Conservation Area; *ENP* Everglades National Park

Richardson 2003; Yamashita et al. 2010; Aiken et al. 2011). Indeed, soil losses in the EAA are on the order of 2 m (Sklar et al. 2005), an enormous store of carbon.

Analysis of the Aquatic Cycling of Mercury in the Everglades (ACME) dataset (available at https://sofia.usgs.gov/exchange/acme/) indicates concentrations of DOM in sediment porewaters show similar spatial trends to that of surface waters. DOM concentrations are highest in WCA2 (e.g., 30–160 mg C/L at site F1 in the northern portion of WCA2A) and lowest in ENP (e.g., 3–15 mg C/L in Taylor Slough Site 7). DOM concentrations tend to increase with depth at most sites, indicating that diffusive flux of DOM from porewaters to overlying surface waters is an important DOM source. Recent characterization of paired porewater and surface water samples using high resolution Fourier Transform Ion Cyclotron Resonance Mass Spectrometry (FT-ICR-MS) revealed that most (59–84%) assigned molecular formulae were common to surface water and porewater (Poulin et al. 2017), indicating that porewaters supply surface water DOM, especially near canals where local vertical hydraulic gradients have increased in response to canal entrenchment, levee construction, and peat subsidence (McCormick et al. 2011).

Stable and radiometric carbon analysis of DOM also support the contention that peat oxidation in the EAA is an important source of DOM to the WCAs. Negative $\delta^{13}C$ for DOM is observed at the inflows to the Stormwater Treatment Areas (STAs; see Chap. 1, Volume I) and WCA-2 receiving runoff from the EAA, but becomes positive down the hydrologic flow path, indicative of an older carbon source in the northern Everglades (Wang et al. 2002; Stern et al. 2007). Waters closest to canal discharges from the EAA also show slightly higher $\delta^{13}C$ due to sugarcane contributions to DOM, as C4 sugarcane undergoes less carbon stable isotope fractionation than C3 wetland macrophytes such as sawgrass or cattails (Wang et al. 2002).

4.2.2 Impact of Nutrient Enrichment

Nutrient enrichment (especially phosphorus; see Chap. 5, this volume) has also altered DOM concentrations. An important change in Everglades ecosystems is the ongoing expansion of cattails in formerly sawgrass dominated freshwater wetlands, a plant community shift driven by cattails' superior P storage capability along with better adaptation to current water levels (Sklar et al. 2005). Because plant production tends to be higher in cattail-dominated regions (Qualls and Richardson 2003) and cattail contains almost twice the leachable OM as does sawgrass (Maie et al. 2006), DOC production is likely to be greater in cattail dominated wetlands. The relative yield of phenolic DOM from cattails vs. sawgrass is also greater (28% of total DOC for cattails vs. 20% for sawgrass; Maie et al. 2006) suggesting that this plant community shift may help to explain the observation (Yamashita et al. 2010; Aiken et al. 2011) of more aromatic DOM in the northern Everglades.

Sulfate enrichment (due to sulfur application in the EAA and inputs via canal drainage (Bates et al. 2002; see Chap. 2, this volume) has also likely altered DOM quality. A comparative study of DOM from three of the world's largest subtropical freshwater wetlands (the Florida Everglades, the Brazilian Pantanal, and the Botswana Okavano Delta) found that Everglades DOM was appreciably enriched in sulfur compared to DOM from these other wetland complexes (Hertkorn et al. 2016). Employing high resolution FT-ICR MS, Hertkorn et al. (2016) discovered that a remarkable 13–16% of all peaks in the Everglades DOM mass spectra could be assigned to CHOS or CHNOS formulas (compared to just 2–3% for Pantanal and Okavano Delta samples). Sulfide produced from dissimilatory sulfate reduction adds to electrophilic centers in DOM via S_N2, Michael addition, and S_NAr reactions to produce sulfidized DOM (Heitmann and Blodau 2006; Hoffmann et al. 2012; Sleighter et al. 2014; Yu et al. 2015; Poulin et al. 2017). These reactions are rapid, occurring on a timescale of hours under environmentally relevant concentrations of sulfide and DOC (Yu et al. 2015). Poulin et al. (2017) investigated the importance of this process in the Everglades, and found that the degree of sulfur incorporation in surface water and porewater DOM increased linearly with porewater sulfide concentration along a trophic gradient, with more highly sulfidized DOM found in the highly sulfate enriched portions of WCA-2. Most (50–80%) of the sulfur incorporated into DOM is as reduced exocyclic (e.g., thiols) and heterocyclic (e.g., thiophenes) species, with the proportion of exocyclic S increasing with increasing degree of DOM sulfidization (Poulin et al. 2017), reaching the upper end of the range (10–70% of total S) previously reported for DOM (Waples et al. 2005; Skyllberg 2008; Manceau and Nagy 2012). Reduced S species in DOM, especially thiols, play a critical role in inorganic Hg(II) and MeHg coordination chemistry, inhibition of HgS precipitation/aggregation, enhancement of microbial Hg methylation, and Hg photodegradation. Thus, sulfate enrichment of the Everglades, followed by sulfidization of DOM, has likely profoundly altered the reactivity of DOM toward Hg.

4.2.3 Hydrologic Controls on DOM Fluxes and Quality

Pronounced seasonal variations in DOC concentration and quality reflect variation in hydrologic inputs/outputs, soil type (peat vs. marl), and primary productivity. These dynamics are most well understood in the southern Everglades, where extensive research has been conducted through the Florida Coastal Everglades Long Term Ecological Research (FCE LTER) Program. In the peat-based Shark River Slough (SRS) in the western portion of ENP, DOC concentrations peak during the dry season (December–May). At the Taylor Slough (TS) sites in the eastern portion of ENP, DOC peaks during the wet season (May–October) (Chen et al. 2013). Seasonally-dependent changes in DOC concentration were accompanied by shifts in DOM quality. At both groups of sampling sites, FI was higher during dry months compared to wet months (Chen et al. 2013). Specific UV-absorbance ($SUVA_{254}$), a measure correlated to DOM aromaticity and molecular weight (MW; Weishaar et al. 2003), was slightly lower during dry months compared to wet months (Chen et al. 2013). These findings indicate that inputs of terrestrially derived organic matter are strongly tied to water delivery. Taking the TS sites as an example, DOM aromaticity declines along a north-south salinity gradient, reflecting the transition from terrestrial-like (soil-derived) to marine-like (microbial- and seagrass-derived) DOM. During the dry season, when freshwater flows are reduced, northward intrusion of water from Florida Bay dilutes the terrestrial, aromatic DOM with more aliphatic, lower MW DOM. Restoration efforts focused on increasing freshwater flows to the southern Everglades are likely to increase delivery of high MW, aromatic, highly-sulfidized, peat-derived DOM to more pristine portions of the southern Everglades. As discussed below, laboratory studies have demonstrated that DOM aromaticity and sulfur content play important roles in controlling DOM-dependent enhancement of microbial Hg methylation in sulfidic waters (Graham et al. 2012, 2013, 2017), so shifts in DOM quality in response to water management have the potential to alter *in situ* rates of microbial Hg methylation. On the other hand, MeHg biouptake by phytoplankton has been reported to be negatively correlated with DOM concentration, including in the Everglades (Gorski et al. 2008; Luengen et al. 2012; Pollman and Axelrad 2014). The impact of DOM character on MeHg uptake by phytoplankton is not well understood, but DOM with greater reduced S functionality should be superior at binding MeHg and rendering it less bioavailable for uptake. The net effect of increasing DOM and shifting DOM character toward more aromatic and sulfidized DOM on fish Hg concentrations is thus difficult to predict, and warrants further attention.

To summarize, surface water DOM concentrations decline along a north-south trophic and salinity gradient in the Florida Everglades. Water management and nutrient enrichment have profoundly altered DOM concentration and quality, resulting in high concentrations of high MW, aromatic, peat-derived DOM exhibiting substantial sulfidization in the northern Everglades. The implications of these perturbations to carbon cycling on the biogeochemical cycling of Hg are discussed in the following sections.

4.3 Hg Complexation by DOM

4.3.1 Hg-DOM and MeHg-DOM Complexes and Competition with Other Ligands

As a Class B "soft" cation, Hg has a strong preference for S-donor ligands (Riccardi et al. 2013). In oxic surface waters, the dominant S-donor ligands for inorganic Hg (II) (henceforth, $Hg(II)_i$) and MeHg are DOM thiols. In anoxic bottom waters or porewaters, DOM thiols are likely outcompeted by inorganic sulfide (H_2S_T) (Skyllberg 2008). Extended X-ray Absorption Fine Structure (EXAFS) analysis of Hg complexes with soil and aquatic humic substances demonstrate that Hg is linearly coordinated with two S atoms with bond lengths similar to that reported for Hg-S in Hg complexes with low molecular weight thiols (Skyllberg et al. 2006; Gerbig et al. 2011).

The thermodynamics and kinetics of $Hg(II)_i$ binding to DOM have been studied extensively. Kinetic investigations suggest that $Hg(II)_i$ initially binds to more abundant and weaker O- and N-donor ligands, followed by slower transfer to strong, but less abundant S-donor ligands, ultimately forming highly stable complexes on the timescale of hours (Hsu and Sedlak 2003; Lamborg et al. 2003; Miller et al. 2009). Studies of the thermodynamics of $Hg(II)_i$ complexation by DOM support the notion of a two site-type model, as conditional stability constants for $Hg(II)_i$ binding by DOM are inversely related to the Hg/DOM ratio (Haitzer et al. 2002). At low Hg/DOM ratios typical of natural environments, $Hg(II)_i$ binding is dominated by coordination to strong S-donor ligands, whereas carboxylic and phenolic O-donor ligands control $Hg(II)_i$ binding at high Hg/DOM ratios. Reported log K values for Hg complexation with DOM (including DOM isolated from the Everglades (Drexel et al. 2002; Haitzer et al. 2002, 2003)) at low, environmentally relevant Hg/DOM ratios range from log K = 23–30 L/kg DOM (Drexel et al. 2002; Haitzer et al. 2002, 2003; Khwaja et al. 2006, 2010; Gasper et al. 2007). In a critical review of $Hg(II)_i$-DOM binding, Skyllberg (2008) recommended a value of log K = 42.0 for the pH-independent equilibrium constant for $Hg(II)_i$ binding to two fully deprotonated DOM thiols (RS^-):

$$Hg^{2+} + 2RS^- = Hg(SR)_2 \tag{4.1}$$

Given DOC concentrations, and DOM thiol content (typically 0.1–1.5 mass percent (Waples et al. 2005)), it is possible to predict the concentration of DOM-bound Hg in natural waters, but we must also consider Hg complexation to inorganic sulfide and β-HgS(s) (metacinnabar) precipitation for anoxic soils and sediments where sulfate reduction occurs. Using Benoit et al. (1999b) equilibrium constants for $Hg(II)_i$-S complexation, a log K of 38.0 for metacinnabar precipitation (National Institute of Standards and Technology 2013) and Skyllberg's log K for Hg-DOM thiol complexation (Skyllberg 2008), β-HgS(s) solubility can be calculated at pH 7.4 for various concentrations of DOM thiols (RSH_T). $[RSH]_T$ can be calculated based on porewater DOC concentration (20–120 mg C/L), knowledge of total S content (typically 1.2–1.6 mass % for northern Everglades DOM (Waples et al. 2005)), and fraction of total S in reduced form (typically 50–70% for northern Everglades DOM (Waples et al. 2005; Poulin et al. 2017)). Overlain on Fig. 4.2 are measured porewater concentrations of filterable total Hg (Hg_T) and sulfide for several sites in the northern Everglades (data from the ACME database).

A few observations can be made based on Fig. 4.2. First, DOM thiols only significantly impact β-HgS(s) solubility at total dissolved sulfide concentrations less than 1–10 μM. Above 10 μM dissolved sulfide (e.g., site WCA2A F1), DOM thiols are outcompeted by inorganic sulfide, and β-HgS(s) solubility increases with increasing sulfide concentration due to $Hg(II)_i$-sulfide complexation. At Everglades sites less impacted by sulfate inputs (e.g. WCA3A 15), however, DOM thiols likely play an important role in $Hg(II)_i$ binding even in anoxic sediment porewaters. Second, Hg_T can be predicted reasonably well assuming equilibrium with metacinnabar and binding to both DOM thiols and inorganic sulfide (with the exception of the WCA2A F1 site for which the model over-predicts observed Hg concentrations). Hg_T can remain relatively invariant over a wide range of sulfide concentrations, as first observed for the Everglades by Benoit et al. (1999a) if DOM

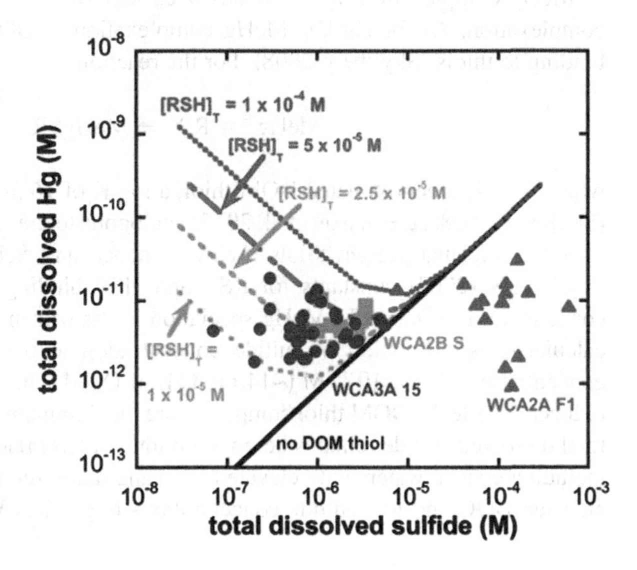

Fig. 4.2 Predicted (lines) total dissolved Hg concentration in equilibrium with metacinnabar (β-HgS (s)) at pH 7.40 and 0.01 M NaCl as a function of porewater total dissolved sulfide concentration at several representative concentrations of DOM thiols ($[RSH]_T$). Symbols are observed porewater sulfide and 0.45 μm filter-passing total Hg for three sites in the Everglades (all data from ACME database)

thiol concentrations change simultaneously with sulfide concentrations. To illustrate this point, a Hg_T of ~10^{-11} M (~2 ng/L) can be maintained as sulfide is increased from 0.1 to 10 μM simply by increasing DOM thiol concentration from ~1×10^{-5} to 5×10^{-5} M. As noted above, both DOM concentration and degree of sulfidization are greater in more sulfidic northern Everglades sites, and the relatively invariant porewater Hg_T across a spatial sulfide gradient may be partially attributable to changing DOM thiol concentrations across this same gradient. Earlier efforts to model Hg_T across a sulfide gradient included the neutral, sulfide-concentration independent species HgS^0 (really HOHgSH; Benoit et al. 1999a), a species now suspected to be misidentified nanoparticulate metacinnabar (Deonarine and Hsu-Kim 2009), and Hg adsorption to surface thiols (particulate organic matter and metal sulfides) rather than precipitation/dissolution of pure metacinnabar. While Hg adsorption cannot be ruled out, the calculations shown here suggest that $Hg(II)_i$-binding by DOM thiols may be an underappreciated aspect of $Hg(II)_i$ speciation in low sulfide Everglades porewaters. The hypothesis that $Hg(II)_i$-DOM thiol complexes are significant in low sulfide porewaters is especially germane in light of recent studies demonstrating exceptional bioavailability of $Hg(II)_i$-thiol complexes Hg-methylating bacteria (Schaefer and Morel 2009; Schaefer et al. 2011; Graham et al. 2017). Whether a shift from $Hg(II)_i$-DOM thiol to $Hg(II)_i$-sulfide complexes as dominant aqueous $Hg(II)_i$ species is partially responsible for well-documented declines in MeHg concentrations with increasing sulfide concentration deserves further attention. In support of this hypothesis, a recent study found that DOM-complexed Hg was more available for Hg methylation than Hg complexed to inorganic ligands (principally Cl^-) when added to marine sediments, and argued that direct uptake of Hg-DOM thiol complexes or shuttling of Hg to microbial metal transporters was the main cause for the enhanced bioavailability of Hg-DOM complexes (Mazrui et al. 2016).

MeHg complexation by DOM has been less well studied compared to $Hg(II)_i$ complexation. As for $Hg(II)_i$, MeHg complexation to DOM is controlled through binding to thiols (Skyllberg 2008). For the reaction:

$$MeHg^+ + RS^- = MeHgSR \tag{4.2}$$

where RS^- is a deprotonated DOM thiol, a log K of 16.5–16.7 has been suggested (Skyllberg 2008; Jeremiason et al. 2015) analogous to the strength of MeHg binding by low molecular weight thiols (Schwarzenbach and Schellenberg 1965). Using Skyllberg's (2008) constants for RS^- and HS^- binding to MeHg and a MeHg concentration of 0.2 ng/L, MeHg speciation in the presence of DOM thiols can be calculated as a function of sulfide concentration at pH 7.4. At a modest thiol concentration of 1×10^{-5} M (~14 mg C/L of DOM with 1.6% S and 70% of S as reduced S), MeHg-DOM thiol complexes are the dominant MeHg species as long as total dissolved sulfide remains below ~3.6 μM. In Everglades' surface waters (even including coastal waters with elevated Cl^-) and many sedimentary porewaters with elevated DOC and low sulfide concentrations (e.g., sites WCA2B S and WCA3A

15 shown in Fig. 4.2), MeHg-DOM thiol complexes will predominate. As discussed below, formation of strong MeHg-DOM complexes may limit the efficiency of MeHg biouptake at lower tropic levels (Luengen et al. 2012), thus ameliorating MeHg biomagnification in aquatic ecosystems.

4.3.2 Implications for Hg and MeHg Transport

What role does complexation with DOM play in the transport of $Hg(II)_i$ and MeHg through the Everglades? Studies of terrestrial ecosystems have noted strong positive correlations between DOC concentrations and both filterable Hg_T and MeHg (Driscoll et al. 1995; Babiarz et al. 1998; Balogh et al. 2004; Hall et al. 2008; Brigham et al. 2009; Dittman et al. 2010; Schuster et al. 2011). Robust positive correlations between DOC and both Hg_T and MeHg have been observed for the mangrove-dominated Shark River estuary in the Southwest Everglades (Bergamaschi et al. 2012), indicating potential for tight coupling of Hg/MeHg and carbon export to coastal areas. However, the relationship between carbon and Hg/MeHg export for the northern Everglades is less clear. An early study found no correlation between DOC and either filterable Hg_T or MeHg for northern Everglades canal and marsh surface waters (Hurley et al. 1998). It is important to note, however, that some northern Everglades surface waters (e.g., the Stormwater Treatment Areas (STAs), site WCA2A F1) have sulfide concentrations in the 1–100 μM range and DOM thus may not be the principal ligand for $Hg(II)_i$ or MeHg. At the landscape scale, there are statistically significant, although weak, correlations between DOC and filterable Hg_T and MeHg. Figure 4.3 shows the relationship between surface water DOC and filterable Hg_T for five of the core ACME stations

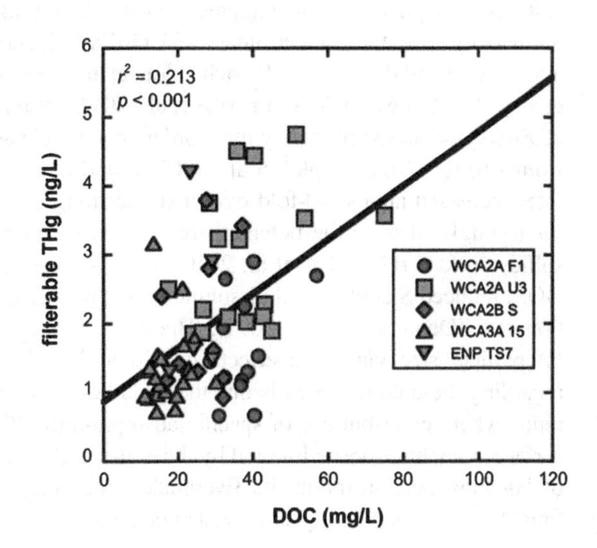

Fig. 4.3 Relationship between dissolved organic carbon (DOC) concentration and filterable (0.45 μm) total Hg (THg) in surface waters sampled from 1995 to 2007 as part of the Aquatic Cycling of Mercury in the Everglades (ACME) project

spanning the north—south trophic gradient in the Everglades over the period 1995–2007. A much weaker ($r^2 = 0.078$), but significant ($p = 0.020$) correlation was observed between DOC and filterable MeHg for these same five sites (data not shown). These findings are similar to that of Liu et al. (2008, 2009) who reported significant, but weak ($r^2 < 0.30$) correlations between log-transformed DOC concentrations and filterable Hg_T, filterable MeHg, and measures of water-particle partitioning in an analysis of the EPA Region 4 Regional Environmental Monitoring and Assessment Program (R-EMAP) data collected at ~250 stations across the Everglades in 2005. To summarize, it appears that filterable Hg_T and MeHg concentrations are weakly related to DOC; thus natural and anthropogenic controls on DOM production and transport have the potential to impact Hg_T and MeHg transport through the Everglades.

4.4 Interactions of DOM with Hg-Sulfide Species

In addition to directly complexing $Hg(II)_i$, DOM may interact with HgS(s) surfaces, altering kinetics of HgS(s) precipitation, dissolution, growth, and aggregation (Ravichandran et al. 1998, 1999; Waples et al. 2005; Deonarine and Hsu-Kim 2009; Slowey 2010; Gerbig et al. 2011). Metacinnabar particles formed in the presence of DOM are very small, probably <20 nm in diameter under conditions of high DOM: $Hg(II)_i$ ratios (Deonarine and Hsu-Kim 2009; Gerbig et al. 2011). Nanoparticles with a high proportion of coordinatively unsaturated atoms exhibit substantially different thermodynamic (e.g., solubility product) and kinetic properties (e.g., surface-area normalized dissolution rates) compared to bulk minerals owing to the important contribution of the surface free energy to the overall thermodynamic stability (Madden and Hochella 2005; Navrotsky et al. 2008). Studies of HgS(s) (and related metal sulfides) precipitation/growth/aggregation and dissolution have shown that DOM size/aromaticity is a strong determinant of DOM's influence on HgS(s) particle growth/aggregation and dissolution (Ravichandran et al. 1998; Waples et al. 2005; Deonarine et al. 2011). For example, Deonarine et al. (2011) found that growth/aggregation rates of ZnS(s) decreased by nearly three orders magnitude as DOM aromaticity increased from ~10 to ~40%. Waples et al. (2005) found that cinnabar (α-HgS(s)) dissolution rates increased almost 40-fold over a similar range of increasing DOM aromaticity. Interestingly, despite the potential for specific binding of DOM thiols to the metal sulfide surface (Gondikas et al. 2010), these studies report no correlations between DOM reduced S content and dissolution rate (Waples et al. 2005) or growth/aggregation rate (Deonarine et al. 2011). These results suggest that DOM controls HgS (s) particle size via steric effects related to DOM size/aromaticity. One caution regarding these conclusions is that these studies were conducted at high metal:DOM ratios where contributions of specific adsorption of DOM thiols to the metal sulfide surfaces may have been obscured by the non-specific steric effects. Within the context of Hg biogeochemistry in the Everglades, relatively aromatic ($SUVA_{254} > 3.0$ L (mg C)$^{-1}$ m^{-1}) and highly sulfidized (Wagner et al. 2015; Hertkorn et al. 2016; Poulin et al. 2017) DOM in the northern Everglades should be especially effective at capping β-HgS(s) nanoparticles, preventing growth and aggregation, and likely increasing the pool of $Hg(II)_i$ available for microbial methylation as discussed below.

4.5 DOM and Microbial Hg Methylation

DOM has the potential to impact microbial Hg methylation in at least three ways. Firstly, as a carbon source, electron donor, or electron acceptor (or shuttle to electron acceptors such as iron oxyhydroxides), changes in DOM concentration and/or quality (lability) can impact the activity of Hg-methylating microorganisms (King et al. 2000; Drott et al. 2007; Zhang et al. 2014). For example, addition of an external carbon source (pyruvate) to sediment microcosms stimulated Hg methylation three to tenfold in estuarine sediments with high porewater sulfate (Zhang et al. 2014). Nutrient enrichment and accompanying shifts in plant communities in the Everglades (Sect. 4.2) have likely led to changes in DOM concentration and quality that have impacted microbial activity (Maie et al. 2006; Davis et al. 2006). Sulfate reduction and methanogenesis rates are greatest in the eutrophied northern wetlands, and microbial diversity amongst sulfate reducing bacteria (SRB) even appears to be dependent upon trophic status (Castro et al. 2002); significant given that SRB and methanogens are important Hg-methylators (Gilmour et al. 2013). While these observations are undoubtedly tied to sulfate enrichment of the northern Everglades, a role for changes in DOM concentration and quality in determining microbial community composition and activity cannot be discounted. An open question is whether substantial changes in DOM production/quality in the Everglades have influenced the abundance and/or activity of Hg-methylating microorganisms. Recent identification of the *hgcA* and *hgcB* genes (Parks et al. 2013) and development of new molecular tools for assessing Hg-methylator abundance and activity (Christensen et al. 2016) should aid in improved understanding of the response of Hg-methylators to environmental change and management decision making.

Secondly, complexation of $Hg(II)_i$ by DOM may alter $Hg(II)_i$ bioavailability for methylation by microorganisms, though it is not wholly resolved whether complexation by DOM reduces or increases $Hg(II)_I$ bioavailability. Complexation of $Hg(II)_i$ by DOM may limit $Hg(II)_i$ bioavailability to phytoplankton (Gorski et al. 2008) and some bacteria (Barkay et al. 1997). However, $Hg(II)_i$ complexed with low molecular weight thiols exhibits exceptionally high bioavailability to bacteria, including Hg-methylating bacteria (Schaefer and Morel 2009; Schaefer et al. 2011; Thomas et al. 2014; Graham et al. 2017). Concentrations of select low molecular weight thiols (e.g., cysteine, thioacetic acid, glutathione) in wetland and estuarine sediment porewaters can individually reach the 10^{-8}–10^{-5} M range (MacCrehan and Shea 1995; Zhang et al. 2004; Liem-Nguyen et al. 2015). Including unidentified DOM-thiols (Poulin et al. 2017), $Hg(II)_i$ complexation to thiols in the DOM pool has the potential to enhance microbial Hg methylation in the Everglades, especially in low-sulfide soils. Indeed, DOM-complexed Hg isotope tracer has been shown to be more bioavailable than an inorganic Hg tracer under conditions of undersaturation with respect to β-HgS(s) (Mazrui et al. 2016).

Thirdly, interactions of DOM with β-HgS(s) surfaces can enhance $Hg(II)_i$ bioavailability to Hg-methylating microorganisms. Nanoparticles of β-HgS(s) have been demonstrated to be more bioavailable to Hg-methylating microorganisms in both pure culture studies (Zhang et al. 2012) and in sediment microcosm experiments (Zhang et al. 2014; Kucharzyk et al. 2015; Mazrui et al. 2016). Laboratory experiments with a model Hg-methylator (*Desulfovibrio desulfuricans* ND132) have shown that DOM can substantially increase Hg-methylation rates under conditions of supersaturation with respect to β-HgS(s) (Graham et al. 2012, 2013, 2017)—a finding attributed DOM's propensity to inhibit β-HgS(s) growth/aggregation, keeping β-HgS(s) nanosized and highly bioavailable. While the exact mechanisms for increased bioavailability of nanoscale β-HgS(s) are unknown, greater intrinsic solubility for nanosized β-HgS(s) and/or more rapid ligand exchange at the cell surface by metal transporters hypothesized to be involved in $Hg(II)_i$ uptake (Szczuka et al. 2015) are plausible explanations. DOM quality strongly modulates the magnitude of this DOM enhancement effect as shown in Fig. 4.4. Highly aromatic, high molecular weight DOM (as inferred from high $SUVA_{254}$ or low slope ratio, S_R, the ratio of spectral slopes in wavelength range of 275–295 nm and 350–400 nm) and highly sulfidized DOM are especially effective at enhancing Hg methylation. The work of Moreau et al. (2015) corroborates these findings, as they found greater enhancement of Hg methylation by higher molecular weight, more aromatic hydrophobic acid fractions of DOM compared to transphilic acid fractions. Of special note are the DOM isolates from the Everglades; 20 mg C/L of these isolates (F1HPoA, F1TPiA, and 2BSHPoA) increased Hg methylation efficiency ~7–15X above that observed in DOM-free controls. The exceptional enhancement in Hg methylation in the presence of northern Everglades DOM is attributable to both the relatively high aromatic character/high molecular weight and the high reduced sulfur content (see Sect. 4.2), attributes which stabilize β-HgS(s) against growth/aggregation, enhancing Hg bioavailability. It is thus hypothesized that historical land use changes and agricultural practices that impacted DOM quality have increased *in situ* Hg methylation rates in the Everglades, especially in wetland sediments where precipitation of β-HgS(s) is predicted. It should also be pointed out that all three means by which DOM enhances microbial Hg methylation (stimulation of microbial activity; increased $Hg(II)_i$ biouptake by formation of $Hg(II)_i$-thiol complexes, and stabilization of nanoscale β-HgS(s)) may be operative simultaneously.

There is strong observational evidence that increased DOM inputs can spur increased in situ Hg methylation rates in the Everglades. Mesocosms at site WCA3A 15 were amended with DOM isolated from surface water and isotopically labeled $Hg(II)_i$. DOM addition stimulated Hg methylation from both "new" enriched isotope pools and "old" ambient Hg pools three to fivefold above DOM-free controls (Gilmour et al. 2004). While a number of studies have demonstrated that "new" tracer Hg is more bioavailable than ambient Hg for methylation (Hintelmann et al. 2000; Jonsson et al. 2014; see also Chap. 3, Volume III), the observation that DOM

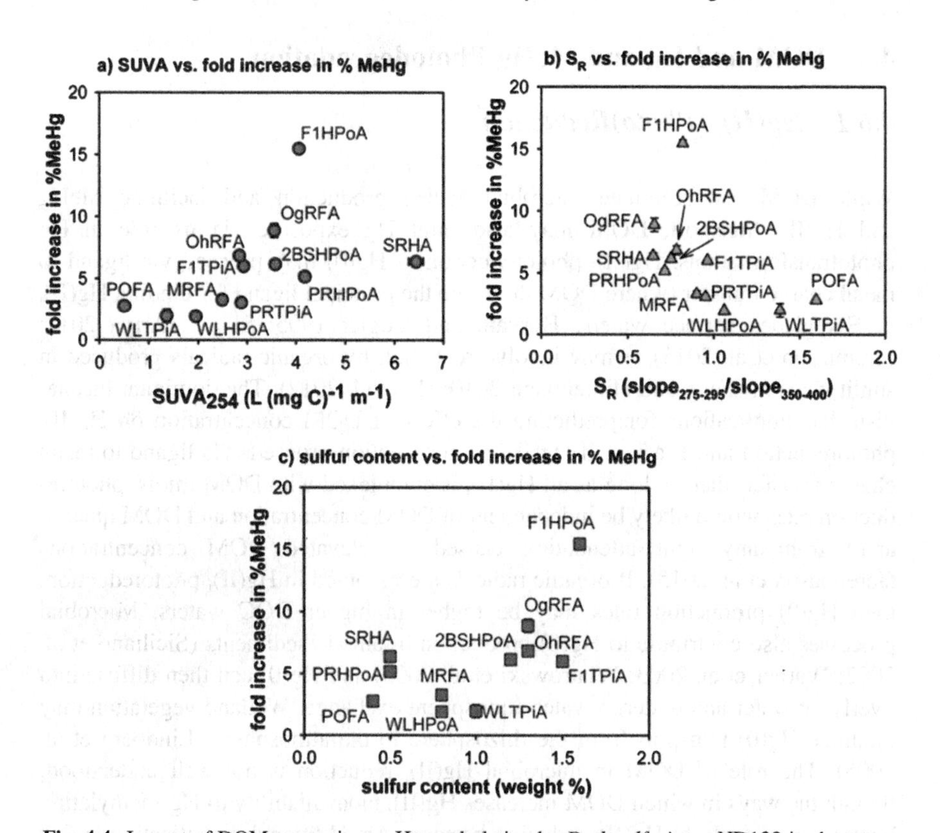

Fig. 4.4 Impact of DOM properties on Hg methylation by *D. desulfuricans* ND132 in short-term (3 h) washed cell assays under conditions of slight supersaturation with respected to β-HgS(s). Scale for MeHg production is fold increase in % Hg methylated relative to DOM-free controls. Experiments were performed with washed cells of ND132 in solutions containing minimal media at pH 7.4 ± 0.1, 20 mg C/L of each DOM isolate, and 1.5 ± 0.2 μM total sulfide. $SUVA_{254}$ = specific UV absorbance at 254 nm. S_R = slope ratio (ratio of spectral slopes in wavelength range of 275–295 nm and 350–400 nm). *2BSHPoA* hydrophobic acid from WCA2B S site in FL Everglades; *F1HPoA* hydrophobic acid from WCA2A F1 site in FL Everglades; *F1TPiA* transphilic acid from WCA2A F1 site in FL Everglades; *MRFA* Missouri River fulvic acid; *OgRFA* Ogeechee River fulvic acid; *OhRFA* Ohio River fulvic acid; *POFA* Pacific Ocean fulvic acid; *PRHPoA* Penobscot River hydrophobic acid; *PRTPiA* Penobscot River transphilic acid; *SRHA* Suwannee River humic acid; *WLHPoA* Williams Lake (MN) hydrophobic acid; *WLTPiA* Williams Lake transphilic acid. Data originally reported in alternate form in Graham et al. (2013); for further details regarding the DOM samples, consult the original paper. Adapted with permission from Graham et al. (2013). Copyright American Chemical Society

addition stimulated Hg methylation from the ambient Hg pool suggests that continued inputs of highly reactive (highly aromatic, highly sulfidized) DOM could continue to mobilize "legacy" Hg for microbial methylation even after more restrictive national and global Hg pollution prevention and control measures are enacted. This observation has implications for ecosystem management and is worthy of further study.

4.6 DOM and Hg and MeHg Photodegradation

4.6.1 $Hg(II)_i$ (Photo)Reduction

While DOM can stimulate microbial MeHg production and facilitate MeHg and $Hg(II)_i$ transport, DOM may also limit Hg exposure via its role in Hg phototransformations. $Hg(II)_i$ photoreduction to Hg(0) may proceed via ligand to metal charge transfer (where DOM thiols are the principal ligand for binding $Hg(II)_i$ in Everglades surface waters; Horvath and Vogler 1993; Si and Ariya 2011; Jeremiason et al. 2015) or may involve reduction by organic radicals produced in sunlit waters (Zheng and Hintelmann 2010; He et al. 2012). The dominant mechanism has implications for predicting the effect of DOM concentration on $Hg(II)_i$ photoreduction and evasion. If $Hg(II)_i$ photoreduction proceeds via ligand to metal charge transfer, then as long as all $Hg(II)_i$ is complexed with DOM thiols, photoreduction rates would likely be independent of DOM concentration and DOM quality, apart from any light attenuation caused by elevated DOM concentrations (Jeremiason et al. 2015). If organic radicals are involved in $Hg(II)_i$ photoreduction, then Hg(0) production rates may be higher in higher DOC waters. Microbial processes also contribute to Hg(0) production in anoxic sediments (Siciliano et al. 2002; Warner et al. 2003; Wiatrowski et al. 2006) and Hg(0) can then diffuse into overlying water and undergo water-atmosphere exchange. Wetland vegetation may facilitate Hg(0) transport from the rhizosphere to the atmosphere (Lindberg et al. 2005). The role of DOM in microbial $Hg(II)_i$ reduction is not well understood, though the ways in which DOM increases $Hg(II)_i$ bioavailability to Hg-methylating bacteria may apply to $Hg(II)_i$-reducing bacteria. An additional complication is that DOM can participate in redox reactions with $Hg(II)_i$ and Hg(0) under dark conditions (Gu et al. 2011; Zheng et al. 2012). At the landscape-scale, Krabbenhoft et al. (1998) measured diel variability of dissolved gaseous mercury (DGM) production (assumed to be mostly Hg(0)) in WCA2A, and estimated total DGM production as ~2.2 μg m^{-2} $year^{-1}$ or about 10% of late 1990s Hg deposition. More recent estimates suggest about 18% of total deposited Hg is returned to the atmosphere via photoreduction and Hg(0) evasion, and that the fraction of recent deposition evaded back to the atmosphere is similar amongst all the WCAs and ENP (Liu et al. 2011). Given uncertainties about the mechanism of $Hg(II)_i$ photoreduction and the relative contributions of microbial vs. photochemical processes as sources of Hg(0), further work is needed to understand the role of DOM in facilitating Hg transfer from water and sediment to the atmosphere in the Everglades.

4.6.2 MeHg Photodegradation

Numerous studies demonstrate that MeHg complexation by DOM (or molecular constituents of DOM) can accelerate rates of MeHg photodegradation (Zhang and

Hsu-Kim 2010; Black et al. 2012; Fernández-Gómez et al. 2013; Tai et al. 2014; Qian et al. 2014; Jeremiason et al. 2015). MeHg photodegradation is also wavelength-dependent, with faster rates of degradation by UVA and UV-B compared to visible light (Lenherr and St. Louis 2009). Our understanding of the mechanism(s) of MeHg photodegradation is rapidly evolving. Quantum-mechanical calculations show that MeHg complexation by strong sulfur-donor ligands weakens the C-Hg bond, and makes MeHg more susceptible to photochemical degradation (Tossell 1998; Zhang and Hsu-Kim 2010) but the exact mechanism(s) of photodecomposition remain to be fully elucidated. Some workers have concluded that singlet oxygen ($^1O_2$) was involved in photodecomposition based on selective inhibitors of reactive oxygen species (ROS) (Zhang and Hsu-Kim 2010). More recent investigations have found either no or only a modest role for ROS in MeHg photodecomposition (Black et al. 2012; Tai et al. 2014; Qian et al. 2014; Jeremiason et al. 2015), and suggested that breaking of the Hg-C bond proceeds via an intramolecular charge transfer from an excited triplet state ($^3DOM^*$) to the Hg-C bond, with coordination of MeHg to a thiol necessary for charge transfer to occur (Qian et al. 2014). An implication of this proposed mechanism is that MeHg photodecomposition should be independent of DOM concentration, provided that the aromatic thiols that facilitate photodecomposition are present at concentrations sufficient to complex all MeHg. MeHg photodecomposition independent of DOM concentration has been observed at low MeHg:RSH ratios (RSH is a DOM thiol; Fernández-Gómez et al. 2013; Tai et al. 2014; Jeremiason et al. 2015), including a study with Everglades DOM isolated by ultrafiltration (Tai et al. 2014). Because DOM significantly increases light attenuation, however, photodegradation rates may diminish with increasing DOM concentration. Indeed, a negative correlation between DOC and MeHg photodemethylation potential has been reported for the Everglades (Liu et al. 2011). Interestingly, MeHg photodecomposition may proceed at very similar rates for DOM isolates of appreciably different character (Jeremiason et al. 2015), suggesting that DOM quality may not be an especially important determinant of MeHg photodegradation rates either. Further studies, including studies at the field scale, are needed to evaluate this hypothesis.

4.7 Impact of DOM on Trophic Transfer of MeHg

Lastly, we should consider the impact of DOM on MeHg bioaccumulation. A study of Hg and MeHg bioaccumulation across diverse freshwater stream ecosystems found similar trophic biomagnification across ecosystems as inferred from plots of $\delta^{15}N$ (a measure of trophic position) vs. Hg concentration in biomass (Chasar et al. 2009), implying that MeHg uptake at the base of the foodwebs may control MeHg concentrations in predatory fish. MeHg uptake by plankton will largely be governed by filterable MeHg concentrations and the bioavailability of that MeHg. We have already noted that DOM may enhance MeHg production by a number of mechanisms (Sect. 4.5) as well as MeHg transport (Sect. 4.4), though the relationship between DOM and MeHg can be complicated by photodegradation in surface waters (Sect. 4.6).

How does DOM impact MeHg bioavailability to plankton? Most studies report that complexation of MeHg by DOM decreases MeHg uptake by planktonic species (Gorski et al. 2008; Zhong and Wang 2009; Tsui and Finlay 2011; Luengen et al. 2012), though there are conflicting reports (Pickhardt and Fisher 2007) in the literature as well. Luengen et al. (2012) found greater algal MeHg uptake by live cells vs. dead cells, and suggested that MeHg uptake involves active transport, possibly inadvertent uptake by an essential trace metal transporter. Thus, while DOM may promote increased MeHg bioaccumulation via increased MeHg production/transport, it may also reduce MeHg bioaccumulation via inhibition of planktonic MeHg biouptake. DOM may decrease bioaccumulation factors (fish MeHg/ water MeHg) by both increasing MeHg concentrations in water, whilst reducing MeHg biouptake by plankton. Indeed, several studies report negative correlations between bioaccumulation factors and DOC, including for the FL Everglades (Watras et al. 1998; Liu et al. 2008; Tsui and Finlay 2011; Pollman and Axelrad 2014). It should be further cautioned that *absolute* Hg tissue concentrations may still increase with increasing DOC (as observed in one large comparative study (Chasar et al. 2009)) even if MeHg trophic transfer *efficiency* (bioaccumulation factor) declines with increasing DOC. Factors such as food web structure, fish lifespan, and time lags between MeHg production and transfer into fish may also be important in determining MeHg concentrations in fish (Cleckner et al. 1998).

4.8 Conclusions

Substantial changes in the concentration and quality of DOM in the Florida Everglades have occurred due to alterations in surface water hydrology, nutrient enrichment, and associated shifts in plant communities. Chief among these changes is an increase in the concentration of peat-derived, aromatic, and subsequently highly sulfidized, DOM in the northern Everglades. Given the state of knowledge regarding Hg-DOM interactions, it is hypothesized that increased DOM concentrations and accompanying shifts in DOM character are partially responsible (along with elevated atmospheric inputs of Hg and sulfate from agricultural runoff) for elevated MeHg production in the Everglades. The negative environmental impacts arising from DOM's role in enhancing MeHg production and transport may be partially offset by DOM's role in enhancing MeHg photodegradation and decreasing MeHg uptake by plankton. Quantifying the direction and magnitude of the net effect of increased DOM flux and alteration of DOM quality on Hg concentrations in biota deserves further attention, especially in the context of Everglades' ecosystem restoration. For example, restoration of sheet flow from the northern Everglades into ENP would likely increase delivery of nutrients (including sulfate) and DOM to ENP, resulting in ENP surface water with DOM concentrations and quality becoming more similar to the present-day northern Everglades. As outlined in this review, changes in

DOM concentration and quality have the potential to impact microbial activity, $Hg(II)_i$ bioavailability for methylation, $Hg(II)_i$ and MeHg mobility, MeHg photodegradation, and the efficiency of MeHg trophic transfer.

As another example, efforts to limit phosphorous pollution have resulted in installation of stormwater treatment areas (STAs) for biological phosphorous removal. Water leaving these STAs has high DOC concentrations and elevated $SUVA_{254}$ (Aiken et al. 2011), and MeHg production may be stimulated in areas receiving these STA effluents. A strategically deployed network of sensors for high temporal resolution monitoring of DOM concentration, and perhaps quality (enabled by rapid improvements in sensor technology; Rode et al. 2016), may represent a valuable tool for adaptive management in the Everglades.

Acknowledgements This chapter is dedicated to the memory of George Aiken, who contributed many seminal studies of DOM in the Everglades and elsewhere, and who provided invaluable mentorship to the author.

References

Aiken GR, Gilmour CC, Krabbenhoft DP, Orem W (2011) Dissolved organic matter in the Florida Everglades: implications for ecosystem restoration. Crit Rev Environ Sci Technol 41:217–248

Aquatic Cycling of Mercury in the Everglades Database. https://sofia.usgs.gov/exchange/acme/. Accessed 10 May 2017

Babiarz CL, Benoit JM, Shafer MM, Andren AW (1998) Seasonal influences on partitioning and transport of total and methylmercury in rivers from contrasting watersheds. Biogeochemistry 41:237–257

Balogh SJ, Nollet YH, Swain EB (2004) Redox chemistry in Minnesota streams during episodes of increased methylmercury discharge. Environ Sci Technol 38:4921–4927

Barkay T, Gillman M, Turner R (1997) Effects of dissolved organic carbon and salinity on bioavailability of mercury. Appl Environ Microbiol 63:4267–4271

Bates AL, Orem WH, Harvey JW (2002) Tracing sources of sulfur in the Florida Everglades. J Environ Qual 31:287–299

Benoit J, Gilmour C, Mason R, Heyes A (1999a) Sulfide controls on mercury speciation and bioavailability to methylating bacteria in sediment pore waters. Environ Sci Technol 33:951–957

Benoit J, Mason R, Gilmour C (1999b) Estimation of mercury-sulfide speciation in sediment pore waters using octanol-water partitioning and implications for availability to methylating bacteria. Environ Toxicol Chem 18:2138–2141

Bergamaschi BA, Krabbenhoft DP, Aiken GR et al (2012) Tidally driven export of dissolved organic carbon, total mercury, and methylmercury from a mangrove-dominated estuary. Environ Sci Technol 46:1371–1378

Black FJ, Poulin BA, Flegal AR (2012) Factors controlling the abiotic photo-degradation of monomethylmercury in surface waters. Geochim Cosmochim Acta 84:492–507

Brigham ME, Wentz DA, Aiken GR, Krabbenhoft DP (2009) Mercury cycling in stream ecosystems. 1. Water column chemistry and transport. Environ Sci Technol 43:2720–2725. https://doi.org/10.1021/es802694n

Castro H, Reddy KR, Ogram A (2002) Composition and function of sulfate-reducing prokaryotes in eutrophic and pristine areas of the Florida Everglades. Appl Environ Microbiol 68:6129–6137

Chasar LC, Scudder BC, Stewart AR et al (2009) Mercury cycling in stream ecosystems. 3. Trophic dynamics and methylmercury bioaccumulation. Environ Sci Technol 43:2733–2739

Chen M, Maie N, Parish K, Jaffé R (2013) Spatial and temporal variability of dissolved organic matter quantity and composition in an oligotrophic subtropical coastal wetland. Biogeochemistry 115:167–183

Christensen GA, Wymore AM, King AJ et al (2016) Development and validation of broad-range qualitative and clade-specific quantitative molecular probes for assessing mercury methylation in the environment. Appl Environ Microbiol 82:6068–6078

Cleckner LB, Garrison PJ, Hurley JP, Olson ML (1998) Trophic transfer of methyl mercury in the northern Florida Everglades. Biogeochemistry 40:347–361

Cory R, McKnight D (2005) Fluorescence spectroscopy reveals ubiquitous presence of oxidized and reduced quinones in dissolved organic matter. Environ Sci Technol 39:8142–8149

Davis SE III, Childers DL, Noe GB (2006) The contribution of leaching to the rapid release of nutrients and carbon in the early decay of wetland vegetation. Hydrobiologia 569:87–97

Deonarine A, Hsu-Kim H (2009) Precipitation of mercuric sulfide nanoparticles in NOM-containing water: implications for the natural environment. Environ Sci Technol 43:2368–2373

Deonarine A, Lau BLT, Aiken GR et al (2011) Effects of humic substances on precipitation and aggregation of zinc sulfide nanoparticles. Environ Sci Technol 45:3217–3223

Dittman JA, Shanley JB, Driscoll CT et al (2010) Mercury dynamics in relation to dissolved organic carbon concentration and quality during high flow events in three northeastern U.S. streams. Water Resour Res 46:W07522

Drexel R, Haitzer M, Ryan J et al (2002) Mercury(II) sorption to two Florida Everglades peats: evidence for strong and weak binding and competition by dissolved organic matter released from the peat. Environ Sci Technol 36:4058–4064

Driscoll CT, Blette V, Yan C, Schofield CL (1995) The role of dissolved organic carbon in the chemistry and bioavailability of mercury in remote Adirondack lakes. Water Air Soil Pollut 80:499–508

Drott A, Lambertsson L, Bjorn E, Skyllberg U (2007) Importance of dissolved neutral mercury sulfides for methyl mercury production in contaminated sediments. Environ Sci Technol 41:2270–2276

Fernández-Gómez C, Drott A, Bjorn E et al (2013) Towards universal wavelength-specific photodegradation rate constants for methyl mercury in humic waters, exemplified by a boreal Lake-wetland gradient. Environ Sci Technol 47:6279–6287

Gasper J, Aiken G, Ryan J (2007) A critical review of three methods used for the measurement of mercury (Hg^{2+})-dissolved organic matter stability constants. Appl Geochem 22:1583–1597

Gerbig C, Kim C, Stegemeier J et al (2011) Formation of nanocolloidal metacinnabar in mercury-DOM-sulfide systems. Environ Sci Technol 45:9180–9187

Gilmour C, Krabbenhoft DP, Orem W (2004) Appendix 2B-3: mesocosm studies to quantify how methylmercury in the Everglades responds to changes in mercury, sulfur, and nutrient loading. In: 2004 Everglades consolidated report, South Florida Water Management District and Florida Department of Environmental Protection

Gilmour CC, Podar M, Bullock AL et al (2013) Mercury methylation by novel microorganisms from new environments. Environ Sci Technol 47:11810–11820

Gondikas AP, Jang EK, Hsu-Kim H (2010) Influence of amino acids cysteine and serine on aggregation kinetics of zinc and mercury sulfide colloids. J Colloid Interface Sci 347:167–171

Gorski PR, Armstrong DE, Hurley JP, Krabbenhoft DP (2008) Influence of natural dissolved organic carbon on the bioavailability of mercury to a freshwater alga. Environ Pollut 154:116–123

Graham AM, Aiken GR, Gilmour CC (2012) Dissolved organic matter enhances microbial mercury methylation under sulfidic conditions. Environ Sci Technol 46:2715–2723

Graham AM, Aiken GR, Gilmour CC (2013) Effect of dissolved organic matter source and character on microbial Hg methylation in Hg–S–DOM solutions. Environ Sci Technol 47:5746–5754

Graham AM, Cameron-Burr KT, Hajic HA et al (2017) Sulfurization of dissolved organic matter increases Hg-sulfide-dissolved organic matter bioavailability to a Hg-methylating bacterium. Environ Sci Technol 51:9080–9088

Gu B, Bian Y, Miller CL, Dong W (2011) Mercury reduction and complexation by natural organic matter in anoxic environments. Proc Nat Acad Sci 108:1479–1483

Haitzer M, Aiken G, Ryan J (2002) Binding of mercury(II) to dissolved organic matter: the role of the mercury-to-DOM concentration ratio. Environ Sci Technol 36:3564–3570

Haitzer M, Aiken G, Ryan J (2003) Binding of mercury(II) to aquatic humic substances: influence of pH and source of humic substances. Environ Sci Technol 37:2436–2441

Hall BD, Aiken GR, Krabbenhoft DP et al (2008) Wetlands as principal zones of methylmercury production in southern Louisiana and the Gulf of Mexico region. Environ Poll 154:124–134

He F, Zheng W, Liang L, Gu B (2012) Mercury photolytic transformation affected by low-molecular-weight natural organics in water. Sci Tot Environ 416:429–435

Heitmann T, Blodau C (2006) Oxidation and incorporation of hydrogen sulfide by dissolved organic matter. Chem Geol 235:12–20

Helms JR, Stubbins A, Ritchie JD et al (2008) Absorption spectral slopes and slope ratios as indicators of molecular weight, source, and photobleaching of chromophoric dissolved organic matter. Limnol Oceanogr 53:955–969

Hertkorn N, Harir M, Cawley KM et al (2016) Molecular characterization of dissolved organic matter from subtropical wetlands: a comparative study through the analysis of optical properties, NMR and FTICR/MS. Biogeosciences 13:2257–2277

Hintelmann H, Keppel-Jones K, Evans RD (2000) Constants of mercury methylation and demethylation rates in sediments and comparison of tracer and ambient mercury availability. Environ Toxicol Chem 19:2204–2211

Hoffmann M, Mikutta C, Kretzschmar R (2012) Bisulfide reaction with natural organic matter enhances arsenite sorption: insights from X-ray absorption spectroscopy. Environ Sci Technol 46:11788–11797

Holmes CD, Krishnamurthy NP, Caffrey JM et al (2016) Thunderstorms increase mercury wet deposition. Environ Sci Technol 50:9343–9350

Horvath O, Vogler A (1993) Photoredox chemistry of chloromercurate(II) complexes in acetonitrile. Inorg Chem 32:5485–5489

Hsu H, Sedlak D (2003) Strong Hg(II) complexation in municipal wastewater effluent and surface waters. Environ Sci Technol 37:2743–2749

Hurley JP, Krabbenhoft DP, Cleckner LB, Olson ML (1998) System controls on the aqueous distribution of mercury in the northern Florida Everglades. Biogeochemistry 40:293–310

Jeremiason JD, Portner JC, Aiken GR et al (2015) Photoreduction of Hg(II) and photodemethylation of methylmercury: the key role of thiol sites on dissolved organic matter. Environ Sci Process Impacts 17:1892–1903

Jonsson S, Skyllberg U, Nilsson MB et al (2014) Differentiated availability of geochemical mercury pools controls methylmercury levels in estuarine sediment and biota. Nat Commun 5:4624

Khwaja A, Bloom P, Brezonik P (2006) Binding constants of divalent mercury (Hg^{2+}) in soil humic acids and soil organic matter. Environ Sci Technol 40:844–849

Khwaja AR, Bloom PR, Brezonik PL (2010) Binding strength of methylmercury to aquatic NOM. Environ Sci Technol 44:6151–6156

King J, Kostka JE, Frischer M, Saunders F (2000) Sulfate-reducing bacteria methylate mercury at variable rates in pure culture and in marine sediments. Appl Environ Microbiol 66:2430–2437

Krabbenhoft DP, Hurley JP, Olson ML, Cleckner LB (1998) Diel variability of mercury phase and species distributions in the Florida Everglades. Biogeochemistry 40:311–325

Kucharzyk KH, Deshusses MA, Porter KA, Hsu-Kim H (2015) Relative contributions of mercury bioavailability and microbial growth rate on net methylmercury production by anaerobic mixed cultures. Environ Sci Process Impacts 17:1568–1577

Lamborg C, Tseng C, Fitzgerald W et al (2003) Determination of the mercury complexation characteristics of dissolved organic matter in natural waters with "reducible Hg" titrations. Environ Sci Technol 37:3316–3322

Lenherr I, St. Louis V (2009) Importance of ultraviolet radiation in photodemethylation of methylmercury in freshwater ecosystems. Environ Sci Technol 43:5692–5698
Liem-Nguyen V, Bouchet S, Bjorn E (2015) Determination of sub-nanomolar levels of low molecular mass thiols in natural waters by liquid chromatography tandem mass spectrometry after derivatization with p-(hydroxymercuri) benzoate and online preconcentration. Anal Chem 87:1089–1096
Lindberg S, Dong W, Chanton J et al (2005) A mechanism for bimodal emission of gaseous mercury from aquatic macrophytes. Atmos Environ 39:1289–1301
Liu G, Cai Y, Philippi T et al (2008) Distribution of total and methylmercury in different ecosystem compartments in the Everglades: implications for mercury bioaccumulation. Environ Pollut 153:257–265
Liu G, Cai Y, Mao Y et al (2009) Spatial variability in mercury cycling and relevant biogeochemical controls in the Florida Everglades. Environ Sci Technol 43:4361–4366
Liu G, Naja GM, Kalla P et al (2011) Legacy and fate of mercury and methylmercury in the Florida Everglades. Environ Sci Technol 45:496–501
Luengen AC, Fisher NS, Bergamaschi BA (2012) Dissolved organic matter reduces algal accumulation of methylmercury. Environ Toxicol Chem 31:1712–1719
MacCrehan W, Shea D (1995) Temporal relationship of thiols to inorganic sulfur compounds in anoxic Chesapeake Bay sediment porewater. ACS Symp Ser 612:294–310
Madden AS, Hochella MF Jr (2005) A test of geochemical reactivity as a function of mineral size: manganese oxidation promoted by hematite nanoparticles. Geochim Cosmochim Acta 69:389–398
Maie N, Yang C, Miyoshi T et al (2005) Chemical characteristics of dissolved organic matter in an oligotrophic subtropical wetland/estuarine ecosystem. Limnol Oceanogr 50:23–35
Maie N, Jaffé R, Miyoshi T, Childers DL (2006) Quantitative and qualitative aspects of dissolved organic carbon leached from senescent plants in an oligotrophic wetland. Biogeochemistry 78:285–314
Manceau A, Nagy KL (2012) Quantitative analysis of sulfur functional groups in natural organic matter by XANES spectroscopy. Geochim Cosmochim Acta 99:206–223
Mazrui NM, Jonsson S, Thota S et al (2016) Enhanced availability of mercury bound to dissolved organic matter for methylation in marine sediments. Geochim Cosmochim Acta 194:153–162
McCormick PV, Harvey JW, Crawford ES (2011) Influence of changing water sources and mineral chemistry on the Everglades ecosystem. Crit Rev Environ Sci Technol 41:28–63
Miller CL, Southworth G, Brooks S et al (2009) Kinetic controls on the complexation between mercury and dissolved organic matter in a contaminated environment. Environ Sci Technol 43:8548–8553
Moreau JW, Gionfriddo CM, Krabbenhoft DP et al (2015) The effect of natural organic matter on mercury methylation by *Desulfobulbus propionicus* 1pr3. Front Microbiol 46:292–299
National Institute of Standards and Technology (2013) NIST critically selected stability constants of metal complexes
Navrotsky A, Mazeina L, Majzlan J (2008) Size-driven structural and thermodynamic complexity in iron oxides. Science 319:1635–1638
Parks JM, Johs A, Podar M et al (2013) The genetic basis for bacterial mercury methylation. Science 339:1332–1335
Pickhardt PC, Fisher NS (2007) Accumulation of inorganic and methylmercury by freshwater phytoplankton in two contrasting water bodies. Environ Sci Technol 41:125–131
Pollman CD, Axelrad DM (2014) Mercury bioaccumulation and bioaccumulation factors for Everglades mosquitofish as related to sulfate: a re-analysis of Julian II (2013). Bull Environ Contam Toxicol 93:509–516
Poulin BA, Ryan JN, Nagy KL et al (2017) Spatial dependence of reduced sulfur in Everglades dissolved organic matter controlled by sulfate enrichment. Environ Sci Technol 51:3630–3639
Qian Y, Yin X, Lin H et al (2014) Why dissolved organic matter enhances photodegradation of methylmercury. Environ Sci Technol Lett 1:426–431

Qualls RG, Richardson CJ (2003) Factors controlling concentration, export, and decomposition of dissolved organic nutrients in the Everglades of Florida. Biogeochemistry 62:197–229
Ravichandran M (2004) Interactions between mercury and dissolved organic matter—a review. Chemosphere 55:319–331
Ravichandran M, Aiken G, Reddy M, Ryan J (1998) Enhanced dissolution of cinnabar (mercuric sulfide) by dissolved organic matter isolated from the Florida Everglades. Environ Sci Technol 32:3305–3311
Ravichandran M, Aiken G, Ryan J, Reddy M (1999) Inhibition of precipitation and aggregation of metacinnabar (mercuric sulfide) by dissolved organic matter isolated from the Florida Everglades. Environ Sci Technol 33:1418–1423
Riccardi D, Guo H-B, Parks JM et al (2013) Why mercury prefers soft ligands. J Phys Chem Lett 4:2317–2322
Rode M, Wade AJ, Cohen MJ et al (2016) Sensors in the stream: the high-frequency wave of the present. Environ Sci Technol 50:10297–10307
Rumbold DG, Lange TR, Axelrad DM, Atkeson TD (2008) Ecological risk of methylmercury in Everglades National Park, Florida, USA. Ecotoxicology 17:632–641
Schaefer JK, Morel FMM (2009) High methylation rates of mercury bound to cysteine by Geobacter sulfurreducens. Nat Geosci 2:123–126
Schaefer JK, Rocks SS, Zheng W et al (2011) Active transport, substrate specificity, and methylation of Hg(II) in anaerobic bacteria. Proc Natl Acad Sci USA 108:8714–8719
Scheidt DJ, Kalla PI (2007) EPA Everglades ecosystem assessment: water management and quality, eutrophication, mercury contamination, soils and habitat monitoring for adaptive management: a R-EMAP status report. US EPA Region 4, Athens, GA
Schuster PF, Striegl RG, Aiken GR et al (2011) Mercury export from the Yukon River basin and potential response to a changing climate. Environ Sci Technol 45:9262–9267
Schwarzenbach G, Schellenberg M (1965) Die komplexchemie des methylquecksilber-kations. Helv Chim Acta 48:28–46
Si L, Ariya PA (2011) Aqueous photoreduction of oxidized mercury species in presence of selected alkanethiols. Chemosphere 84:1079–1084
Siciliano SD, O'Driscol NJ, Lean DRS (2002) Microbial reduction and oxidation of mercury in freshwater lakes. Environ Sci Technol 36:3064–3068
Sklar FH, Chimney MJ, Newman S et al (2005) The ecological–societal underpinnings of Everglades restoration. Front Ecol 3:161–169
Skyllberg U (2008) Competition among thiols and inorganic sulfides and polysulfides for Hg and MeHg in wetland soils and sediments under suboxic conditions: illumination of controversies and implications for MeHg net production. J Geophys Res Biogeosci 113:G00C03
Skyllberg U, Bloom P, Qian J et al (2006) Complexation of mercury(II) in soil organic matter: EXAFS evidence for linear two-coordination with reduced sulfur groups. Environ Sci Technol 40:4174–4180
Sleighter RL, Chin Y-P, Arnold WA et al (2014) Evidence of incorporation of abiotic S and N into prairie wetland dissolved organic matter. Environ Sci Technol Lett 1:345–350
Slowey AJ (2010) Rate of formation and dissolution of mercury sulfide nanoparticles: the dual role of natural organic matter. Geochim Cosmochim Acta 74:4693–4708
Stern J, Wang Y, Gu B, Newman J (2007) Distribution and turnover of carbon in natural and constructed wetlands in the Florida Everglades. Appl Geochem 22:1936–1948
Szczuka A, Morel FMM, Schaefer JK (2015) Effect of thiols, zinc, and redox conditions on Hg uptake in *Shewanella oneidensis*. Environ Sci Technol 49:7432–7438
Tai C, Li Y, Yin Y et al (2014) Methylmercury photodegradation in surface water of the Florida Everglades: importance of dissolved organic matter-methylmercury complexation. Environ Sci Technol 48:7333–7340
Thomas SA, Tong T, Gaillard J-F (2014) Hg(II) bacterial biouptake: the role of anthropogenic and biogenic ligands present in solution and spectroscopic evidence of ligand exchange reactions at the cell surface. Metallomics 6:2213–2222

Tossell JA (1998) Theoretical study of the photodecomposition of methyl Hg complexes. J Phys Chem A 102:3587–3591

Tsui MTK, Finlay JC (2011) Influence of dissolved organic carbon on methylmercury bioavailability across Minnesota stream ecosystems. Environ Sci Technol 45:5981–5987

Wagner S, Jaffé R, Cawley K et al (2015) Associations between the molecular and optical properties of dissolved organic matter in the Florida Everglades, a model coastal wetland system. Front Chem 3:155

Wang Y, Hsieh YP, Landing WM et al (2002) Chemical and carbon isotopic evidence for the source and fate of dissolved organic matter in the northern Everglades. Biogeochemistry 61:269–289

Waples J, Nagy K, Aiken G, Ryan J (2005) Dissolution of cinnabar (HgS) in the presence of natural organic matter. Geochim Cosmochim Acta 69:1575–1588

Warner KA, Roden EE, Bonzongo J-C (2003) Microbial mercury transformation in anoxic freshwater sediments under iron-reducing and other electron-accepting conditions. Environ Sci Technol 37:2159–2165

Watras CJ, Back RC, Halvorsen S et al (1998) Bioaccumulation of mercury in pelagic freshwater food webs. Sci Total Environ 219:183–208

Weishaar JL, Aiken GR, Bergamaschi BA et al (2003) Evaluation of specific ultraviolet absorbance as an indicator of the chemical composition and reactivity of dissolved organic carbon. Environ Sci Technol 37:4702–4708

Wiatrowski HA, Ward PM, Barkay T (2006) Novel reduction of mercury(II) by mercury-sensitive dissimilatory metal reducing bacteria. Environ Sci Technol 40:6690–6696

Yamashita Y, Scinto LJ, Maie N, Jaffé R (2010) Dissolved organic matter characteristics across a subtropical wetland's landscape: application of optical properties in the assessment of environmental dynamics. Ecosystems 13:1006–1019

Yu Z-G, Peiffer S, Göttlicher J, Knorr K-H (2015) Electron transfer budgets and kinetics of abiotic oxidation and incorporation of aqueous sulfide by dissolved organic matter. Environ Sci Technol 49:5441–5449

Zhang T, Hsu-Kim H (2010) Photolytic degradation of methylmercury enhanced by binding to natural organic ligands. Nat Geosci 3:473–476

Zhang J, Wang F, House J, Page B (2004) Thiols in wetland interstitial waters and their role in mercury and methylmercury speciation. Limnol Oceanogr 49:2276–2286

Zhang T, Kim B, Levard C et al (2012) Methylation of mercury by bacteria exposed to dissolved, nanoparticulate, and microparticulate mercuric sulfides. Environ Sci Technol 46:6950–6958

Zhang T, Kucharzyk KH, Kim B et al (2014) Net methylation of mercury in estuarine sediment microcosms amended with dissolved, nanoparticulate, and microparticulate mercuric sulfides. Environ Sci Technol 48:9133–9141

Zheng W, Hintelmann H (2010) Isotope fractionation of mercury during its photochemical reduction by low-molecular-weight organic compounds. J Phys Chem A 114:4246–4253

Zheng W, Liang L, Gu B (2012) Mercury reduction and oxidation by reduced natural organic matter in anoxic environments. Environ Sci Technol 46:292–299

Zhong H, Wang W-X (2009) Controls of dissolved organic matter and chloride on mercury uptake by a marine diatom. Environ Sci Technol 43:8998–9003

Chapter 5
Phosphorus in the Everglades and Its Effects on Oxidation-Reduction Dynamics

Sara A. Phelps and Todd Z. Osborne

Abstract Nutrient enrichment—particularly with respect to phosphorus—has long been a major concern in the Everglades (see Chap. 2, Volume I). This perturbation is of keen interest with respect to the Everglades mercury (Hg) problem because the biogeochemical cycling of Hg in aquatic ecosystems is intrinsically linked to trophic state through multiple pathways, including the effects of nutrient status on food web structure and dynamics, in situ particle production, and redox dynamics in surficial sediments (see Fig. 1.1, Chap. 1, this volume). As a result, decision makers charged with the responsibility of restoring the Everglades must also consider the resultant impacts of management strategies on not just trophic state dynamics, but also the linked effects of those strategies on Hg biogeochemical cycling and trophic transfer. This chapter thus reviews phosphorus enrichment in the Everglades and its effects on Hg biogeochemical cycling, including its effects on methyl mercury production related to perturbations in redox dynamics in particular.

Keywords Phosphorus · Redox · Nutrient enrichment · Trophic dynamics · Legacy phosphorus

5.1 Introduction

Eutrophication in aquatic systems is dictated by the principles of limiting nutrient theory. A nutrient is considered limiting if additions of that nutrient to the system result in an increase in net primary production. In most instances, phosphorus is the limiting nutrient controlling eutrophication in freshwater systems, and nitrogen is the limiting nutrient in marine systems. The differences in nutrient limitation between these two systems lie within various ecological and biogeochemical mechanisms. In

S. A. Phelps (✉) · T. Z. Osborne
Wetland Biogeochemistry Laboratory, University of Florida, Gainesville, FL, USA

Whitney Laboratory for Marine Bioscience, University of Florida, St. Augustine, FL, USA
e-mail: saraphelps@ufl.edu

© Springer Nature Switzerland AG 2019

D. G. Rumbold et al. (eds.), *Mercury and the Everglades. A Synthesis and Model for Complex Ecosystem Restoration*, https://doi.org/10.1007/978-3-030-32057-7_5

the historically oligotrophic Everglades, research towards restoration has focused on the effects of surplus phosphorus in the system (Davis and Ogden 1994). The Everglades developed under severe phosphorus-limited conditions, resulting in unique vegetation, biota (e.g. periphyton), and associated biogeochemical conditions. Under these conditions, atmospheric deposition was the most significant phosphorus source, with few alternative phosphorus inputs available besides internal recycling of phosphorus from soils to plants (Osborne et al. 2015; Scheidt et al. 2000; Noe et al. 2001). The tight biogeochemical cycling of phosphorus forms in the ecosystem resulting from the limited availability of phosphorus results in the majority of phosphorus taken up by plant or microbial tissues as well as particulate detrital material. Due to limited phosphorus availability, plant communities adapted to growing in low phosphorus conditions became dominant in the Everglades (Osborne et al. 2011c; Davis 1991; Miao and DeBusk 1999; Miao and Sklar 1998).

Multiple studies agree that even small amounts of additional phosphorus to the Everglades ecosystem can result in significant changes in ecosystem productivity and functioning (Osborne et al. 2011c, 2015; Gaiser et al. 2005; Childers et al. 2003; Chiang et al. 2000). Perhaps where these effects are most readily observed are in changes in vegetation community structure. The historical oligotrophic condition of the Everglades resulted in native sawgrass (*C. jamaicense*) dominating much of the region. With the addition of allochthonous nutrient inputs, large monocultures of cattail (*Typha domingensis*) have become increasingly common (Davis 1994; Newman et al. 1998; Richardson et al. 2008). Cattails grow rapidly in the presence of increased available phosphorus, thereby effectively outcompeting vegetation adapted for nutrient-limited conditions such as sawgrass (Osborne et al. 2011c; Davis 1991, 1994; Miao and Sklar 1998; Miao and DeBusk 1999). Total phosphorus (TP) concentrations between 500 and 600 mg kg^{-1} in soil have been used by researchers to indicate areas of enrichment in the Everglades (Craft and Richardson 1993; Reddy et al. 1991).

The shift in vegetation community structure in the Everglades has far-reaching implications. As the hub of soil and water biogeochemistry, the carbon source in a system impacts nutrient cycling and resulting concentrations in both soil and water. As such, eutrophication can influence other ecological processes of concern, including mercury (Hg) methylation.

5.2 Phosphorus in the Everglades

5.2.1 Sources of Everglades Phosphorus Enrichment

Significant declines of the Everglades ecosystem seen in the 1970s through the 1990s including cattail encroachment and loss of calcareous periphyton mats have been linked to phosphorus enrichment (Sklar et al. 2005; Wu et al. 1997; Gaiser et al. 2004), although the initial research and the conclusions ascribing cause and effect were highly controversial (Reddy et al. 1993; Koch and Reddy 1992; Rader and Richardson 1992; Craft et al. 1995). Extensive research in the Water Conservation Areas (WCAs) began in the 1990s to better understand phosphorus loading and ecosystem changes resulting from increased phosphorus availability, including vegetation community shifts and habitat degradation (SFWMD 1992; Davis 1994; Noe et al. 2001; McCormick et al. 2002). Additional studies were conducted to characterize phosphorus enrichment in soils following the tremendous evidence of phosphorus enrichment in the northern Everglades provided by these initial investigations (Koch and Reddy 1992; Amador and Jones 1993, 1995; Craft and Richardson 1993; Reddy et al. 1993; DeBusk et al. 1994; Qualls and Richardson 1995; Newman et al. 1996, 1997, 1998; Miao and Sklar 1998; Noe et al. 2002, 2003; Daoust and Childers 2004; Chambers and Pederson 2006). As the long term consequences of phosphorus enrichment in soils were recognized, additional studies were undertaken to understand changes to soil condition over time (Childers et al. 2003; Reddy et al. 2005; Bruland et al. 2006; Rivero et al. 2009; Newman et al. 2009).

Results from these studies concur that waters entering wetlands from agriculture operations upstream have caused phosphorus enrichment of soils in many areas across the Everglades landscape (Scheidt and Kalla 2007; Hagerthey et al. 2008; Osborne et al. 2011a, c). Data from the past 30 years suggest that drainage canals have shunted runoff to the Everglades with TP concentrations ranging from 100 to 1000 μg/L (Sklar et al. 2005). These concentrations are likely associated with the liberal use of fertilizers in the EAA (SFWMD 2007), necessitating stormwater treatment areas (STAs) to remove and store nutrients from agricultural inflows before water enters the lower portion of the Everglades (Pietro et al. 2007). The association of eutrophication across the northern Everglades with runoff from the EAA was one of the significant catalysts in Everglades restoration, and has since been incorporated into numerous ecological models and performance measures to assist with restoration efforts (Ogden 2005; RECOVER 2006).

Surface runoff from the EAA is not the only source of phosphorus loading to the Everglades. Soil subsidence, a result of hydrologic alterations to wetlands, also plays a role in phosphorus enrichment. Rich organic Histosol soils of the Lake Okeechobee area are formed under saturated soils but are well suited as agricultural soils when drained. As such, the northern Everglades were engineered with extensive ditching to drain the wetlands for agricultural purposes. These alterations compromise the unique conditions that support the Everglades ecosystem and the peat accretion that characterizes Histosols (Davis 1994). Drainage results in primary subsidence and secondary subsidence as a result of two distinct processes. Primary subsidence is a result of soil sinking under its own weight due to a loss of buoyant force provided by saturation. More important to phosphorus enrichment is secondary subsidence, where organic matter is oxidized as a result of aerobic conditions associated with drainage. Since the advent of draining circa 1900–1910, widespread soil subsidence within the EAA has been investigated and documented (Davis 1946; Jones 1948; Stephens and Johnson 1951; Gleason and Stone 1994; Snyder and Davidson 1994; Snyder 2005). Drainage of these wetland soils has resulted not only in soil losses of up to 3 m, but has also contributed to elevated nutrient runoff to the Everglades system downstream (WCAs, Everglades National Park) via mineralization (Osborne et al. 2011c).

5.2.2 *Spatial Distribution of Soil Phosphorus*

5.2.2.1 Northern Everglades

Ecological conditions in the northern Everglades are clearly documented by a large body of research conducted over the past few decades (LOTAC II 1990; Scheidt et al. 2000; Noe et al. 2001; McCormick et al. 2002; Childers et al. 2003; Hagerthey et al. 2008). Several studies have implemented monitoring at a sufficient spatial and temporal scale to identify trends in phosphorus enrichment with time. These studies are important for determining the source of phosphorus to northern Everglades, as well as for estimating timelines for restoration. Over the 10-year period from 1995–1996 to 2005, the proportion of soils exceeding 500 or 400 mg/kg thresholds both increased by approximately 50% (Scheidt and Kalla 2007). In light of continued decreases in phosphorus loading, the mechanisms (and spatial variability) behind these increases in soil phosphorus enrichment must be considered.

In general, peat soils located in WCA-3A north of Alligator Alley, northern WCA-2A, and the edges of Loxahatchee National Wildlife Refuge close to the rim canal have the highest soil TP concentration (Osborne et al. 2011c, 2015). On a volume basis (μg cm^{-3}), the peat soils in WCA-2A and at the edges of Loxahatchee NWR have the highest TP content (Osborne et al. 2011c, 2015).

The magnitude of TP enrichment throughout the northern Everglades is largely mediated by hydrology. For example, TP enrichment of soils in WCA-1 is concentrated in fringe areas where EAA runoff-derived inflows interact more directly with the marsh. In WCA-3A, consistent temporal trends in TP concentrations across the region are not apparent due to multiple drivers related to hydrology at work. A comparison of 2003 data (Bruland et al. 2006) to 1992 data (Reddy et al. 1994) found that TP concentrations in the upper portion of soils (0–10 cm) increased 53% (Bruland et al. 2007), with the proportion of soils classified as enriched increasing by approximately 50%. In some areas, the likely cause of increased TP concentrations is surface water inputs, since peripheral areas of WCA-3 receiving surface water inputs from canals exhibit elevated TP concentrations. In other regions, TP on a volumetric basis increased from 1992 to 2003, with concurrent increases in bulk density. Investigations aimed at determining the mechanism for phosphorus increases in this area have suggested oxidation of organic matter as a result of subsidence and potentially fire as possible drivers (Scheidt et al. 2000; Bruland et al. 2007; Scheidt and Kalla 2007). As previously discussed, significant soil subsidence in this region supports this theory (Scheidt and Kalla 2007).

Internal processes within the Northern Everglades wetlands also control the degree of observed TP enrichment. In WCA-2, comparisons of 2003 data (Rivero et al. 2007) with data from 1990 (DeBusk et al. 1994) and 1998 (DeBusk et al. 2001) identified spatial variation in floc TP concentrations corresponding to sinks and sources of TP in the area (Grunwald et al. 2008). While inflow TP concentrations were lower for the 5-year time period from 1999 to 2003, there is evidence that soils in portions of WCA-2 are saturated with phosphorus and result in a net flux of TP to the water column (Grunwald et al. 2008). This pattern suggests that while phosphorus loads to WCA-2A have been reduced, internal cycling of phosphorus in certain portions of WCA-2 has continued to contribute to elevated water column-TP concentrations. It is likely that these trends will continue over time as surface soils with elevated TP concentrations are either buried or transported following flow patterns within the wetland.

5.2.2.2 Southern Everglades

With the exception of the Everglades Ecosystem Assessment Regional Environmental Monitoring and Assessment Program (Scheidt and Kalla 2007; Kalla and Scheidt 2017) and Osborne et al. (2011b), the edaphic properties of Everglades National Park (ENP) in its spatial entirety, the landscape end member of the Everglades ecosystem, have been largely ignored. The unique eco-type diversity of the ENP makes generalizations concerning soil biogeochemical properties difficult. Perhaps for this reason, many studies characterizing ENP soils (Chen et al. 2000; Childers et al. 2003; Daoust and Childers 2004; Chambers and Pederson 2006) use small numbers of sample sites focusing on one or two of the five main ENP ecotypes (Shark River Slough, eastern marl prairies, western marl and wet prairies, Taylor Slough, and the mangrove interface). While at times more practical, these sampling approaches render landscape-level inferences of ENP soil biogeochemistry problematic.

Recent efforts to characterize the spatial diversity of ENP have indicated a wide range of soil properties across the region, with considerable variability in bulk density (0.01–1.39 g/cm) and loss on ignition (4.1–94.6%) when compared to northern Everglades soils. Also unique to ENP soils is the storage of TP across various soil compartments. In contrast to all other published studies of Everglades soils, TP in ENP soils tends to be lowest in the floc layer and highest in the soil surface layer. Despite this difference in phosphorus storage pools, only 16% of ENP sites are considered phosphorus enriched based on TP concentrations in the soil surface layer (Osborne et al. 2011b). When compared to 21% in WCA-1 (Corstanje et al. 2006), 39% in WCA-2A (Rivero et al. 2007), and 25% in WCA-3A (Bruland et al. 2006), ENP is the least impacted region with respect to soil phosphorus enrichment. A comparison of soil measures from 2006 (Osborne et al. 2014), 2012 (Osborne et al. 2012), and 2018 (August 2018) suggest a downward trend in soil TP enrichment within the organic soils of Taylor Slough. This is correlated with improved water quality from restoration measures in the northeast portions of ENP.

Soil TP spatial patterns in ENP do not suggest gradient-driven enrichment associated with inflows as seen in the north Everglades (Koch and Reddy 1992; DeBusk et al. 1994; Qualls and Richardson 1995; Newman et al. 1997), with the exception of soils in the main channel of upper Taylor Slough (Osborne et al. 2014; August 2018). Instead, TP patterns are likely the result of biogeochemical conditions unique to the southern Everglades. For example, elevated soil TP concentrations in the southeastern mangrove interface region are thought to reflect the natural accumulation and burial of phosphorus from mangrove primary production (Rivera-Monroy et al. 2011).

5.2.3 *Ecological Effects*

Phosphorus enrichment and resulting eutrophic conditions have caused measurable shifts in key biological communities that are known to be highly nutrient-sensitive (Osborne et al. 2011c). These communities are linked to soil characteristics through biogeochemical processes and can therefore influence phosphorus enrichment and processing in both soils and the water column. As a result of the tight biogeochemical cycling that occurs between nutrients in the water column, soils, and plants, surface water quality and vegetative communities are complexly interconnected with Everglades soil characteristics (Daoust and Childers 2004; Davis et al. 2005; Ogden 2005; Hagerthey et al. 2008). Surface water with high nutrient concentrations has resulted in altered vegetation and periphyton communities, thereby altering litter quality and soil processes including microbial activity (Wright and Reddy 2001), mineral precipitation, and accretion (Osborne et al. 2011c).

5.2.3.1 Vegetation

Vegetation communities in wetlands are largely dictated by hydroperiod and nutrient availability. Long hydroperiods (8–12 months/year) yield emergent graminoid or floating-leaved aquatic macrophytes as the primary producers (Osborne et al. 2011c). In these systems, extended durations of anaerobic conditions result in the slow decomposition of roots and detrital materials which ultimately dictates peat soil formation. Shorter hydroperiods (3–8 months/year) result in lower macrophyte productivity and the dominance of periphytic and benthic algal communities (Osborne et al. 2011c; Davis et al. 2005; Ewe et al. 2006).

Variation of vegetation communities within hydroperiods is controlled by nutrient availability. Historically, sawgrass and other low phosphorus-adapted plants dominated the Everglades (Osborne et al. 2011c). As discussed previously, increases in phosphorus concentrations have resulted in a distinctive shift in vegetation communities. Native marsh communities dominated by sawgrass are often outcompeted by cattails as soil TP concentrations exceed 650 mg/kg (Wu et al. 1997).

Shifts in vegetation have significant impacts on the Everglades ecosystem through alterations to the heterotrophic food web. By changing the carbon source, organic matter quality is also altered. This is evidenced in vegetation decomposition rates, where cattail detritus has been found to decompose at a faster rate compared to sawgrass (Craft and Richardson 1997; DeBusk and Reddy 1998, 2005). This rapid organic matter turnover is also linked to phosphorus concentrations, as nutrients within plant material are released during decomposition. As such, the lower lignin content of cattails results in a positive feedback loop that drives further cattail expansion as a result of increased phosphorus turnover within the system. Furthermore, more phosphorus is deposited in the detrital pool as phosphorus re-absorption during senescence is less efficient in cattails versus sawgrass (Miao and Sklar 1998; Osborne et al. 2008).

Eutrophication also leads to higher gross and net sedimentation rates in cattail-dominated systems as compared to sawgrass systems. This is because cattail exhibits extremely high rates of primary production, especially under eutrophic conditions. These higher sedimentation rates result in increases in organic soil accumulation. However, the fractional amounts of carbon permanently sequestered through settling and burial is lower compared to the amount of carbon fixed because of the increased lability of cattail compared to sawgrass. While net sedimentation rates may be higher in cattail systems, the opportunity for permanent storage of carbon is comparatively reduced.

Cattail expansion also negatively influences dissolved oxygen content through increased demand for oxygen as a terminal electron acceptor (TEA) associated with rapid organic matter turnover. This results in poor fish and wading bird habitat by low dissolved oxygen levels limiting the availability and diversity of invertebrates. Rapid organic matter turnover also increases the rates of biogeochemical cycling resulting in accelerated availability of harmful nutrients and metals of concern such as Hg (Osborne et al. 2011b). In these ways, a phosphorus enriched environment results in vegetation shifts which gives rise to negative feedback loops as a result of inefficient phosphorus cycling, ultimately leading to negative impacts to water quality and decreased habitat quality.

5.2.3.2 Biota

In addition to sawgrass, another unique low phosphorus-adapted organism is calcareous periphyton (McCormick and Stevenson 1998). Found throughout the Everglades, periphyton consists of filamentous cyanobacteria and diatoms that form a laminar sheet that can be attached to the soil surface, the surfaces of plants below the water column, or floating on the water surface (Osborne et al. 2011c). The calcium carbonate precipitation associated with the process of marl formation co-precipitates phosphorus, thereby becoming part of the soil (McCormick et al. 2001; Gleason et al. 1974). Periphyton is of particular importance in the Everglades as it is responsible for significant primary production, phosphorus and carbon sequestration, and actively regulating other biogeochemical processes in the water column (Osborne et al. 2011c, 2015; McCormick and O'Dell 1996; Noe et al. 2001; Gaiser et al. 2006). Marls form in Everglades marshes when macrophyte vegetation production is low while calcareous periphyton production is high, a plentiful source of calcium carbonate is available, light can adequately penetrate the water column, and a significant lowering of the water table occurs during most years (Gleason et al. 1974; Browder et al. 1994). As such, the presence of periphyton suggests unspoiled wetland conditions and a system vulnerable to ecosystem alterations.

Periphyton mats are dominated by blue-green algae that are not tolerant of high phosphorus concentrations, and therefore periphyton is not formed in phosphorus-enriched wetlands (Gaiser et al. 2004). Intermediate concentrations of water column phosphorus have been shown to result in relatively high periphyton cover and higher nutrient concentrations, since periphyton growth is nutrient limited (King and

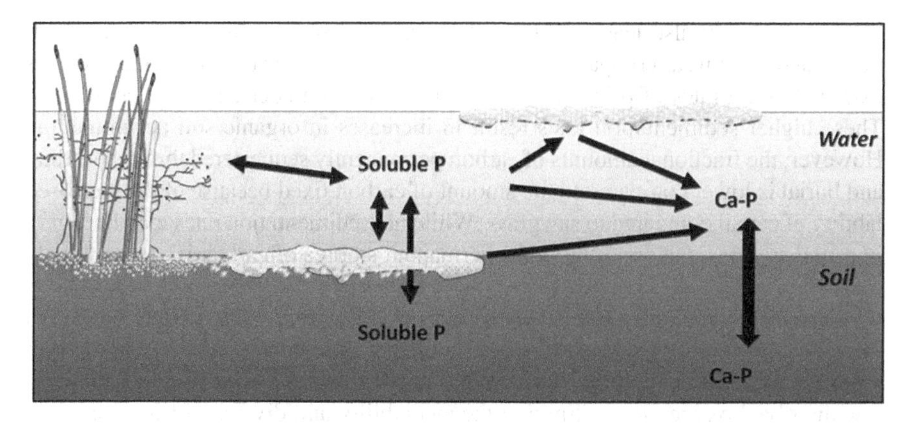

Fig. 5.1 Schematic depicting periphyton cycling phosphorus. Adapted from artwork by P. Inglett from Reddy and DeLaune (2008)

Richardson 2007). However, as resource limitation is overcome and phosphorus enrichment occurs, cyanobacteria begins to dominate over filamentous green algae (McCormick et al. 2001; Gaiser et al. 2005, 2006) and peripohyton cover decreases (King and Richardson 2007). When this occurs, the resulting loss of calcium carbonate precipitation is detrimental to marl formation and accretion, and water column phosphorus concentrations increase (Fig. 5.1). Furthermore, cattail-dominated systems resulting from phosphorus enrichment can result in shading that limits periphyton abundance (Goldsborough and Robinson 1996; Grimshaw et al. 1997).

In accordance with the subsidy-stress model (Odum et al. 1979), nutrient enrichment is often hypothesized to eventually result in the simplification of food webs. Phosphorus enrichment can affect the structure and function of Everglades aquatic food webs through impacts to vegetation and algal species (Chase 2003; King et al. 2004). Since periphyton has been identified as an important component of the Everglades food web, extensive phosphorus enrichment and the attendant loss of calcareous periphyton can result in a loss of taxa that rely heavily on this food resource (King and Richardson 2007). As such, a loss of periphyton results in a decrease in invertebrate trophic diversity (Sergeant et al. 2010). This simplification of the food web carries through to higher trophic levels, as fish density has been shown to ultimately decline at peak phosphorus levels with the disappearance of periphyton (King and Richardson 2007; Liston et al. 2008).

5.2.3.3 Soil Processes

The physical and biogeochemical properties of soil are strong determinants of plant community structure and help regulate nutrient cycling in the Everglades (Osborne et al. 2015). As organic matter accumulates, soils can serve as a sink for elements (DeBusk et al. 1994; Newman et al. 1997; Bruland et al. 2006; Scheidt and Kalla 2007) or as a source of nutrients in biogeochemical cycling (Reddy et al. 2005;

Reddy and DeLaune 2008). In the context of eutrophication, whether soils serve as a source or sink is related to organic matter turnover. As living plants take up nutrients from the water or soil, these nutrients are stored in plant tissues. Soil accretes when plants senesce and plant litter simultaneously contributes to both the organic matter pool and to the available nutrient pool via the breakdown of organic matter and subsequent release of nutrients stored in plant organic tissues. Soil accretion in wetlands is largely dependent on hydrologic regime and vegetation (Reddy and DeLaune 2008).

While the recycling of nutrients between plants and soil is closely coupled, some quantity of nutrients can be exported outside the plant cycle, usually through mineralization of organic matter. As a result, nutrient concentrations in the water column increase. This nutrient release can occur during soil oxidation by microbial activity or through fire. As vegetation shifts from sawgrass to cattails, organic matter typically becomes more labile. This means that organic matter is more readily oxidized as microbes utilize organic matter as an electron donor in order to produce energy. Thus, eutrophication in the Everglades indirectly influences soil processes through alteration of carbon sources. This in turn becomes a positive feedback loop via accelerated mineralization of organic matter, thereby further contributing to eutrophic conditions. It is important to note when considering the net balance of this process that nutrients can also be incorporated into organic matter by microbial uptake and plant growth (Osborne et al. 2011c, 2015). While the processes are complex, it is important to remember that the drivers of soil enrichment of nutrients and ultimate export to the water column can also result from a range of interactions, including simple concentration gradient driven mass transfer.

These soil processes and the resulting concentrations of nutrients in soil are important indicators of nutrient loading in the Everglades and as a measurement of ecosystem change (Reddy et al. 2005; Scheidt and Kalla 2007). The regulation of water column nutrient concentrations by soils means that we cannot understand the extent of ecosystem impact associated with changes in nutrient loading by solely studying water column nutrient concentrations (Reddy and DeLaune 2008). For this reason, decreasing nutrient concentration loadings from inflows does not necessarily result in corresponding water column nutrient concentrations, at least not over short time scales (<5 years[1]). Nonetheless, over long time scales, sediment phosphorus concentrations should generally correlate with overlying water concentrations. Thus, as a general rule, when sediment phosphorus concentrations range from 500 to 600 mg/kg, TP concentrations in the water column tend to exceed 10 ppb (Payne et al. 2003). Nutrient fluxes to and from sediments are the result of dynamic processes that buffer water column nutrient concentrations over potentially extensive

[1]For example, the likely depth of surficial sediments actively involved in exchanging and methylating Hg is perhaps on the order of several centimeters (see Chap. 3, Volume III) while bioturbation can result in exchange depths approximating 10 cm in lacustrine and estuarine sediments for other constituents, including nutrients (Teal et al. 2008). If we assume an average sediment burial rate of 2.33 mm/year for relatively undisturbed sites in the Everglades (Craft and Richardson 1993), these values equate to sediment nutrient turnover rates as long as ~40 years.

periods of time. As such, examining nutrient concentrations in soils provides a better understanding of both the long-term implications (legacy effects) of nutrient enrichment in Everglades wetlands on water quality, and our ability to predict the effects of achieving target nutrient load concentrations.

5.3 Phosphorus Dynamics Effects on Reduction-Oxidation Reactions in Wetland Soils

5.3.1 Overview of Reduction-Oxidation Reactions in Wetland Soils

All other factors being equal, methyl Hg (MeHg) concentrations in Everglades waters increase with increased rates of microbial methylation (Chap. 1, this volume). Microbial methylation depends on anaerobic conditions, with methylating agents commonly including obligate anaerobes like sulfate reducing bacteria (see Chaps. 1 and 3, this volume). Microbial Hg demethylation can occur under aerobic conditions where Hg concentrations are high (reductive demethylation), or under anaerobic conditions where Hg concentrations are low (oxidative demethylation). It is important to note that the net balance of Hg methylation and demethylation is reflected in methylated Hg measurements; however, net MeHg concentrations in most ecosystems studies (including the Everglades) seem to be driven by controls on MeHg production rather than degradation (Chap. 1, this volume). Therefore, understanding the effects of phosphorus dynamics on reduction-oxidation (redox) reactions in wetlands soils is an important component to understanding Hg methylation and resulting MeHg concentrations.

Wetland microorganisms use redox reactions in order to sustain life. In redox reactions, electron donors (commonly organic matter) are oxidized terminal while terminal electron acceptors (TEAs) are reduced in order to produce energy. The electron acceptors used in this process depends on their availability, which is ultimately dictated by hydrology. Since aerobic respiration yields the greatest amount of energy, oxygen is rapidly consumed. In saturated wetland soils, oxygen is quickly depleted as the rate of oxygen consumption outpaces the rate of supply (Mitsch and Gosselink 2000). Once oxygen is depleted, alternative electron acceptors for respiration are sequentially utilized in order of energy yields. The depletion of TEAs is related to redox state and energy yield (Reddy et al. 1986; Reddy and D'Angelo 1996), but soil respiratory pathways in peat wetlands are not strictly dictated solely by TEA availability (Dettling et al. 2006).

While anthropogenic impacts in the Everglades are associated with nutrient loading, it is likely that other elements are simultaneously added to the ecosystem as a result of the same activities (e.g. fertilizer and soil amendment applications). Alternative TEAs can be added to the system not only through these direct anthropogenic sources but also through saltwater intrusion as a result of water table manipulation (Osborne et al. 2011c). An increased availability of TEAs means that soil respiration rates can increase, which affects the stability of soil carbon and can result in subsidence over long periods of time.

Redox reactions drive biogeochemical processes in wetlands by the utilization of organic matter as well as elements functioning as TEAs. This process results in the conversion of elements from one redox state to another as they are reduced or oxidized. For example, in wetlands, denitrification can be observed when nitrate is utilized as a TEA. In natural wetlands, nitrate is derived primarily from precipitation or from the chemoautotrophic oxidation of ammonium to nitrite and ultimately to nitrate. In impacted regions, additional inorganic nitrogen enters the system from agricultural canals or through increased soil nitrogen mineralization where hydroperiods have been reduced. As oxygen becomes depleted, denitrification becomes energetically favored as a respiratory pathway. The utilization of nitrate as a TEA depends on the sufficient supply of organic matter as well as NO_3^-, an oxidized inorganic form of nitrogen.

In order of decreasing energy yield, primary alternative TEAs in natural waters include nitrate, manganese, ferric iron, sulfate, and methane. In the Everglades, sulfate is the primary TEA. Sulfur contamination from agricultural activities in the Everglades Agricultural Area has resulted in high sulfate concentrations throughout much of the Everglades (Chap. 2, this volume). Hg methylation in the Everglades is primarily related to the activity of sulfate-reducing bacteria (Chaps. 1 and 3, this volume).

5.3.2 Phosphorus Effects on Redox Dynamics in the Everglades

In eutrophic settings, net primary production is higher as a result of increased nutrient availability. Increased net primary production influences macrophyte growth and detritus accumulation rates, as well as the proliferation of phytoplankton in the water column. The organic matter derived from macrophytes favored under eutrophic conditions (cattail) is typically more labile and readily oxidized by microbes as an electron donor in order to produce energy. More labile fractions (e.g. lignin) readily yield smaller compounds which can be utilized by microbes in respiration. Increased detritus as well as changes in the quality of substrates due to shifts in vegetation community structure as a result of eutrophication have been implicated in increased enzyme activity and carbon metabolism (Wright and Reddy 2001; Corstanje and Reddy 2004), as well as altered organic matter decomposition dynamics (DeBusk and Reddy 2003). In fact, Wright and Snyder (2009) found that microbial respiration is higher in detritus than soil in eutrophic systems as compared to oligotrophic systems since detritus quantities are increased and the detrital environment is more conducive for microbial activity. Accelerated detrital turnover rates in eutrophic systems are therefore the result of changes in vegetation associated with increased nutrient availability as well as increased microbial activity associated with soil and plant conditions (e.g. organic matter quality) (Wright and Snyder 2009).

During the process of organic matter mineralization, positive feedback loops are evident in the processing of phosphorus. Nutrient enrichment has been found to increase the mineralization of phosphorus, with a positive relationship between increases in microbial biomass and increases in phosphorus mineralization rates (Corstanje et al. 2006). The release of additional nutrients associated with phosphorus mineralization favors continued increases in net primary productivity, carbon accumulation, and increasing microbial biomass. Increased nutrient concentrations therefore directly contribute to rapid organic matter turnover.

Increased microbial respiration and organic matter oxidation requires an associated increase in TEAs. As such, rapid organic matter turnover is associated with an increased demand for oxygen as a TEA, with subsequent demand for other less energetically favorable TEAs. In this way, eutrophication results in decreased oxygen concentrations in affected systems as oxygen consumption exceeds the rate of supply. The availability of electrons from oxygen or other TEAs in solution can be measured by redox potential (E_h). When organic matter is oxidized, E_h drops as TEAs are reduced in order of energetic favorability. As oxygen is used up as a TEA and E_h falls below +400 mV, conditions become increasingly anaerobic or more reduced.

Under very reduced conditions (−100 to −200 mV), sulfate is reduced to sulfide. While this pathway commonly exists under natural conditions, the enrichment of TEAs in Everglades soils can lead to accelerated carbon mineralization and associated nutrient or contaminant release. Increased sulfate concentrations result in an increase in TEAs and thereby an increase in organic matter decomposition rates and sulfide concentrations. When organic matter is decomposed, nutrients including phosphorus are released. Given the immense amount of carbon found in Everglades peat soils, the process of carbon mineralization is especially important to preventing soil losses and maintaining the balance between Histosol production and decomposition. Furthermore, in light of eutrophication issues facing the Everglades, controlling nutrient release associated with mineralization is critical for preventing negative water quality impacts since Everglades soils represent the major pool of nutrients (particularly phosphorus) in the system. As nutrients are released, organic matter decomposition can simultaneously release other elements that affect the availability of TEAs. Oxidation of organic matter results in the mineralization of sulfur and nitrogen bound in the soil, which can then remain available in the soil, taken up by plants, or transported downstream. A buildup of reduced sulfur can also influence phosphorus concentrations in the water column. Sulfide oxidation results in the release of hydronium ions, which ultimately lowers the pH of the system. Under these conditions, the calcium in marls can become soluble and its associated constituents will be released to the water column. If phosphorus is retained in these marls, it will also become available within the water column.

5.3.3 Prevalence and Spatial Distribution of Redox Reactions in the Everglades

The enormous area of organic soils contained within the Everglades ecosystem is one of its most significant attributes (Davis 1946; Stephens 1956; Davis and Ogden 1994; Bruland et al. 2006). These peat deposits represent over 5000 years of soil accretion and, as such, store nutrients accumulated over that time period. Not only do these soils function as a storage pool of nutrients, but they also act as a storage pool of carbon. In the northern and central Everglades, the mean total carbon content of peat soils is approximately 47% (dominated by organic carbon), with total carbon for marl soils of ENP being lower due to lack of organic carbon content of marls (Reddy et al. 2005; Osborne et al. 2011b). The formation of immense stores of carbon such as those found in the Everglades is only possible under unique conditions where TEAs are limited, thereby limiting organic matter oxidation. These conditions ultimately result in slow decomposition rates which allow primary production to outpace decomposition, resulting in peat accretion and burial.

The balance of these processes and the associated environmental implications (e.g. storage of nutrients and contaminants) vary spatially across the Everglades landscape (Fig. 5.2). The decomposition of organic matter in wetland soils depends on many factors, including the type of organic matter, the microbial community, available nutrients, pH, temperature, and redox potential (D'Angelo and Reddy 1999; Wright and Reddy 2007). Several of these factors are often dictated by hydrologic regimes. Changes over the last century to Everglades hydrology (Light and Dineen 1994; Steinman et al. 2002; see also Chap. 2, Volume I) have been found to influence biological changes evident in vegetation and fauna (McIvor et al. 1994; Sklar et al. 2002; Willard and Cronin 2007; Bernhardt and Willard 2009). These studies have found that altered hydrology has resulted in significant impacts to soils through changes in soil redox potential, organic matter losses in regions with reduced hydroperiods, and increases in organic matter (peat accumulation) in regions with increased hydroperiods – especially when combined with increases in phosphorus loading (Craft and Richardson 2008). Carbon flux resulting from lowered water levels has been clearly documented (DeBusk and Reddy 2003), and is enhanced by phosphorus enrichment.

The majority of organic soils found in the Everglades are located in the northern and central Everglades (Reddy et al. 2005), with the highest organic matter soils found in the peat soils of WCA-1 where organic content then trends downward towards the southern portion of WCA-3 (Osborne et al. 2015) (Fig. 5.2). The soils in the southern Everglades are less organic due to the prevalence of shallow wet prairies and marl prairies. Peat soils also can be found in Shark River Slough, where longer hydroperiods are present (Reddy et al. 2005; Scheidt et al. 2000; Scheidt and Kalla 2007). As such, organic matter in soil generally decreases across the Everglades from north to south. An exception to this trend, and a testament to the influence of hydrology and soil redox processes, is found in the northwestern corner of WCA-3A (Fig. 5.3). Here, soil subsidence due to drainage (Scheidt et al. 2000) has resulted in losses of soil depth over the period from 1946 to 2005. Over the entire Everglades Protection Area (EvPA), up to 28% of peat soils have been lost between 1946 and 2005 (Osborne et al. 2011a, b, c; Scheidt et al. 2000). The hysteresis in soil creation and loss is a critical determinant in the present-day Everglades ecosystem (DeAngelis 1994), and is associated with the susceptible and unstable nature of the ecosystem (Maltby and Dugan 1994). Fluctuations in hydrology can undo decades, even centuries of organic matter burial and storage in deep peat soils resulting in dramatic implications to water quality, vegetation communities, and the carbon budget of the system.

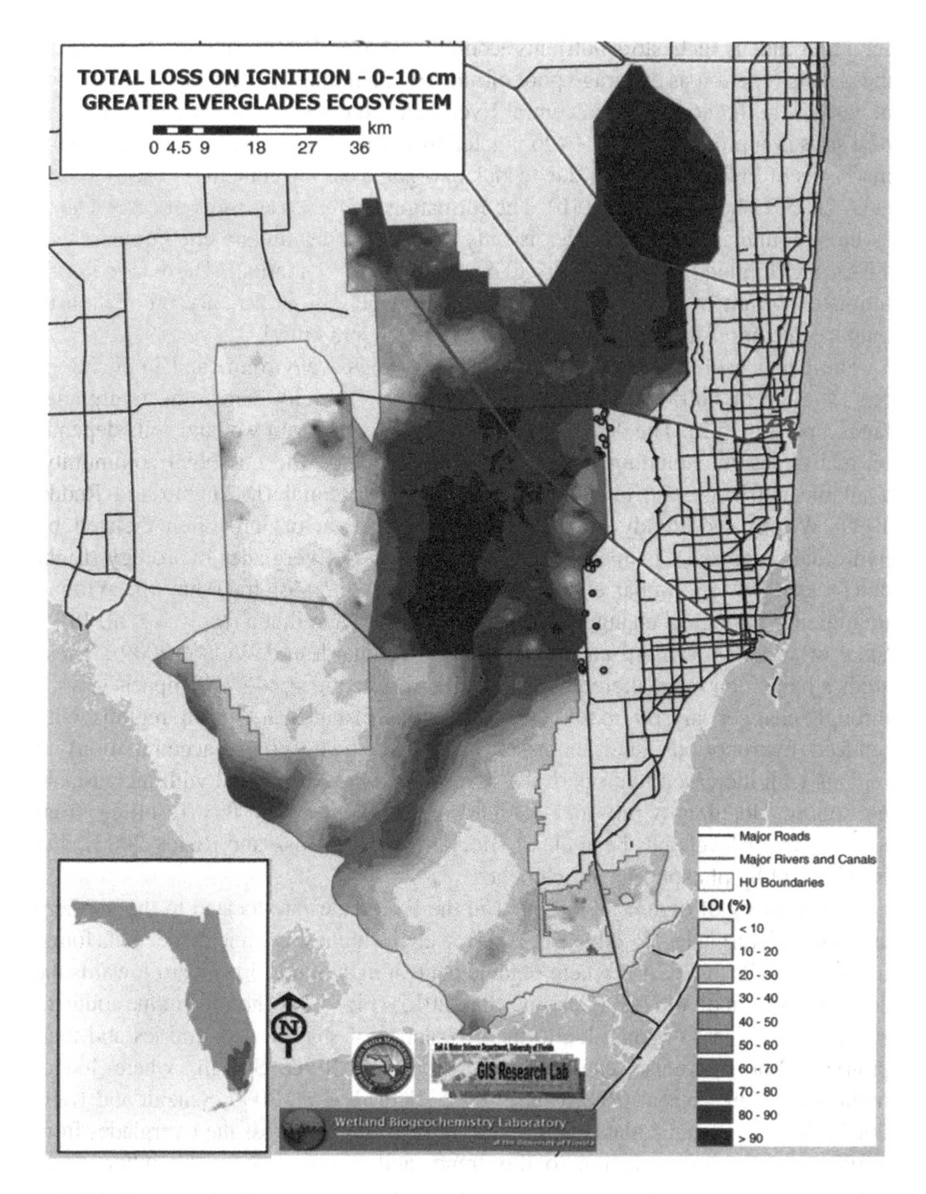

Fig. 5.2 Patterns of soil organic matter across the Greater Everglades landscape characterized by loss on ignition (LOI). Figure adapted from Reddy et al. (2005)

Re-flooding disturbed soils can release phosphate and ammonium and further exacerbate eutrophication problems (Corstanje and Reddy 2004; Aldous et al. 2005). The extent of nutrient release upon re-wetting depends on the mechanism of nutrient storage in soils. For example, if phosphorus is retained by iron oxides, redox-driven reduction of these precipitates (*i.e.*, the reduction of Fe(III), which is insoluble, to Fe

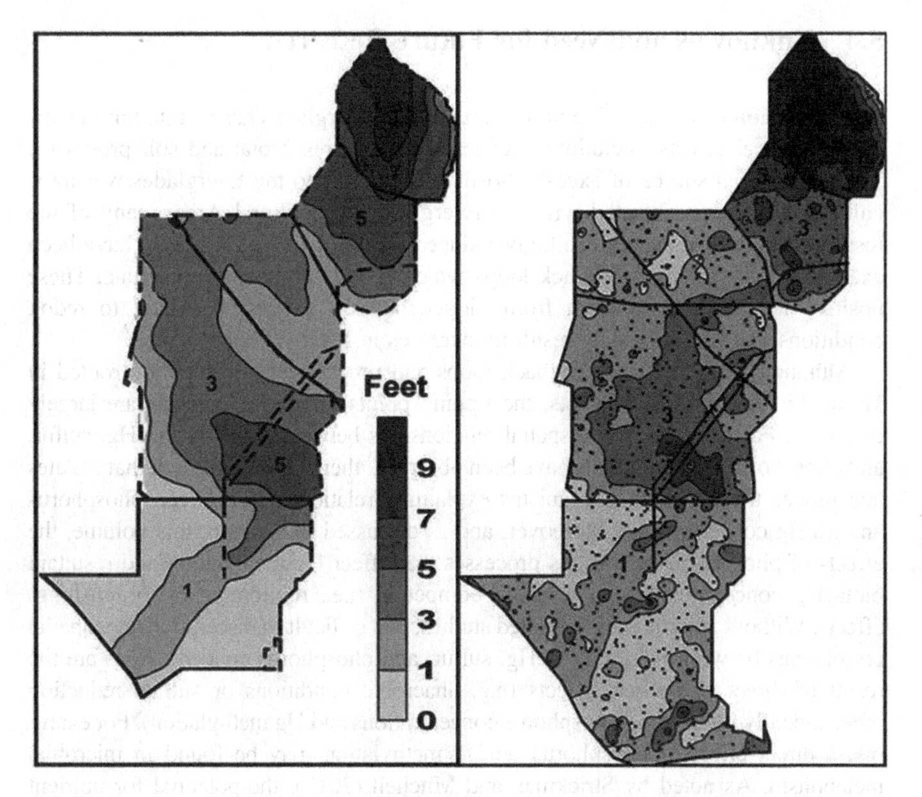

Fig. 5.3 A comparison of soil depths in 1946 (left panel) and 1995–2003 (right panel), indicating soil loss throughout the Everglades. Maps created by D.J. Scheidt from Davis (1946) and Scheidt and Kalla (2007)

(II), which is soluble) upon re-wetting will release bound phosphorus into the water column. In the highly organic peat soils or carbonate-rich marl soils of the Everglades, oxidation of organic matter is more likely to contribute to phosphorus efflux upon re-wetting (Pant and Reddy 2003; Aldous et al. 2005). In the western and eastern ENP where marls dominate, hydrology can influence both periphyton species composition and production; species composition in turn can affect periphyton mat accretion (Davis et al. 2005), and periphyton chemical composition and biomass (Gottlieb et al. 2005).

Lastly, fire in the Everglades, often resulting from reduced hydroperiods (Lodge 2010; see also Chaps. 1 and 2; Volume I), also poses significant implications to the biogeochemical balance of the system through redox reactions. Under very dry conditions, fire can result in the rapid oxidation (combustion) of organic rich soils, in turn influencing plant community structure (Lodge 2010). Additionally, muck fires convert less available organic phosphorus to more available inorganic phosphorus, directly effecting water column concentrations when saturated conditions return (Leeds et al. 2009).

5.4 Unknowns and Need for Future Research

Eutrophication from excessive phosphorus in the Everglades has resulted in numerous ecological effects, including changes in vegetation, biota, and soil processes. While the initial source of excess phosphorus loading to the Everglades was agricultural activities associated with the Everglades Agricultural Area, many of the resulting ecological changes initially induced by the increased loadings have been exacerbated by positive feedback loops which reinforce the eutrophic state. These positive feedback loops stem from biogeochemical processes related to redox conditions that can ultimately result in increases in MeHg concentrations.

Although many of these feedback loops and process relationships are rooted in known biogeochemical concepts, the tipping points or precise controls are largely unknown. For example, while spatial relationships between total Hg, MeHg, sulfur, and phosphorus concentrations have been observed, there is little research that isolates and proves the potential mechanisms explaining relationships between phosphorus and MeHg concentrations. Moreover, and as discussed in Chap. 6, this volume, the effects of phosphorus on various processes that affect Hg methylation and resultant biota Hg concentrations can result in competing (i.e., reinforcing or diminishing) effects. Without more process-oriented studies, it is difficult to discern whether spatial correlations between total Hg, MeHg, sulfur, and phosphorus concentrations are the result of direct or indirect effects (e.g. anaerobic conditions or sulfate reduction independently influencing phosphorus concentrations and Hg methylation). For example, a direct effect of phosphorus on Hg methylation may be found in microbial metabolism. As noted by Strickman and Mitchell (2017), the potential for nutrient limitation or nutrient enhancement within microbial communities responsible for methylation is not well understood. Conversely, excessive phosphorus concentrations can lead to the development of high porewater sulfide concentrations, particularly in areas that also are exposed to high sulfate concentrations. This latter scenario can lead to lower rates of Hg methylation because of the strong binding or sequestration of inorganic Hg (a necessary precursor to support Hg methylation) with sulfide (Chap. 1, this volume). As result, each component that links eutrophication with MeHg production must be considered individually as well as in concert with the other components to adequately understand the consequences of increasing trophic state, including altered redox conditions, on Hg cycling and trophic transfer in the Everglades.

References

Aldous A, McCormick P, Ferguson C, Graham S, Craft C (2005) Hydrologic regime controls soils phosphorus fluxes in restoration and undisturbed wetlands. Restor Ecol 13(2):341–347

Amador JA, Jones RD (1993) Nutrient limitation on microbial respiration in peat soils with different total phosphorus content. Soil Biol Biochem 25:793–801

Amador JA, Jones RD (1995) Carbon mineralization in pristine and phosphorus enriched peat soils of the Florida Everglades. Soil Sci 159:129–141

August KR (2018) Reduced soil nutrient enrichment and Typha expansion due to restoration efforts: a temporal analysis of Taylor Slough in Everglades National Park. Masters Thesis. University of Florida, Gainesville, FL

Bernhardt CE, Willard DA (2009) Response of the Everglades ridge and slough landscapes to climate variability and 20th-century water management. Ecol Appl 19(7):1723–1738

Browder JA, Gleason PJ, Swift DR (1994) Periphyton in the Everglades: spatial variation, environmental correlates, and ecological implications. In: Davis SM, Ogden JC (eds) Everglades, the ecosystem and its restoration. St. Lucie Press, Delray Beach, FL, pp 379–418

Bruland GL, Grunwald S, Osborne TZ, Reddy KR, Newman S (2006) Spatial distribution of soil properties in water conservation area 3 of the Everglades. Soil Sci Soc Am J 70:1662–1676

Bruland GL, Osborne TZ, Reddy KR, Grunwald S, Newman S, DeBusk WF (2007) Recent changes in soil total phosphorus in the Everglades: water conservation area 3A. Environ Monit Assess 129:379–395

Chambers RM, Pederson KA (2006) Variation in soil phosphorus, sulfur, and iron pools among South Florida wetlands. Hydrobiologia 569:63–70

Chase JM (2003) Strong and weak trophic cascades along a productivity gradient. Oikos 101:187–195

Chen M, Ma LQ, Li YC (2000) Concentrations of P, K, Al, Fe, Mn, Cu, Zn, and As in marl soils from South Florida. Proc Soil Crop Sci Soc Fla 59:124–129

Chiang C, Craft CB, Rogers DW, Richardson CJ (2000) Effects of 4 years of nitrogen and phosphorus additions on Everglades plant communities. Aquat Bot 68:61–78

Childers DL, Doren RF, Jones R, Noe GB, Rugge M, Scinto LJ (2003) Decadal change in vegetation and soil phosphorus pattern across the Everglades landscape. J Environ Qual 32:344–362

Corstanje R, Reddy KR (2004) Response of biogeochemical indicators to a drawdown and subsequent reflood. J Env Qual 33:2357–2366

Corstanje R, Grunwald S, Reddy KR, Osborne TZ, Newman S (2006) Assessment of the spatial distribution of soil properties in a northern Everglades marsh. J Environ Qual 35:938–949

Craft CB, Richardson CJ (1993) Peat accretion and phosphorus accumulation along a eutrophic gradient in northern Everglades. Biogeochem 22:133–156

Craft CB, Richardson CJ (1997) Relationships between soil nutrients and plant species composition in Everglades peatlands. J Env Qual 26:224–232

Craft CB, Richardson CJ (2008) Soil characteristics of the Everglades peatland. In: Richardson CJ (ed) The Everglades experiments. Springer, pp 59–74

Craft CB, Vymazal J, Richardson CJ (1995) Response of Everglades plant communities to nitrogen and phosphorus additions. Wetlands 15(3):258–271

D'Angelo EM, Reddy KR (1999) Regulators of heterotrophic microbial potentials in wetland soils. Soil Biol Biochem 31:815–830

Daoust RJ, Childers DL (2004) Ecological effects of low level phosphorus additions on two plant communities in a neotropical freshwater wetland. Oecologia 141:672–686

Davis JH (1946) The peat deposits of Florida: Florida geologic survey. Geol Bull 30:247

Davis SM (1991) Growth, decomposition, and nutrient retention of *Cladium jamaicense* Crantz and *Typha domingensis* Pers. in the Florida Everglades. Aquat Bot 40:203–224

Davis SM (1994) Phosphorus inputs and vegetation sensitivity in the Everglades. In: Davis SM, Ogden JC (eds) Everglades: the ecosystem and its restoration. St. Lucy Press, Delray Beach, FL, p 826

Davis SM, Ogden JC (eds) (1994) Everglades: the ecosystem and its restoration. St. Lucy Press, Delray Beach, FL, p 826

Davis SM, Childers DL, Lorenz JL, Wanless HR, Hopkins TE (2005) A conceptual ecological model of ecological interactions in the mangrove estuaries of the Florida Everglades. Wetlands 25:832–842

DeAngelis DL (1994) Synthesis: spatial and temporal characteristics of the environment. In: Davis SM, Ogden JC (eds) Everglades: the ecosystem and its restoration. St. Lucie Press, Delray Beach, FL

DeBusk WF, Reddy KR (1998) Turnover of detrital organic carbon in a nutrient-impacted Everglades marsh. Soil Sci Soc Am J 62:1460–1468
DeBusk WF, Reddy KR (2003) Nutrient and hydrology effects on soil respiration in a northern Everglades marsh. J Environ Qual 32:702–710
DeBusk WF, Reddy KR (2005) Litter decomposition and nutrient dynamics in a phosphorus enriched Everglades marsh. Biogeochemistry 75:217–240
DeBusk WF, Reddy KR, Koch MS, Wang Y (1994) Spatial patterns of soil phosphorus in Everglades water conservation area 2A. Soil Sci Soc Am J 58:543–552
DeBusk WF, Newman S, Reddy KR (2001) Spatio-temporal patterns of soil phosphorus enrichment in Everglades water conservation area 2A. J Environ Qual 30:1348–1446
Dettling MD, Yavitt JB, Zinder SH (2006) Control of organic carbon mineralization by alternative electron acceptors in four peatlands, Central New York State, USA. Wetlands 26(4):917–927
Ewe SML, Gaiser EE, Childers DL, Rivera-Monroy VH, Iwaniec D, Fourquerean J, Twilley RR (2006) Spatial and temporal patterns of aboveground net primary productivity (ANPP) in the Florida Coastal Everglades LTER (2001–2004). Hydrobiologia 569:459–474
Gaiser EE, Scinto LJ, Richards JH, Jayachandran K, Childers DL, Trexler JC, Jones RD (2004) Phosphorus in periphyton mats provides the best metric for detecting low-level P enrichment in an oligotrophic wetland. Water Res 38:507–516
Gaiser EE, Trexler JC, Richards JH, Childers DL, Lee D, Edwards AL, Scinto LJ, Jayachandran K, Noe GB, Jones RD (2005) Cascading ecological effects of low-level phosphorus enrichment in the Florida Everglades. J Environ Qual 34:717–723
Gaiser EE, Childers DL, Jones RD, Richards JH, Scinto LJ, Trexler JC (2006) Periphyton responses to eutrophication in the Florida Everglades. Cross-system patterns of structural and compositional change. Limnol Oceanogr 51:617–630
Gleason PJ, Stone P (1994) Age, origin, and landscape evolution of the Everglades peatland. In: Davis SM, Ogden JC (eds) Everglades: the ecosystem and its restoration. St. Lucy Press, Delray Beach, FL, p 826
Gleason PJ, Cohen AD, Stone P, Smith WG, Brooks HK, Goodrick R, Spackman W Jr (1974) The environmental significance of Holocene sediments from the Everglades and saline tidal plains. In: Gleason PJ (ed) Environments of South Florida. Present and past. Miami Geological Society, Coral Gables, FL, pp 297–351
Goldsborough LG, Robinson GGC (1996) Pattern in wetlands. In: Stevenson RJ, Bothwell ML, Lowe RL (eds) Algal ecology. Academic Press, San Diego, CA, pp 77–117
Gottlieb A, Richards J, Gaiser E (2005) Effects of desiccation duration on the community structure and nutrient retention of short and long-hydroperiod Everglades periphyton mats. Aquat Bot 82:99–112
Grimshaw HJ, Wetzel RG, Brandenburg M, Segerblom K, Wenkert LJ, Marsh GA, Charnetzky W, Haky JE, Carraher C (1997) Shading of periphyton communities by wetland emergent macrophytes: decoupling of algal photosynthesis from microbial nutrient retention. Arch Fur Hydro 139:17–27
Grunwald S, Osborne TZ, Reddy KR (2008) Temporal trajectories of phosphorus and pedo-patterns mapped in water conservation area 2A, Everglades, Florida, USA. Geoderma 146:1–13
Hagerthey SE, Newman S, Rutchey K, Smith EP, Godin J (2008) Multiple regime shifts in a subtropical peatland: community-specific thresholds to eutrophication. Ecol Monogr 78:547–565
Jones LA (1948) Soils, geology, and water control in the Everglades region. Bulletin 442. University of Florida Agricultural Experiment Station and Soil Conservation Service, Gainesville, FL
Kalla PI, Scheidt DJ (2017) Everglades ecosystem assessment – Phase IV, 2014: data reduction and initial synthesis. United States Environmental Protection Agency, Science and Ecosystem Support Division. SESD Project 14-0380. Athens, Georgia

King RS, Richardson CJ (2007) Subsidy–stress response of macroinvertebrate community biomass to a phosphorus gradient in an oligotrophic wetland ecosystem. J N Am Benthol Soc 26:491–508
King RS, Richardson CJ, Urban DL, Romanowicz EA (2004) Spatial dependency of vegetation-environment linkages in an anthropogenically influenced wetland ecosystem. Ecosystems 7:75–97
Koch MS, Reddy KR (1992) Distribution of soil and plant nutrients along a trophic gradient in the Florida Everglades. Soil Sci Soc Am J 56:1492–1499
Lake Okeechobee Technical Advisory Council (LOTAC) II (1990) Final report to the Governor, State of Florida, Secretary, Department of Environmental Regulation, Governing Board, South Florida Water Management District, West Palm Beach, Florida. 64 pp
Leeds JA, Garrett PB, Newman JM (2009) Assessing impacts of hydropattern restoration of an overdrained wetland on soil nutrients, vegetation, and fire. Restor Ecol 17(4):460–469
Light SS, Dineen JW (1994) Water control in the Everglades: a historical perspective. In: Davis S, Ogden J (eds) Everglades: the ecosystems and its restoration. St. Lucie Press, Delray Beach, FL, pp 47–84
Liston SE, Newman S, Trexler JC (2008) Macroinvertebrate community response to eutrophication in an oligotrophic wetland: an in situ mesocosm experiment. Wetlands 28:686–694
Lodge TE (2010) The Everglades handbook: understanding the ecosystem, 3rd edn. CRC, Boca Raton, FL
Maltby E, Dugan PJ (1994) Wetland ecosystem protection, management, and restoration: an international perspective. In: Davis SM, Ogden JC (eds) Everglades: the ecosystem and its restoration. St. Lucie Press, Delray Beach, FL
McCormick PV, O'Dell MB (1996) Quantifying periphyton responses to phosphorus in the Florida Everglades: a synoptic-experimental approach. J N Am Benthol Soc 15(4):450–468
McCormick PV, Stevenson RJ (1998) Periphyton as a tool for ecological assessment and management in the Florida Everglades. J Phycol 34:726–733
McCormick PV, O'Dell MB, Shuford RBE III, Backus JG, Kennedy WC (2001) Periphyton responses to experimental phosphorus enrichment in a subtropical wetland. Aquat Bot 71:119–139
McCormick PV, Newman S, Miao S, Gawlik DE, Marley D, Reddy KR, Fontaine TD (2002) Effects of anthropogenic phosphorus inputs on the Everglades. Chapter 3. In: Porter JW, Porter KG (eds) The Everglades, Florida Bay and coral reefs of the Florida keys: an ecosystem sourcebook. CRC, Boca Raton, FL, pp 123–146
McIvor CC, Ley JA, Bjork RD (1994) Changes in freshwater inflow from the Everglades to Florida Bay including effects on biota and biotic processes: a review. In: Davis SM, Odgen JC (eds) Everglades: the ecosystem and its restoration. St. Lucie Press, Delray Beach, FL
Miao SL, DeBusk WF (1999) Effects of phosphorus enrichment on structure and function of sawgrass and cattail communities in Florida wetlands. In: Reddy KR, O'Connor GA, Schelske CL (eds) Phosphorus biogeochemistry of subtropical ecosystems. Lewis Publishers, Boca Raton, FL, 275 pp
Miao SL, Sklar FH (1998) Biomass and nutrient allocation of sawgrass and cattail along a nutrient gradient in the Florida Everglades. Wetlands Ecosyst Manag 5:245–264
Mitsch WJ, Gosselink JG (2000) Wetlands, 3rd edn. Wiley, New York, NY, 920 pp
Newman S, Grace JB, Kobel JW (1996) Effects of nutrients and hydroperiod on *Typha*, *Cladium*, and *Eleocharis*: implications for Everglades restoration. Ecol Appl 6:774–783
Newman S, Reddy KR, DeBusk WF, Wang Y (1997) Spatial distribution of soil nutrients in a northern Everglades marsh: water conservation area 1. Soil Sci Soc Am J 61:1275–1283
Newman S, Schuette J, Grace JB, Rutchey K, Fontaine T, Reddy KR, Pietrucha M (1998) Factors influencing cattail abundance in the northern Everglades. Aquat Bot 60:265–280
Newman S, Osborne TZ, Rutchey K, Reddy KR, Hagerthey SE (2009) Anthropogenic influences drive wetland landscape evolution: the Everglades trajectory. Ecol Monogr

Noe GB, Childers DL, Jones RD (2001) Phosphorus biogeochemistry and the impact of phosphorus enrichment: why is the Everglades so unique? Ecosystems 4:603–624

Noe GB, Childers DL, Edwards AL, Gaiser E, Jayachandran K, Lee D, Meeder J, Richards J, Scinto LJ, Trexler JC, Jones RD (2002) Short-term changes in phosphorus storage in an oligotrophic Everglades wetland ecosystem receiving experimental nutrient enrichment. Biogeochemistry 59:239–267

Noe GB, Scinto LJ, Taylor J, Childers DL, Jones RD (2003) Phosphorus cycling and partitioning in an oligotrophic Everglades wetland ecosystem: a radioisotope tracing study. Freshw Biol 48:1993–2008

Odum EP, Finn JT, Franz EH (1979) Perturbation theory and the subsidy-stress gradient. Bioscience 29:349–352

Ogden JC (2005) Everglades ridge and slough conceptual ecological model. Wetlands 25:810–831

Osborne TZ, Newman S, Reddy KR (2008) Spatial distribution of total sulfur in the soils of the northern and southern Everglades. Final report # 45000-12699/14856. South Florida Water Management District, West Palm Beach, FL

Osborne TZ, Newman S, Kalla P, Scheidt DJ, Bruland GL, Cohen MJ, Scinto LJ, Ellis LR (2011a) Landscape patterns of significant soil nutrients and contaminants in the greater Everglades ecosystem: past, present, and future. Crit Rev Environ Sci Technol 41:121–148

Osborne TZ, Bruland GL, Newman S, Reddy KR, Grunwald S (2011b) Spatial distributions and eco-partitioning of soil biogeochemical properties in Everglades National Park. Environ Monit Assess 183:395–408

Osborne TZ, Davis SE, Naja GM, Rivero RG, Ross MS (2011c) Report of the soils subgroup. In the SERES project: review of Everglades science, tools and needs related to key science management questions. Available at http://everglades-seres.org/SERES-_Everglades_Foundation/Products_files/SERES_Soils_Review%20copy.pdf

Osborne TZ, Ellis LR, Castro J, Sadle J (2012) Monitoring of phosphorus storage in park marsh land sediments: an assessment of the C-111 spreader canal project. Final report. South Florida Natural Resources Center, National Park Service, Homestead, FL, 117p

Osborne TZ, Reddy KR, Ellis LR, Aumen N, Surratt DD, Zimmerman MS, Sadle J (2014) Evidence of recent phosphorus enrichment in surface soils of Taylor Slough and Northeast Everglades National Park. Wetlands 34(1):37–45

Osborne TZ, Newman S, Reddy KR, Ellis LR, Ross M (2015) Spatial distribution of soil nutrients in the Everglades protection area. In: Entry JA, Gottlieb AD, Jayachandran K, Ogram A (eds) Microbiology of the Everglades ecosystem. CRC, Boca Raton, FL, pp 38–67

Pant HK, Reddy KR (2003) Potential internal loading of phosphorus in constructed wetlands. Water Res 37:965–972

Payne G, Weaver K, Bennett T (2003) Chapter 5: Development of a numeric phosphorus criterion for the Everglades protection area. In SFWMD (Eds.) Everglades consolidated report, South Florida Water Management District, West Palm Beach, FL

Pietro K, Bearzotti R, Chimney M, Germain G, Iricanin N, Piccone T (2007) STA performance, compliance and optimization. Chapter 5, Vol 1. In: South Florida environmental report. South Florida Water Management District, West Palm Beach, FL

Qualls RG, Richardson CJ (1995) Forms of soil phosphorus along a nutrient enrichment gradient in the northern Everglades. Soil Sci 160:183–197

Rader RB, Richardson CJ (1992) The effects of nutrient enrichment on algae and macroinvertebrates in the Everglades: a review. Wetlands 12(2):121–135

RECOVER (2006) CERP monitoring and assessment plan: part 2, 2006 Assessment strategy for the MAP, restoration, coordination and verification. US Army Corps of Engineers, Jacksonville District, Jacksonville, FL and South Florida Water Management District, West Palm Beach, FL

Reddy KR, D'Angelo EM (1996) Biogeochemical indicators to evaluate pollutant removal efficiency in constructed wetlands. Water Sci Technol 35:1–10

Reddy KR, DeLaune RL (2008) Biogeochemistry of wetlands: science and applications. CRC, Boca Raton, FL, p 774

Reddy KR, Feijtel TC, Patrick WH Jr (1986) Effect of soil redox conditions oon microbial oxidation of organic matter. In: Chen Y, Avnimelch Y (eds) The role of organic matter in modern agriculture. Dev Plant Sci Martinus Nijhoff, Dordrecht, pp 117–148

Reddy KR, DeBusk WF, Wang Y, DeLaune R, Koch M (1991) Physico-chemical properties of soils in the water conservation area 2 of the Everglades. Soil Science Department, Institute of Food and Agricultural Sciences, Gainesville, FL

Reddy KR, DeLaune R, DeBusk WF, Koch MS (1993) Long term nutrient accumulation rates in the Everglades. Soil Sci Soc Am J 57:1147–1155

Reddy KR, Wang Y, DeBusk WF, Newman S (1994) Physico-chemical properties of soils in water conservation area 3 (WCA-3) of the Everglades. Final report. South Florida Water Management District, West Palm Beach, FL

Reddy KR, Newman S, Grunwald S, Osborne TZ, Corstanje R, Bruland GL, Rivero RG (2005) Everglades soil mapping final report. South Florida Water Management District, West Palm Beach, FL

Richardson CJ, King RS, Qian SS, Vaithiyanathan P, Qualls RG, Stowe CA (2008) An ecological basis for establishment of a phosphorus threshold for the Everglades ecosystem. Chapter 25. In: Richardson CJ (ed) The Everglades experiments: lessons for ecosystem restoration. Springer, New York

Rivera-Monroy VH, Twilley RR, Davis SE, Childers DL, Simrad M, Chambers R (2011) The role of the Everglades mangrove ecotone region (EMER) in regulating nutrient cycling and wetland productivity in South Florida. Crit Rev Environ Sci Technol 41(S1)

Rivero RG, Grunwald S, Osborne TZ, Reddy KR, Newman S (2007) Characterization of the spatial distribution of soil properties in water conservation area 2A, Everglades, Florida. Soil Sci 172:149–166

Rivero RG, Grunwald S, Binford MW, Osborne TZ (2009) Integrating spectral indices into prediction models of soil phosphorus in a subtropical wetland. Remote Sens Environ 113:2389–2402

Scheidt DJ, Kalla PI (2007) Everglades ecosystem assessment: water management and quality, eutrophication, mercury contamination, soils and habitat: monitoring for adaptive management: a R-EMAP status report. USEPA Region 4, Athens, GA. EPA 904-R-07-001. 98 pp

Scheidt D, Stober J, Jones R, Thornton K (2000) South Florida ecosystem assessment: Everglades water management, soil loss, eutrophication and habitat. Report no. 904-R-00-003. US Environmental Protection Agency, Athens, GA

Sergeant BL, Gaiser EE, Trexler JC (2010) Biotic and abiotic determinants of intermediate-consumer trophic diversity in the Florida everglades. Mar Freshw Res 61:11–22

SFWMD (1992) Surface water improvement and management plan for the Everglades. Supporting information document. South Florida Water Management District, West Palm Beach, FL

SFWMD (South Florida Water Management District) (2007) South Florida environmental report. South Florida Water Management District, West Palm Beach, FL. www.sfwmd.gov/sfer

Sklar F, McVoy C, VanZee R, Gawlik DE, Tarbonton K, Rudnick D, Miao S, Armentano T (2002) The effects of altered hydrology on the ecology of the Everglades. In: Porter J, Porter K (eds) The Everglades, Florida Bay, and coral reefs of the Florida keys: an ecosystem sourcebook. CRC, Boca Raton, FL

Sklar FH, Chimney MJ, Newman S, McCormick P, Gawlik D, Miao S, McVoy C, Said W, Newman J, Coronado C, Crozier G, Korvela M, Rutchey K (2005) The ecological-societal underpinnings of Everglades restoration. Front Ecol Environ 3(3):161–169

Snyder GH (2005) Everglades agricultural area soil subsidence and land use projections. Proc Soil Crop Sci Soc Fla 64:44–51

Snyder GH, Davidson JM (1994) Everglades agriculture: past, present, and future. In: Davis SM, Ogden JC (eds) Everglades: the ecosystem and its restoration. St. Lucy Press, Delray Beach, FL, p 826

Steinman AD, Havens KE, Carrick HJ, Van Zee R (2002) The past, present, and future hydrology and ecology of Lake Okeechobee and its watersheds. In: Porter JW, Porter KG (eds) The

Everglades, Florida Bay, and coral reefs of the Florida keys: an ecosystem sourcebook. CRC, Boca Raton, FL

Stephens JC (1956) Subsidence of organic soils in the Florida Everglades. Soil Sci Soc Am Proc 20:77–80

Stephens JC, Johnson L (1951) Subsidence of organic soils in the upper Everglades region of Florida. Proc Soil Sci Soc Fla 57:20–29

Strickman RJ, Mitchell CPJ (2017) Accumulation and translocation of methylmercury and inorganic mercury in Oryza sativa: an enriched isotope tracer study. Sci Total Environ 574:1415–1423

Teal LR, Bulling MT, Parker ER, Solan M (2008) Global patterns of bioturbation intensity and mixed depth of marine soft sediments. Aquat Biol 2:207–218

Willard DW, Cronin TM (2007) Paleoecology and ecosystem restoration: case studies from Chesapeake bay and the Florida Everglades. Front Ecol Environ 5(9):491–498

Wright AL, Reddy KR (2001) Heterotrophic microbial activity in northern Everglades wetland soils. Soil Sci Soc Am J 65(6):1856–1864

Wright AL, Reddy KR (2007) Substrate-induced respiration for phosphorus-enriched and oligotrophic peat soils in an Everglades wetland. Soil Sci Soc Am J 71:1579–1583

Wright AL, Snyder GH (2009) Soil subsidence in the Everglades agricultural area. Soil and Water Science Department, Florida Cooperative Extension Service, Institute of Food and Agricultural Sciences, University of Florida. Publication #SL 311

Wu Y, Sklar FH, Rutchey K (1997) Analysis and simulations of fragmentation patterns in the Everglades. Ecol Appl 7:268–276

Chapter 6
Major Drivers of Mercury Methylation and Cycling in the Everglades: A Synthesis

Curtis D. Pollman

Abstract This chapter synthesizes information presented in Chaps. 1 through 5 of this volume on the processes that govern the biogeochemical cycling of mercury (Hg) in the Everglades. Particular emphasis is devoted to the processes that influence Hg methylation. Key variables include inorganic Hg, dissolved organic matter (DOM) and sulfate. The role of phosphorus which, during the early years of the South Florida Mercury Science Program was vigorously ascribed by different scientists to have two widely divergent and essentially unidimensional effects on Hg methylation and trophic transfer, also is explored. The role of each these latter variables with respect to Hg methylation (DOC, sulfate and phosphorus) and trophic transfer (DOM and phosphorus) is complex and needs to be considered in its entirety. Given the key role of sulfate and the controversy it has engendered with some Everglades stakeholders, the chapter also includes a detailed discussion of the statistical basis—both for and against—this key role.

Keywords Methylmercury · Sulfate · Dissolved organic matter · Phosphorus · Eutrophication

6.1 Mercury Cycling in the Everglades

At the heart of the mercury (Hg) problem in all aquatic ecosystems is the fact that it would not exist without the presence of methylmercury (MeHg). For most aquatic ecosystems—including the Everglades—MeHg is produced largely in situ with only

C. D. Pollman (✉)
Aqua Lux Lucis Inc., Gainesville, FL, USA
e-mail: cpollman@aqualuxlucis.org

© Springer Nature Switzerland AG 2019

D. G. Rumbold et al. (eds.), *Mercury and the Everglades. A Synthesis and Model for Complex Ecosystem Restoration*, https://doi.org/10.1007/978-3-030-32057-7_6

a small fraction derived from external inputs via atmospheric deposition.[1] The occurrence of MeHg in aquatic environments thus generally requires both a source of inorganic Hg to serve as its precursor and favorable biogeochemical conditions to promote methylation.

Atmospheric deposition is the predominant input pathway for inorganic Hg entering the Everglades (>95%, Chap. 6, Volume I). Despite its presence as a necessary precondition for methylation to occur, spatial and temporal variations in atmospheric inputs of inorganic Hg alone cannot account for the wide variations in the occurrence of both water column concentrations of MeHg and biota Hg concentrations across the Everglades landscape. For example, based on USEPA R-EMAP data (Cycles 0–7; 1995–2005), the coefficient of variation (CV) in MeHg and *Gambusia* Hg concentrations across the Everglades ranges from 84.9 to 163% and 61.7 to 83.1%, respectively; based on measurements conducted by the Mercury Deposition network (MDN 2019), variations in annual deposition fluxes of total Hg *across* Florida (three sites with concomitant deposition data for at least 50 weeks in each year; N years = 9) ranges from only 5.23 to 31.8%; deposition fluxes across the Everglades are likely even more uniform. In contrast, the CVs related to the spatial occurrence of total or dissolved organic carbon (TOC/DOC) and sulfate in surface water more closely approximate (TOC/DOC) or exceed (sulfate) the CVs for both MeHg and *Gambusia* Hg concentrations across the Everglades (Fig. 6.1). It is thus clear that spatial variations in biogeochemistry are more important in governing spatial variations in MeHg and its resultant trophic transfer at any given point in time than spatial variations in atmospheric deposition fluxes of Hg, although *temporal* trends in biota Hg concentrations at a given location can be related at least in part to temporal changes in atmospheric deposition (see Chap. 4, Volume III).

6.2 The Role of Organic Carbon

6.2.1 DOC Influence on Hg Methylation

In his 2002 State of the Union speech, then president George H.W. Bush labelled three countries—Iran, Iraq, and North Korea—as comprising an "axis of evil" primarily responsible for state-sponsored terrorism throughout the world (Glass 2019). Shortly thereafter, USGS geochemist George Aiken (2004) similarly

[1]Direct releases of MeHg into aquatic environments are rare but, when they occur, can be devastating. Perhaps the most significant example is the Minamata Bay, Japan disaster which became recognized as an incipient and growing problem in the mid-1950s (Harada 1995). In this case, industrial releases of MeHg directly into the bay resulted in very high exposure levels of Hg in fish and shellfish, with 2252 individuals diagnosed with a range of neurological and visual symptoms, including fetal developmental problems, categorized as "Minamata disease."

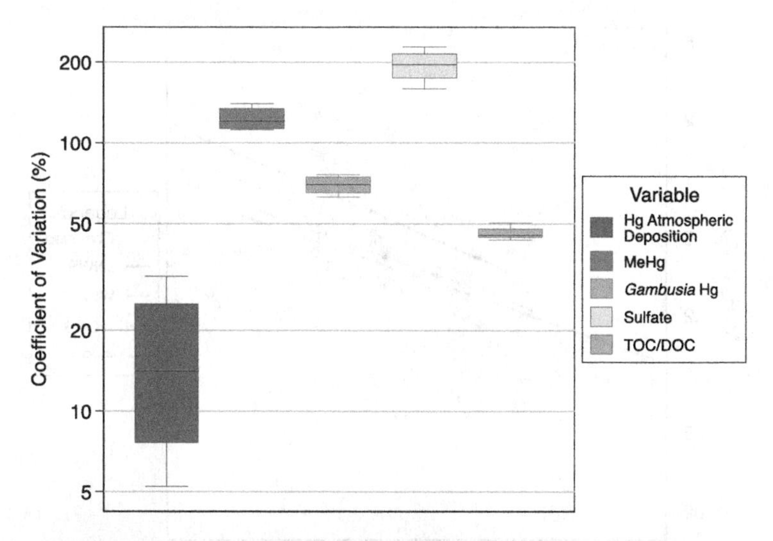

Fig. 6.1 Variability in annual coefficients of variation (CV) for key variables involved in the biogeochemical cycling of Hg in the Everglades. Variables shown as part of the analysis include annual variations in CV for wet deposition Hg fluxes measured across three sites in Florida (MDN sites FL05, FL11 and FL97; N years = 9; MDN 2019). Also included are annual variations in CV measured for MeHg, *Gambusia* Hg, sulfate and TOC/DOC concentrations measured across the EvPA during R-EMAP Cycles 0–7 (N = 4 years; Scheidt and Kalla 2007)

identified a "biogeochemical axis of evil" of three key variables that, acting in concert, were responsible for the globally endemic aquatic Hg problem. The three "bad actors" in this biogeochemical *troika* included Hg, sulfur and carbon, with Aiken using this symbolism as a plea to convince fellow colleagues "of the important role of dissolved organic matter (DOM) in the methylation of mercury."

In general, DOM can influence or affect microbial methylation of Hg in at least three ways.

1. DOM can stimulate activity of methylating bacteria by serving as a substrate in support of metabolism;
2. Complexation by DOM may alter Hg(II) availability for microbially-mediated methylation; and
3. In reducing environments where sulfide production can lead to sequestration of Hg(II) (the necessary Hg precursor for forming MeHg) through the formation of the precipitate cinnabar, DOM interactions with cinnabar surfaces can enhance the bioavailability of Hg(II) to Hg-methylating microorganisms that is otherwise essentially unavailable.

Whether complexation with DOM reduces or increases availability is not fully understood. Despite this uncertainty regarding the effects of complexation, based on cross-sectional survey data and related statistical modeling, it is clear that the overall

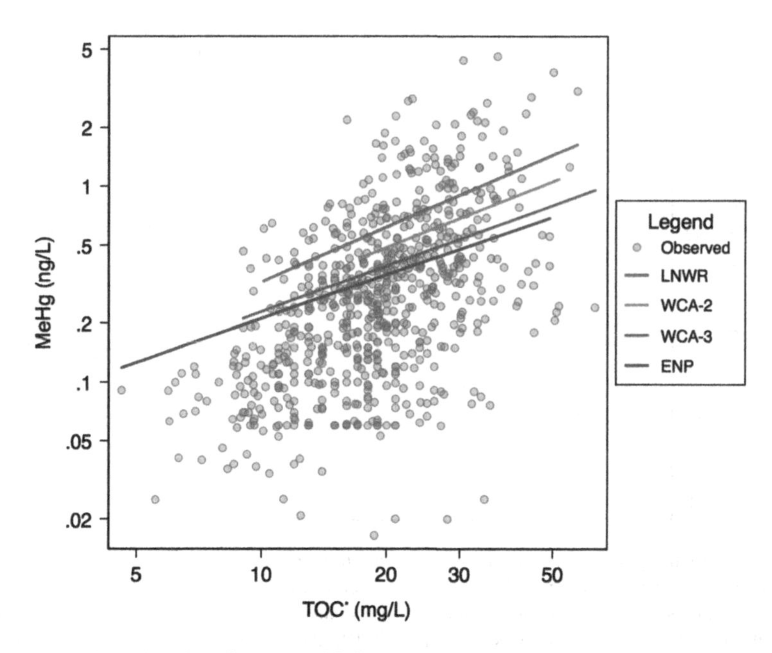

Fig. 6.2 Bivariate plot of surface water MeHg concentrations in the Everglades as a function of TOC*, where TOC* refers to either DOC or TOC concentrations measured by the USEPA as part of R-EMAP, 1995–2014 (N = 696; see Chap. 6, Volume III for a description of the data). Also shown are the fitted slopes for the MeHg-TOC relationship specific for each major hydrologic unit based on mixed effects generalized linear modeling

net effect of DOM with respect to Hg methylation in the Everglades is strongly positive (Fig. 6.2; Pollman 2014; Chap. 6, Volume III).[2]

The nature or quality of DOM can also affect how DOM influences the microbial community composition and activity, including sulfate reducers. Two characteristics of DOM—aromaticity and the degree to which reduced sulfur components (and thiols in particular) have been integrated as part of the DOM molecular structure (i.e., sulfidization)—are particularly important. Laboratory studies have shown that increasing aromaticity and DOM molecular weight results in a greater enhancement effect of DOM on Hg methylation. Sulfidization influences a number of important aspects related to inorganic Hg(II)—Hg methylation dynamics, including inorganic Hg(II) and MeHg coordination chemistry, inhibition of HgS precipitation/aggregation, enhancement of microbial Hg methylation, and Hg photodegradation. As Graham notes in Chap. 4, aromaticity and sulfidization are both increased in the

[2]Organic carbon in the water column was measured during R-EMAP as total organic carbon (TOC) during Cycles 0–5, and as dissolved organic carbon (DOC) during Cycles 6–7. Cycle 11 data include both TOC and DOC (which are shown in Chap. 6, Vol. III to be essentially equivalent); the Cycle 11 data included in the plot are for TOC.

northern Everglades, resulting in exceptional enhancement of methylation seen in laboratory studies.

This spatial effect is evident in the R-EMAP data shown in Fig. 6.2 as well. Included in Fig. 6.2 are the slopes for the fitted relationship between TOC* and MeHg concentrations obtained from mixed effects generalized linear modeling (StataCorp 2017). The model includes hydrologic unit as a fixed effect through an interaction term with TOC*; sample year and season (wet or dry) were included in the model as random effects (year nested within season) to isolate the fixed effects of spatial location and determine if the fixed effects vary significantly between hydrologic units. The dependent variable, MeHg, was modeled using a log link function, with the independent variable, TOC*, log-transformed. The modeling results confirm that the slope for the TOC* relationship with MeHg is indeed higher for the northern Everglades; moreover, a *post hoc* analysis (χ^2 test) of the results also indicates that the slopes for the two more southern hydrologic units are both statistically indistinguishable from each other and are significantly lower than the two more northern hydrologic units ($p \leq 0.0135$).

6.2.2 *DOM Influence on MeHg Uptake by Biota and Photodegradation*

Although increasing DOM leads to higher MeHg concentrations, the *direct* effect of DOM on biota MeHg is more complex. For example, most studies indicate that high concentrations of DOC can limit the bioavailability of MeHg for uptake by algae (e.g., Watras et al. 1995, 1998; Gorski et al. 2008). Structural equation modeling of Hg methylation dynamics and the subsequent trophic transfer of MeHg to *Gambusia* in the Everglades indicates that the direct effect of DOM on *Gambusia* Hg concentrations is indeed negative and significant ($p = 0.001$; Pollman 2014; Chap. 6, Volume III). This direct negative effect blunts the indirect positive effect of DOM exerted on *Gambusia* Hg concentrations (through the direct effect of DOM on Hg methylation) such that, while the overall effect of DOM on *Gambusia* Hg is still positive (see Fig. 6.6, Volume III), it is only weakly significant ($p = 0.037$). These results suggest that any negative impacts resulting from DOM's role in enhancing MeHg production and transport thus may be partially or even largely offset by DOM's role in decreasing MeHg uptake by plankton.

DOM may also affect the importance of photodegradation of MeHg. Photodegradation of MeHg is complex and not fully understood; nonetheless it appears that the primary effect of DOM on photodegradation is through the profound effect DOM has on light attenuation in the water column. Several researchers have found that photodegradation rates (k_{PD}) are wavelength-dependent (UVA, UVB, and photosynthetic active radiation [PAR]), with the following sequence of greater rate constants (cf., Black et al. 2012; and Fernandez-Gomez et al. 2013):

$$k_{PD,UVB} \gg k_{PD,UVA} \gg k_{PD,PAR}$$

Light extinction coefficients ($k_{extinction}$) associated with higher concentrations of DOM likewise are wavelength dependent and follow the same sequence as for k_{PD} (Scully and Lean 1994; and Pérez-Fuentetaja et al. 1999).

$$k_{extinction,UVB} > k_{extinction,UVA} \gg k_{extinction,PAR}$$

Increasing DOM thus rapidly attenuates the wavelengths of light that promote the highest rates of photodegradation. Restated, increasing DOM thus should result in lower rates of photodegradation of MeHg.

6.2.3 Effects of Disturbance on DOM Dynamics

The nature and dynamics of DOM in the Everglades and associated impacts on Hg cycling are intrinsically linked to anthropogenic disturbances that have led to both nutrient and sulfate enrichment. Nutrient enrichment has led to expansion of cattails; cattails, which are more readily decomposed than native sawgrass, contain twice the leachable DOM as does sawgrass. In addition, the amount or yield of phenolic DOM is higher in cattails compared to sawgrass. Drainage of the peat soils within the EAA results in both oxidation of the peat to produce DOM with higher aromaticity and greater fluxes of DOM exported from the EAA to the EvPA. This is evidenced by the high concentrations and aromaticity of DOC in canals draining the EAA (Aiken et al. 2011). As a result, DOM concentrations in the northern Everglades are both higher and more aromatic, with soil oxidation in the EAA likely the primary driver. Anthropogenic sulfate enrichment likely has led to alterations in the chemical nature of Everglades DOM including, in particular, increased sulfidization; this sulfidization can help account for the observed enhanced effect that DOM has on Hg methylation in the northern Everglades compared to more southern locations.

Seasonal differences in DOM quality likely occur as well, with higher aromaticity occurring during the wet season when terrestrially (soil) derived DOM fluxes are better able to penetrate the Everglades marsh because of higher rates of water delivery. Based on these results, Graham (Chap. 4) suggests that Everglades restoration efforts designed to increase freshwater delivery to the southern Everglades are likely to increase the flux of aromatic, highly sulfidized DOM to the southern Everglades as well, and thus increased methylation of Hg. This concern extends to water treated to remove elevated phosphorus concentrations exported from the EAA by routing the water through stormwater treatment areas (STAs) before it is released to the EvPA. These waters contain high concentrations of DOM that are highly aromatic.

6.3 The Role of Sulfate and Statistical Red Herrings

In Chap. 3, Rumbold uses the causal analysis/diagnosis decision information system (CADDIS) developed by the USEPA (Suter et al. 2002, http://www.epa.gov/caddis) to evaluate the dominant factor that most likely can account for the extremely high Hg concentrations found in Everglades biota [see Chaps. 8 and 9, for discussions on biota concentrations, including wading birds (Chap. 8) and piscivorous fish (Chap. 9)]. Evidence of causality requires supporting data from a variety of different avenues that ultimately are self-consistent, including mechanistic and empirical studies that consider spatial and temporal relationships between candidate causative factors and the response variable of interest. Based on that analysis, Rumbold concludes that "that it is the balance between sulfate and sulfide that is the key control of net Hg methylation rate and the dominating cause for the extreme temporal and geographic differences in the magnitude of biomagnification in the Everglades." At the same time, Rumbold takes care to acknowledge the fundamental causative role of the very high inputs of atmospheric Hg to the Everglades (see Chap. 4; Volume I). High Hg loadings, however, do not necessarily translate to high concentrations of MeHg produced and cascading through aquatic food webs. Restated, elevated biota Hg concentrations require not only a source of Hg, but also the capability to produce MeHg concentrations in the presence of an inorganic Hg source.

The primary arguments against the causative role of sulfate with respect to the Everglades Hg problem have been largely statistical in nature. These statistical arguments fall into three categories: (1) the relationship between sulfate and Hg concentrations in biota in the Everglades is *negative*—i.e., sulfate has no stimulatory effect on net methylation (Julian 2013); (2) the relationship between sulfate and MeHg production and Hg biota concentrations is not unimodal (and thus defaults to either no relationship or a negative relationship (Julian 2013, 2014; Julian et al. 2015); and (3) the sulfate-MeHg/biota Hg relationships account for small fraction of the variability of the dependent variable (Julian et al. 2017, 2018) and thus the relationship between sulfate and MeHg remains "elusive" and precludes our ability to empirically and robustly model the effects of sulfate on MeHg formation and bioaccumulation (Julian et al. 2018).

6.3.1 Arguments 1 and 2

These two arguments overlap because of the nature of the unimodal sulfate—methylation response curve and the location of the methylation maximum relative to the overall distribution of sulfate concentrations ($SO4_{critical}$) in the Everglades. It should go without saying that the data used to test the unimodal hypothesis must be appropriate for the task. With the goal of testing the whether the relationship between sulfate and "bioaccumulation" in *Gambusia* Hg is indeed unimodal, Julian (2013) used R-EMAP data collected between 1994 and 2005 (654 joint observations for *Gambusia* Hg, water column sulfate and MeHg concentrations within the EvPA) and

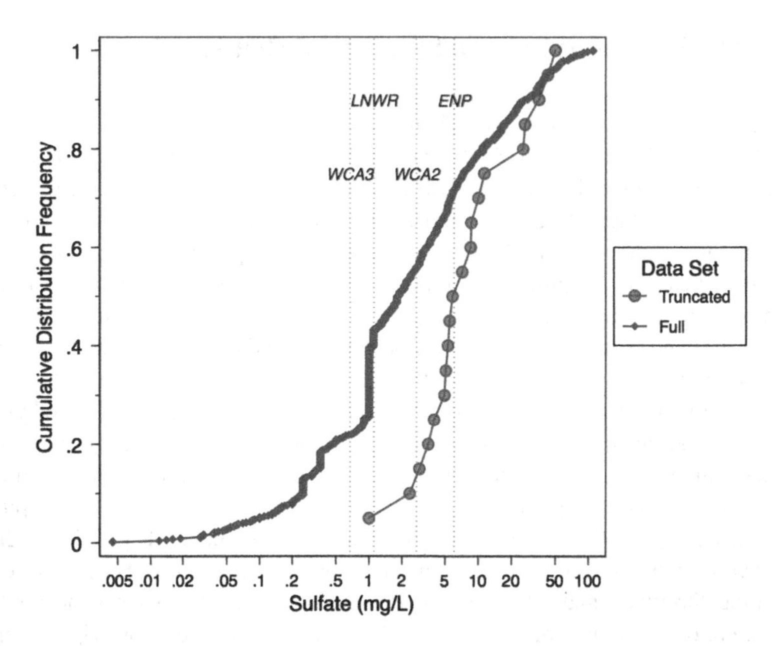

Fig. 6.3 Comparison of the cumulative distribution frequency for sulfate concentrations in the R-EMAP data set for the EvPA for which joint observations of both sulfate and MeHg occur (Cycles 0–7, Scheidt and Kalla 2007; N = 654) and the same data set after truncation (N = 20) by Julian (2013). Vertical lines show the estimated sulfate concentrations that correspond to Hg methylation maxima for each major hydrologic unit (see Chap. 6, Volume III)

then truncated the data by averaging across major hydrologic units and water years to produce a final data set comprising only 20 points. In addition to problems related to inconsistencies in how he aggregated the data, this truncation severely distorts the distribution of the data compared to the original data set (Fig. 6.3), including obscuring the effects of dilute sulfate concentrations on Hg methylation and subsequent trophic transfer. Moreover, by essentially eliminating a large number of data points in the critical low concentration range of sulfate concentrations, Julian's analysis precluded his ability to properly elucidate not only the curvilinear nature of the sulfate-biota Hg relationship but also the locus of the sulfate maximum.[3] This conceptual problem in data analysis is shown through a more complete and statistically robust treatment of the R-EMAP data by Pollman (Chap. 7; Volume III), who is able to both demonstrate the curvilinear relationship and estimate the critical sulfate concentration for each major hydrologic unit in the Everglades that correspond to maximum *Gambusia* Hg concentrations. Those values range from 0.7 to 6.1 mg/L while Julian's truncated data set ranges in concentration from 1 to 50 mg/L, and thus are not adequate to resolve the underlying structure of the relationship

[3]There are other issues inherent in this paper which render its utility in resolving any debate on nature of the sulfate-Hg methylation debate problematic, including inappropriately using *bioaccumulation factors* as a proxy for directly concluding effects of sulfate on biota Hg *concentrations* (see Pollman and Axelrad 2014).

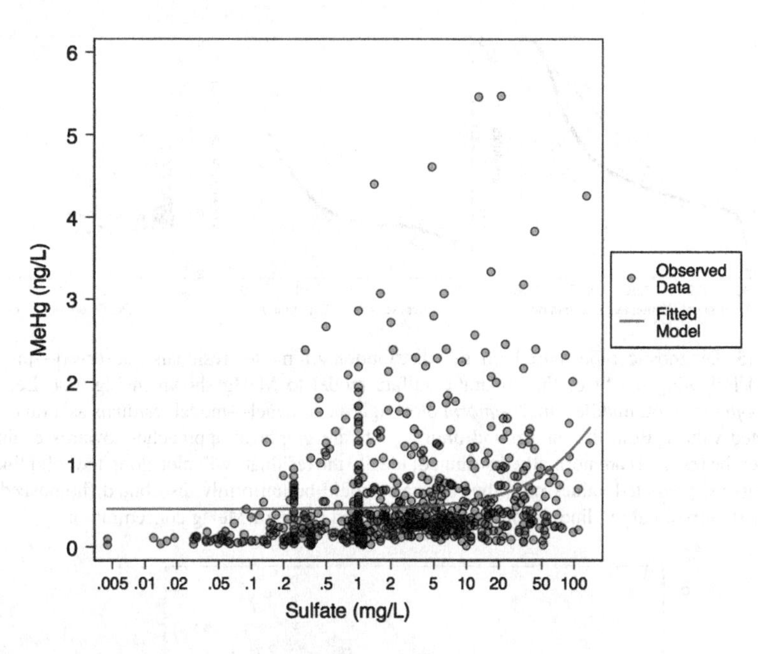

Fig. 6.4 Predicted relationship between sulfate and MeHg fitted through linear regression using sulfate as a quadratic function (after Julian 2014)

because of the obfuscation of a broad range of data points at the critical low end of the distribution (in excess of two orders of magnitude; Fig. 6.3).

A second issue that is problematic with some of the work used to refute or weaken the unimodal sulfate conceptual model relates to statistical model misspecification. Misspecification encompasses a variety of sins, and Rumbold (Chap. 3) discusses how model misspecification in two different forms compromises the modeling conducted by Julian et al. (2014). Another example of misspecification is seen in Julian (2014). In this analysis, Julian modeled the variations in R-EMAP surface water MeHg concentrations as a quadratic function of sulfate (Fig. 6.4). Diagnostic plots of the model residuals (Fig. 6.5) however clearly indicate that the model is misspecified and violates key assumptions of linearity and residual distribution. More specifically, both *pnorm* and *qnorm* plots (StataCorp 2017) demonstrate that the residuals strongly deviate from an expected normal distribution inherent in a well-behaved linear regression model. Likewise, a plot of the model residuals as a function of predicted values shows that the residuals are severely heteroscedastic (non-uniformly distributed). These problems ultimately reflect several issues. First both the dependent and independent variables in this case should be log-transformed. Second, as discussed by Graham (Chap. 4), it is well-established that dissolved organic carbon (DOC) strongly influences methylation dynamics. As a result, we herein revise the Julian (2014) model to address both of these issues. Similar to Julian (2014), our revised model (which uses the same R-EMAP data set used by Julian (2014) and is applied to only those sites within the EvPA) incorporates the

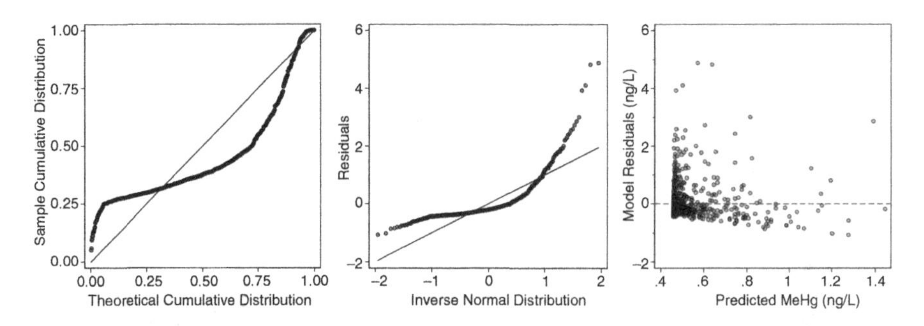

Fig. 6.5 Diagnostic plots based on the distribution of model residuals (observed—predicted values) following the fit of the quadratic sulfate model to MeHg shown in Fig. 6.4. Left-hand panel—*pnorm* plot; middle panel—*qnorm* plot; right-hand panel—model residuals as a function of predicted values. Both the pnorm and qnorm plots are graphical approaches towards evaluating whether the residuals are normally distributed; ideally the residuals will plot along the solid line. For residuals vs. predicted values plot, the residuals should be uniformly distributed (homoscedastic) around 0 (horizontal red line) throughout the predicted range of MeHg concentrations

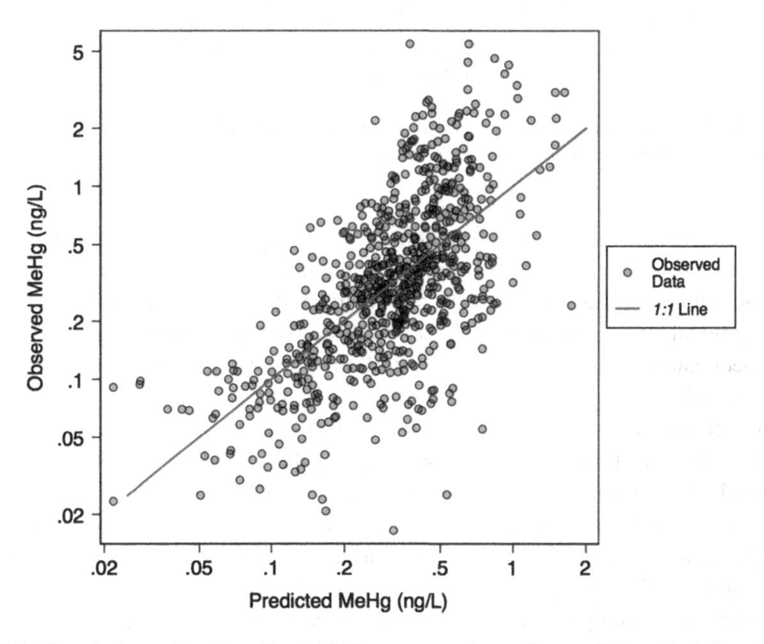

Fig. 6.6 Plot of observed and predicted MeHg concentrations after modifying the Julian (2014) model to (1) use the log-transform of sulfate prior to including it in the model as a quadratic term; and (2) including DOC as a second independent variable

sulfate effect as a quadratic function in an attempt to identify a possible unimodal response; in our case, however sulfate is initially log-transformed. In addition, our model includes DOC (log-transformed) as a second independent variable.

A plot of comparing observed MeHg concentrations as a function of predicted values is shown in Fig. 6.6. The revised model now accounts for approximately 41% of the variance in the dependent variable (compared to only 4% for the Julian (2014)

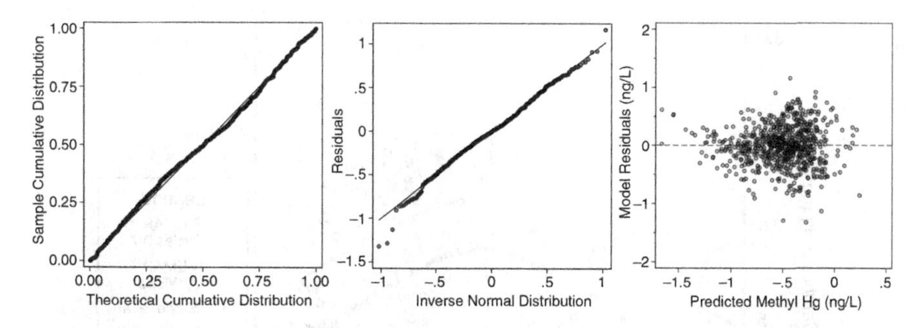

Fig. 6.7 Similar to Fig. 6.5, except that the diagnostic plots are for the residuals obtained from the modified MeHg model shown in Fig. 6.6. Unlike the residuals for the original Julian (2014) model, the residuals for the modified model are far better behaved and more closely approximate normal and homoscedastic distributions

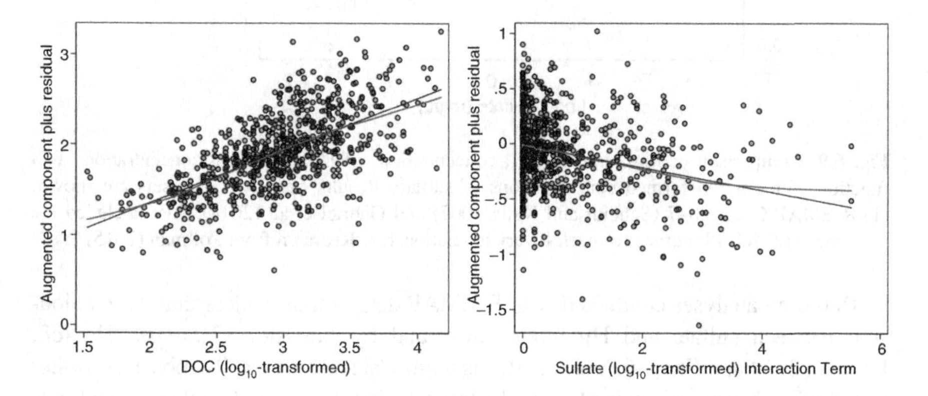

Fig. 6.8 Augmented component plus residuals plots constructed for DOC (log_{10}-transformed) term (left-hand panel) and the sulfate (log_{10}-transformed) interaction (squared) term (right-hand panel) included in the revised MeHg model. Gray straight line depicts the expected linear relationship fitted between the independent and dependent variables. The red lines are smoothed relationships obtained from LOWESS. For both variables the LOWESS and expected linear curves agree quite well, with the minor exception of tailing at the extreme end of the sulfate interaction distribution that is an artifact of two data points. Data from USEPA R-EMAP Cycles 0 through 7 sampled within the EvPA (Scheidt and Kalla 2007; N = 717)

model). Moreover, inspection of the revised model residuals shows that—with the exception of only 3 points and a single point at the lower and upper tails of the residuals distribution—the residuals now are extremely well-behaved and the difficulties with the original model have been largely eliminated (Fig 6.7). The sulfate interaction term in the Julian model shifts from being highly insignificant ($p = 0.728$) to highly significant ($p < 0.001$). Moreover, the effect of DOC and the sulfate interaction term are strongly and robustly linear. This is shown in Fig. 6.8, which includes two augmented components plus residuals plots (Hamilton 2013) that are useful in elucidating any underlying non-linear behavior between an independent variable and the model dependent variable when the effects of other independent variables in the model are accounted for.

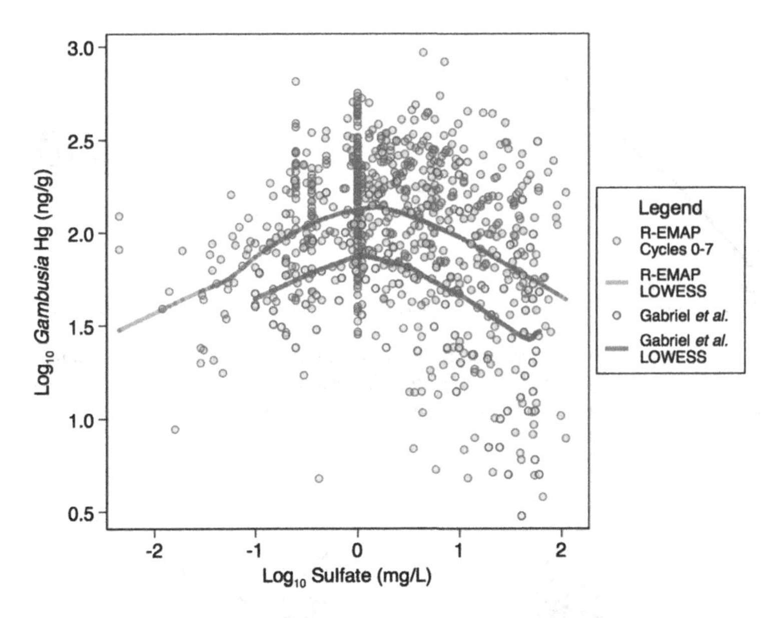

Fig. 6.9 Comparison of $\log_{10}$-transformed concentrations of *Gambusia* Hg concentrations as a function of $\log_{10}$-transformed concentrations of sulfate. Results from two data sets are shown: (1) R-EMAP Cycles 0–7 (Scheidt and Kalla 2007) and Gabriel et al. (2014). Also included are smoothed (LOWESS) curves for each bivariate relationship. Redrawn from Pollman (2015)

Based on analyses conducted with R-EMAP data, it thus is clear that the relationship between sulfate and Hg methylation, and by extension *Gambusia* Hg (cf., Pollman 2014 and Chap. 6, Volume III), is unimodal and statistically robust. A further test of this robustness is gained by evaluating whether the unimodal relationship holds across data sets. For example, Fig. 6.9 compares *Gambusia* Hg concentrations as a function of sulfate concentrations for both the R-EMAP data and data compiled by Gabriel et al. (2014), the latter which includes samples collected between 1998 and 2009 at 11 stations within the Everglades (9 stations within the EvPA). Locally weighted scatterplot smoothing (LOWESS; StataCorp 2017) was applied separately to both data sets. Both data sets show very similar smoothed unimodal curves, with the major difference related to overall higher concentrations observed in the R-EMAP data. This is consistent with the fact that much of the R-EMAP data predates the Gabriel et al. (2014) data, and thus is more strongly influenced by higher concentrations observed during the monotonic decline in fish tissue Hg concentrations that began likely before 1990 and concluded ~2000 (Chap. 2, Volume III).

The unimodal relationship should also hold across species. Because they compiled fish tissue Hg data from sunfish and Largemouth bass (LMB) in addition to *Gambusia*, this hypothesis can also be tested with the Gabriel et al. (2014) data set. Similar to the previous analysis, LOWESS plots were constructed for each individual species. Because each of these three species of fish occupy different niches and thus expectedly biomagnify Hg to differing degrees, the individual Hg concentrations for each observation (after first transforming to $\log_{10}$ values) were standardized to zero mean and unit variance according to species. Standardizing the data in this

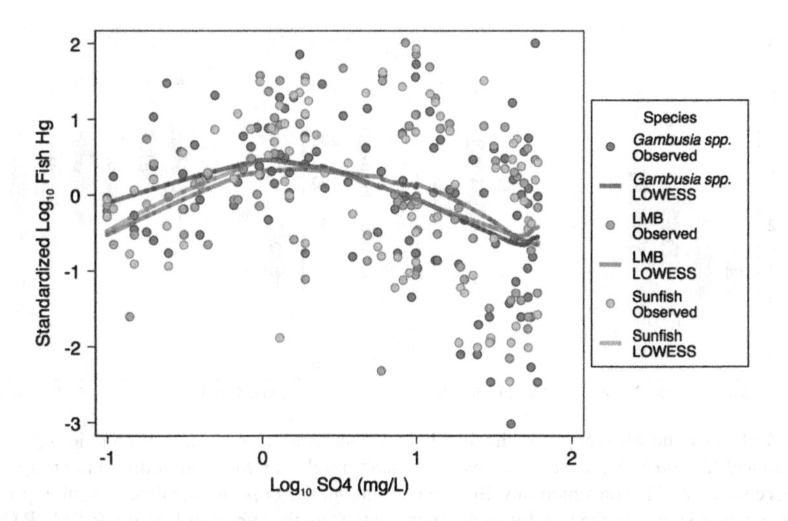

Fig. 6.10 Comparison of log_{10}-transformed fish tissue Hg concentrations as a function of log_{10}-transformed concentrations of sulfate for three species of fish—*Gambusia*, Largemouth bass (LMB), and sunfish. Also included are smoothed (LOWESS) curves for each bivariate relationship. Transformed concentrations have been standardized (0 mean and unit variance) so that the results across species can be compared directly. All data from Gabriel et al. (2014). Redrawn from Pollman (2015)

manner allows a direct comparison of the results across each species. This comparison, which is shown in Fig. 6.10, shows remarkable concordance across the three species to the same unimodal relationship, indicating that the unimodal relationship indeed is self-consistent across different trophic levels of fish in the Everglades.

6.3.2 Argument 3

The argument that, because a statistical relationship that models the effect of a given variable x on a particular response variable y accounts for only a small fraction of the latter variable's overall variance, controlling x can or will not lead to meaningful changes in the values of y is an often used and often specious and misleading argument. Econometricians and epidemiologists have long been using adjusted predictions—also known as margins analysis—to quantify the effect of an independent variable x on a dependent variable y with confidence intervals reflecting the inherent uncertainty throughout the prediction interval—for multivariate models (cf., Long and Freese 2014). The adjusted predictions are conducted by holding the other independent variables in the model at *a priori* fixed levels and allowing x to vary across some range of interest that is also consistent with the characteristics of the data used to construct the model. This type of analysis not only quantitatively isolates the effects of x, but also can be used to identify subgroups of the modeled population where controlling x has predicted lesser or greater benefit. Thus, if the form of the model is well-specified, margins analysis can provide management-relevant and valuable

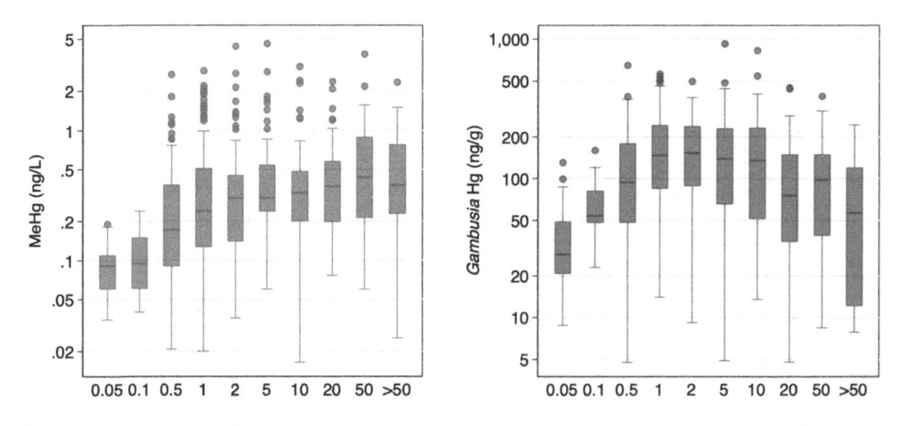

Fig. 6.11 Box and whisker plot of the distribution of surface water MeHg concentrations (left-hand panel) and *Gambusia* Hg concentrations (right-hand panel) as a function of different groupings of surface water sulfate concentrations. Sulfate concentration groups are identified by an upper limit concentration shown in the plot for each group. Data from the subset of USEPA R-EMAP Cycles 0–11 used for structural equation modeling by Pollman (Chap. 6, Volume III). N = 696 and 684 for MeHg and *Gambusia* Hg concentrations, respectively

insight regarding control strategy benefits for different independent variables on a response variable of interest (e.g., the effects of age, sex, diet, smoking, exercise, weight, region and exposure level to a potential carcinogen on the occurrence of particular type of cancer). Chapter 7 (Volume III) uses margins analysis applied to a "sulfate management model" that has been derived from structural equation modeling (Chap. 6, Volume III) and tested to evaluate underlying assumptions of bivariate linearity to predict with associated levels of uncertainty the effects of different control strategies on *Gambusia* Hg concentrations throughout the Everglades.

A second problem with this argument is that it ignores the constraining effects of very low sulfate concentrations on both MeHg and fish Hg concentrations in the EvPA. This is unambiguously illustrated by showing the distribution of surface water MeHg and *Gambusia* Hg for different categories of sulfate concentrations (Fig. 6.11). For example, the average *Gambusia* Hg concentration in the lowest sulfate concentration category (sulfate $<=0.05$ mg/L) is 37.9 ng/g ($N = 42$), with lower and upper 95% confidence limits ranging from 29.8 to 46.0 ng/g. Less than 10% of the observations in the lowest sulfate category had *Gambusia* concentrations in excess of 80 ng/g. At the overall sulfate maximum (between 1 and 2 mg/L), *Gambusia* Hg concentrations are over 4 times higher, averaging 169 ng/g ($N = 69$) and with lower and upper 95% confidence limits ranging from 145 to 193 ng/g.

6.4 The Role of Phosphorus

The Everglades historically developed as an ultra-oligotrophic system, with low ambient phosphorus concentrations driven by atmospheric deposition as inputs. The progressive development of the Everglades Agricultural Area (EAA), which converted nearly 3000 km^2 immediately south of Lake Okeechobee from native custard apple and cypress swamp forests and sawgrass plains south of these forests to agricultural lands, was a major anthropogenic perturbation to the Everglades (see Chaps. 1 and 2, Volume I). This perturbation included a number of impacts, including in particular increased loadings of phosphorus to the remnant Everglades discharged as runoff from the EAA. This additional loading was implicated in observed shifts in macrophyte community structure from native sawgrass to cattail, including shifts associated with a transitional nutrient gradient or advancing nutrient front believed then to be expanding over time (cf. Belanger et al. 1989; Koch and Reddy 2001) in the northeastern region of WCA-2A below the S10 structures, as well as a variety of other trophic state related impacts. In 1988, concerns about further excursions of this nutrient front deeper into the Everglades helped spark federal litigation against both the Florida Departmental of Environmental Regulation (now FDEP) and the South Florida Water Management District (SFWMD) for failing to protect water quality in the Loxahatchee National Wildlife Refuge (LNWR; also known as WCA1) and the Everglades National Park (ENP) (Green and Perko 2001).

As the Hg problem in the Everglades became more apparent, concerns about the impacts of increasing phosphorus concentrations in the Everglades extended to possible effects on Hg cycling and trophic transfer as well. The concern was first expressed by Ron Jones, who hypothesized that phosphorus exported from the EAA controlled "the rate at which mercury is being methylated in the system because it controls the amount of oxygen available in the system" (Jones 1992; cited in Green and Perko 2001). At the other end of the spectrum Green and Perko (2001) argued that "phosphorus enrichment in the Everglades might have the beneficial effect of reducing wildlife mercury contamination." This argument was based on an inverse relationship observed between *Gambusia* Hg concentrations and surface water total phosphorus (TP) concentrations observed along the nutrient gradient below the S10 structures in WCA2 (cf. Fig. 6.12a), presumably through biodilution and enhanced sedimentation rates through increased particle production.

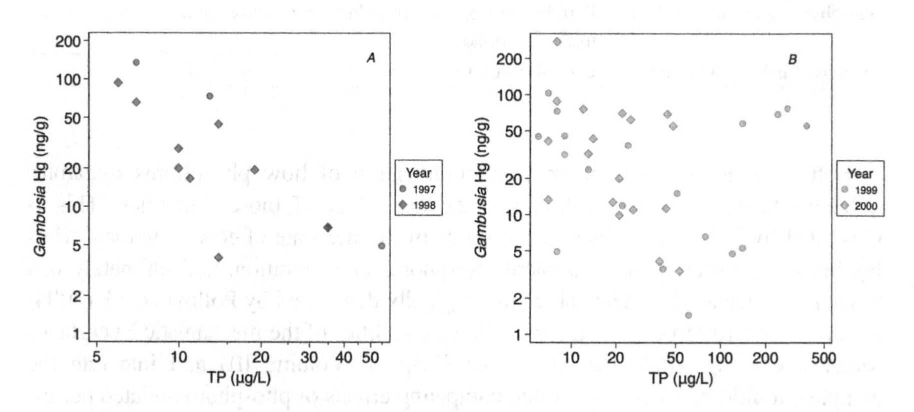

Fig. 6.12 *Gambusia* Hg concentrations vs. surface water TP concentrations along the nutrient gradient below the S10 structures (transect stations F1 through F5 and U3) in WCA-2A. Panel (A): 1997 and 1998; Panel (B): 1999 and 2000. SFWMD data provided by L. Fink (personal communication, 2002)

Table 6.1 List of the effect of increasing phosphorus concentrations on Everglades trophic state related variables, as well as concomitant effects on Hg biogeochemical cycling and whether the phosphorus related effect has a net increasing (+), decreasing (–), or indeterminate (+/–) effect on fish tissue Hg concentrations. Modified from Pollman et al. (2004)

Trophic state effect	Hg biogeochemical cycling effect	Fish Hg concentration effect
Increased particle production	Dilution of Hg(II) and MeHg concentrations due to more particles	–
Increased particle production	Increased sedimentation rates of Hg (II) and MeHg	–
Shifts in trophic structure (simplification of food web)	Change in fish diet	–
Plant coverage of open water	Decreased light transmission, lower rates of photodegradation of MeHg, and Hg (II) photoreduction	+
Plant coverage of open water	Shifts in coverage affect particle production and turnover rates	+/–
Increased decomposition rates	Higher rates of carbon decomposition with higher TP stimulates bacteria to methylate at higher rates	+
Reduced dissolved oxygen	Increased sulfide concentrations (which affects the pool of Hg available to methylate)	–
Reduced dissolved oxygen	Increased rates of methylation due to more favorable conditions for sulfate reducing bacteria	+
Reduced dissolved oxygen	More rapid rates of water transport across gills and thus more gill uptake of MeHg.	+
Macrophyte species shift	Species shifts affect particle production and turnover rates	+/–
Macrophyte species shift	Cattails contain twice the amount of leachable DOM compared to sawgrass; cattails also yield more phenolic (aromatic) DOM	+
Periphyton community shifts	Periphyton is an important locus for methylation in the Everglades	?
Increased fish growth rates	Growth dilution	–

Both arguments represent an oversimplification of how phosphorus dynamics influence biogeochemical cycling and the net effect of those dynamics. This is illustrated by Table 6.1, which lists a series of trophic state effects, how they affect Hg biogeochemical cycling, ambient phosphorus concentration, and ultimately fish tissue Hg concentration. This table was originally developed by Pollman et al. (2004) as part of an effort to expand the modeling capabilities of the mechanistic Everglades Mercury Cycling Model (E-MCM; see Chap. 4, Volume III) and integrate the complex, multidimensional and often competing effects of phosphorus-related perturbations. In addition to changes in redox dynamics and increased particle production mentioned above, other major trophic state effects include shifts in macrophyte community from sawgrass towards cattail (see Chap. 5), shifts in the degree of plant coverage and thus reductions in the relative amount of open water available for light transmission and gas exchange, shifts in organic carbon decomposition rates, shifts in trophic structure (simplification of the food web), and increased rates of fish growth leading to growth dilution of fish tissue Hg concentrations.

A cursory examination of Table 6.1 demonstrates that the overall effect that phosphorus dynamics can exert on Hg cycling and the concomitant effects on fish tissue Hg concentrations is not obvious. For example, reduced dissolved oxygen concentrations can more readily lead to anaerobic conditions necessary for methylation; it also can lead to the generation of higher levels of sulfide that can inhibit methylation by binding Hg(II). Reduced dissolved oxygen concentrations can also force fish to transfer more water across their gills to maintain adequate uptake rates of oxygen; increased uptake of MeHg directly from the water by fish can result. Although the primary uptake pathway for MeHg in fish is dietary (see Chap. 7), simulations conducted with the dynamic Everglades Mercury Cycling Model (E-MCM) simulations suggest that enhanced uptake across the gills can be important as well (Pollman et al. 2004).

Biodilution net impacts also are neither one-dimensional nor necessarily obvious. E-MCM simulations suggest that the effect of increased particle production is primarily indirect, as opposed to directly burying MeHg or diluting it in the long term. Instead, increasing particle production had more impact on predicted inorganic Hg(II) concentrations, and subsequently on methylation rates. Moreover, the inverse relationship suggested by Fig. 6.12a and ascribed to biodilution does not hold up in the presence of more data. This is shown in Fig. 6.12b, which shows results for 1999 and 2000 at the same WCA-2A transect locations shown in Fig. 6.12a, and Fig. 6.13, which also is a bivariate plot of *Gambusia* Hg concentrations as a function of water column TP concentrations, but is constructed using R-EMAP data collected across the Everglades for both wet and dry seasons.

Lastly, it is well documented in lakes that increasing nutrient enrichment generally leads to simplification of the structure of aquatic food webs. More complexity in aquatic webs is manifested in more trophic levels and greater opportunity for increased biomagnification of MeHg through each succeeding trophic level. *Ceteris paribus*, food web simplification resulting from incipient eutrophication should accordingly lead to lower biota Hg concentrations at the apex of the aquatic food web compared to the apex of more complex systems. However, and as discussed by Rumbold (Chap. 7), while evidence supporting food web simplification in lakes in response to eutrophication is evident, such evidence for the Everglades in lacking.

Given the multitude of competing effects imposed by eutrophication on Hg cycling in the Everglades, it is perhaps not surprising that, based on structural equation modeling that included multiple independent variables and both indirect and direct pathways of influence, Pollman (2014) found the overall effects of TP on the occurrence of MeHg and *Gambusia* Hg concentrations to be statistically insignificant.

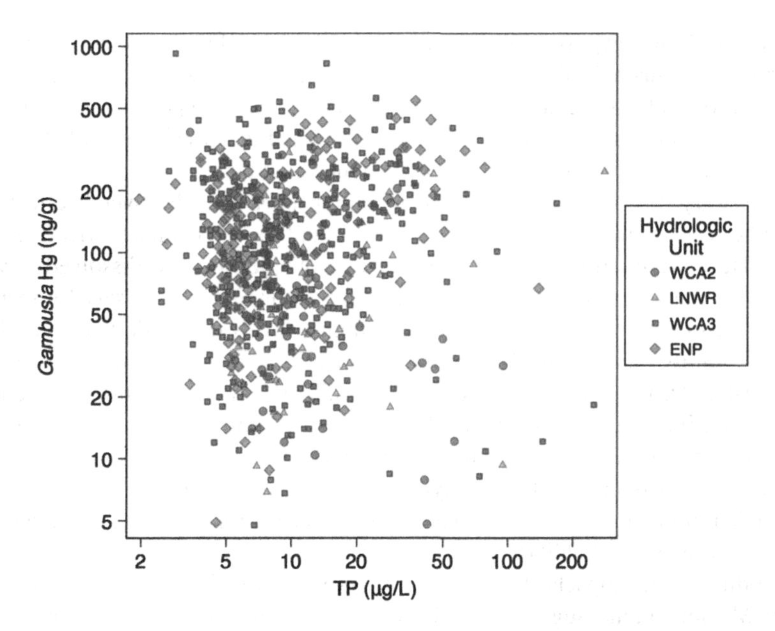

Fig. 6.13 *Gambusia* Hg concentrations vs. surface water TP concentrations plotted using R-EMAP data, Cycles 0–11. Plot includes wet and dry season. Data from the subset of USEPA R-EMAP Cycles 0–11 used for structural equation modeling by Pollman (Chap. 6, Volume III; N = 684)

6.5 Major Loss Mechanisms

There are three major pathways via which total Hg is lost from aquatic ecosystems: (1) photochemical reduction of Hg(II) to form gaseous elemental Hg(0) (or GEM) which in turn leads to either direct evasion of GEM into the atmosphere or losses to the atmosphere via transpiration through the roots of rooted macrophytes; (2) sedimentation of particulate Hg and MeHg and subsequent burial in the sediments; and (3) advective losses associated with water moving through and out of the ecosystem. In addition, MeHg also can be lost via: (1) sedimentation of detritus and net burial following either biota uptake or adsorption to particulates; (2) photodegradation of MeHg; and (3) demethylation.

Dynamic modeling conducted with the E-MCM at the Site 3A-15 within WCA-3A helps provide some perspective on the relative importance of these sinks. Harris et al. (2001) calibrated the E-MCM to Site 3A-15—which was then considered to be a Hg "hot-spot" because of elevated biota Hg concentrations—as part of the overall Florida Hg Pilot Total Maximum Daily Load (TMDL) study conducted by Atkeson et al. (2003). E-MCM is a mechanistic model and part of the model calibration involves evaluating fluxes into and out of the system, as well as changes in storage in different model compartments, to ensure that a reasonable Hg mass balance is achieved. The 3A-15 model results suggest that gross sedimentation or settling is the primary mechanism for removing both Hg(II) and MeHg from the water column. Photodegradation is an important sink for MeHg in the water column as well (equating to ~60% of the gross sedimentation flux). The modeling conducted by Harris et al. (2001), however, also indicated major uncertainties in their ability to parameterize key processes. Model results were highly sensitive to predicted *net* methylation rates, which in turn reflect the difference between gross methylation rates and demethylation rates. Uncertainties in the magnitude of both these rates translate to more magnified uncertainties in resultant net methylation rates. Moreover, as discussed in Chap. 6, Volume I), field measurements of GEM evasion/transpiration rates from within portions of the Everglades are much higher than overall Hg mass balances can support. Lastly, sediment burial rates in the Everglades are not well characterized, and this uncertainty adversely affected the ability of Harris et al. (2001) to develop a more highly constrained mass balance. Further research regarding all of these uncertainties would be useful to help better understand Hg cycling in the Everglades from a quantitative perspective.

6.6 Conclusions

The occurrence of high Hg concentrations in the biota of most aquatic ecosystems, including the Everglades, is a complex process that first requires the conversion of dissolved inorganic Hg(II) into MeHg. This methylation process is microbially driven and, in addition to requiring inputs of biologically available inorganic Hg as the precursor to forming its, also requires suitable biogeochemical conditions to support methylation and its incorporation into aquatic food webs. As discussed in the next chapter (Chap. 7), once methylation occurs the critical step in the trophic transfer process is its incorporation into the base of the aquatic food web where bioconcentration can lead to increases in MeHg concentrations in phytoplankton relative to dissolved concentrations in the water column by factors in excess of 10^6.

In addition to requiring inorganic Hg as a precursor, Hg methylation requires both a substrate to support metabolic activity (DOM) and a suitable terminal electron acceptor. A variety of microbial groups have the capacity to methylate Hg, including sulfate reducing bacteria, iron reducing bacteria, manganese reducing bacteria, and methanogens. Regardless of the mediating microbial group responsible, methylation generally requires anaerobic conditions to proceed.

The overall effect of DOM on Hg methylation in the Everglades is positive—i.e., increasing concentrations of DOM generally correlate with increasing concentrations of MeHg. The nature and quality of the DOM also influences how DOM influences methylation. Increasing aromaticity and increased incorporation of reduced sulfur into the structure of DOM (sulfidization) both accentuate the positive effects of DOM on methylation, resulting in enhanced methylation rates observed in the northern Everglades. The exacerbating effects of DOM enhancing MeHg production and transport may be partially offset by DOM's role in enhancing MeHg photodegradation and decreasing MeHg uptake by plankton. These results thus underscore the necessity of further research on the effects of increased DOM flux and alteration of DOM quality on Hg concentrations in biota, especially in the context of Everglades' ecosystem restoration.

Despite the presence of other methylating groups of bacteria in the Everglades, methylation driven by sulfate reducers appears to be the dominant Hg methylation pathway. Moreover, the fact that MeHg in the Everglades is constrained to substantially lower concentrations at low ($\leq$0.1 mg/L) sulfate concentrations indicates that methylation by other groups of bacteria are not sufficient to fully compensate for or overcome this constraint. As we shall see in Chap. 7, Volume III, this has important policy/management implications for efforts attempting to mitigate the Everglades Hg problem.

When considering the linkages between anthropogenically driven perturbations to the Everglades and the dynamics of both DOM and sulfate, it becomes clear that these perturbations have concomitantly resulted in profound effects on Hg biogeochemical cycling. Hydrologic disturbance has resulted in increased releases of terrestrial DOM from the EAA comparatively enriched in aromaticity and is primarily responsible for the observed gradients in both DOC concentration and aromaticity decreasing from north to south in the Everglades. Eutrophication has led to changes to incursions of cattails displacing native sawgrass along disturbance gradients and pathways; because cattail detritus is more readily decomposed, this displacement has increased the release of DOM into the marsh. DOM derived from cattails is comparatively more aromatic as well, which further exacerbates the effects of cattail encroachment on Hg methylation.

Disturbance largely in the form of using elemental sulfur as a soil conditioning amendment with in the EAA has also led to large scale (orders of magnitude; see Chap. 7, Volume III) increases in surface water sulfate concentrations throughout much of the Everglades. This of course results in sulfidization of DOM and thus, in the absence of any other sulfate related effects, increasing rates of Hg methylation. Increasing sulfate concentrations also leads to increasing activity of sulfate reducing bacteria activity, particularly if ambient concentrations in the absence of any enrichment are expectedly low (as is the case for entirety of the Everglades not influenced by tidal inputs of marine waters). How increased sulfate concentrations translate to changes in MeHg however is complex and inherently unimodal. This is because very high sulfate concentrations can result in the development of high concentrations of sulfide in anaerobic environments where Hg methylation occurs. This sulfide in turn sequesters Hg(II) as an insoluble precipitate (cinnabar), thus limiting methylation. The directional effect of reducing sulfate concentrations on Hg methylation (positive or negative) within the Everglades thus is site specific and must consider the starting and ending sulfate concentrations.

References

Aiken GR (2004) Carbon, sulfur, and mercury—A biogeochemical axis of evil. Proceedings of the 2004 CALFED science conference, "Getting results: integrating science and management to achieve system-level responses", Sacramento, CA, October 4–6, 2004. http://pubs.er.usgs.gov/publication/70195438

Aiken GR, Gilmour CC, Krabbenhoft DP, Orem WH (2011) Dissolved organic matter in the Florida Everglades: Implications for ecosystem restoration. Rev Environ Sci Technol 41 (S1):217–248

Atkeson T, Axelrad D, Pollman C, Keller G (2003) Integrating atmospheric mercury deposition and aquatic cycling in the Florida Everglades: an approach for conducting a Total Maximum Daily Load analysis for an atmospherically derived pollutant. In: Integrated summary. Final report prepared for the United States Environmental Protection Agency. Florida Department of Environmental Protection, Tallahassee, FL

Belanger TV, Scheidt DJ, Platko JR II (1989) Effects of nutrient enrichment on the Florida Everglades. Lake and Reservoir Manage 5(1):101–111

Black FJ, Poulin BA, Flegal AR (2012) Factors controlling the abiotic photo-degradation of monomethylmercury in surface waters. Geochim Cosmochim Acta 84:492–507

Fernandez-Gomez C, Drott A, Björn A, Díez S, Bayona JM, Tesfalidet S, Lindfors A, Skyllberg U (2013) Towards universal wavelength-specific photodegradation rate constants for methyl mercury in humic waters, exemplified by a boreal lake-wetland gradient. Environ Sci Technol 47:6279–6287

Gabriel MC, Howard N, Osborne TZ (2014) Fish mercury and surface water sulfate relationships in the Everglades Protection Area. Environ Manag 53:583–593

Glass A (2019) President Bush cites 'axis of evil,' Jan. 29, 2002. Politico. https://www.politico.com/story/2019/01/29/bush-axis-of-evil-2002-1127725

Gorski PR, Armstrong DE, Hurley JP, Krabbenhoft DP (2008) Influence of natural dissolved organic carbon on the bioavailability of mercury to a freshwater alga. Environ Pollut 154:116–123

Green WH, Perko GV (2001) Good science or myopia: will the 1991 Everglades settlement lead to an optimal restoration or will phosphorus reductions be taken too far? St Thomas Law Rev 13 (3):697–728

Hamilton LC (2013) Statistics with STATA: updated for Version 12, 8th edn. Brooks/Cole, Boston, MA

Harada M (1995) Minamata disease: methylmercury poisoning in Japan caused by environmental pollution. Crit Rev Toxicol 25(1):1–24

Harris R, Pollman CD, Hutchinson D, Beals D (2001) Florida Pilot Mercury Total Maximum Daily Load (TMDL) Study: application of the Everglades Mercury Cycling Model (E-MCM) to site WCA 3A-15. Report to the Florida Department of Environmental Protection, Tallahassee, FL. Submitted by Tetra Tech, Inc., Lafayette, CA

Jones RD (1992) Hearing transcript, at 183, Sugar Cane Growers Cooperative of Fla. v South Fla. Water management dist., DOAH case nos. 92-3038, 92-3039, and 92-3040 (Fla. Div. of Admin. Hearings, Oct. 16, 1992)

Julian P (2013) Mercury bio-concentration factor in mosquito fish (*Gambusia spp.)* in the Florida Everglades. Bull Environ Contam Toxicol 90:329–332

Julian P (2014) Reply to "Mercury bioaccumulation and bioaccumulation factors for everglades mosquitofish as related to sulfate: a re-analysis of Julian II (2013)". Bull Environ Contam Toxicol 93:517

Julian P, Gu B, Frydenborg R, Lange T, Wright A, Mabry Macray J (2014) Chapter 3B: mercury and sulfur environmental assessment for the Everglades. In: 2014 South Florida environmental report. South Florida Water Management District, West Palm Beach, FL

Julian P, Gu B, Redfield G (2015) Comment on and reinterpretation of Gabriel *et al.* (2014) 'Fish mercury and surface water sulfate relationships in the Everglades Protection Area'. Environ Manag 55:1–5

Julian II P, Gu B, Freitag A (2017) Limiting factors in mercury methylation hotspot development: the tangled web. Presented at Greater Everglades Ecosystem Restoration Conference (GEER) 2017, April 17–20, 2017. Coral Springs, FL

Julian P, Gu B, Weaver K, Jerauld M, Dierberg FE, Debusk TA, Potts JA, Larson NR, Hileman K, Sierer Finn D (2018) Chapter 3B: mercury and sulfur environmental assessment for the Everglades. In: 2018 South Florida environmental report. South Florida Water Management District, West Palm Beach, FL

Koch MS, Reddy KR (2001) Distribution of soil and plant nutrients along a trophic gradient in the Florida Everglades. Soil Sci Soc Am J 56(5):1492–1499

Long JS, Freese J (2014) Regression models for categorical dependent variables using Stata, 3rd edn. Stata Press, College Station, TX

Mercury Deposition Network (MDN) (2019). http://nadp.slh.wisc.edu/data/sites/list/?net=MDN. Data accessed April 30, 2019

Pérez-Fuentetaja A, Dillon PJ, Yan ND, McQueen DJ (1999) Significance of dissolved organic carbon in the prediction of thermocline depth in small Canadian shield lakes. Aquatic Ecol 33:127–133

Pollman CD (2014) Mercury cycling and trophic state in aquatic ecosystems: implications from structural equation modeling. Sci Tot Env 499:62–73

Pollman CD (2015) The role of sulfate as a driver for mercury methylation in the Everglades—what does statistics really have to say? Invited paper. Greater Everglades Ecosystem Restoration Conference 2015, Coral Springs, FL, April 21–23, 2015

Pollman CD, Axelrad DM (2014) Mercury bioaccumulation and bioaccumulation factors for Everglades mosquitofish as related to sulfate: a re-analysis of Julian II (2013). Bull Environ Contam Toxicol 93:509–516

Pollman CD, Harris R, Beals D, Axelrad D (2004) Effects of trophic factors on fish mercury concentrations in the Florida Everglades: a sensitivity analysis using the E-MCM model. Paper presented at the Seventh International Conference on Mercury as a Global Pollutant, June 27—July2, 2004. Lubljana, Slovenia

Scheidt DJ, Kalla PI (2007) Everglades ecosystem assessment: Water management and quality, eutrophication, mercury contamination, soils and habitat: monitoring for adaptive management: A R-EMAP status report. USEPA region 4, Athens, GA. EPA 904-R-07-001. 98 pp. http://www.epa.gov/region4/sesd/reports/epa904r07001/epa904r07001.pdf

Scully NM, Lean DRS (1994) The attenuation of ultraviolet radiation in temperate lakes. Arch Hydrobiol Beih 43:135–144

StataCorp (2017) Stata statistical software: release 15. StataCorp LLC, College Station, TX

Suter GW, Norton SB, Cormier SM (2002) A methodology for inferring the causes of observed impairments in aquatic ecosystems. Environ Toxicol Chem 21:1101–1111

Watras CJ, Morrison KA, Host JS, Bloom NS (1995) Concentration of mercury species in relationship to other site-specific factors in the surface waters of northern Wisconsin lakes. Limnol Oceanogr 40:553–565

Watras CJ, Back RC, Halvorsen S et al (1998) Bioaccumulation of mercury in pelagic freshwater food webs. Sci Total Environ 219:183–208

Chapter 7
Primer on Methylmercury Biomagnification in the Everglades

Darren G. Rumbold

Abstract The purpose of this chapter is to provide background information on the entry of methylmercury into the food web and the process by which it biomagnifies through food webs.

Keywords Methylmercury · Bioconcentration · Bioaccumulation · Biomagnification

7.1 Introduction

The complex process of bioconversion of inorganic mercury (Hg) to methylmercury (MeHg) was described in Chap. 1, this volume. While the inorganic form of Hg is toxic, we have long known that MeHg (principally monomethylmercury) is more toxic, more readily accumulated and is the form that is biomagnified (Westöö 1966; Watras et al. 1998; for review, see Wiener et al. 2003).

Biomagnification is defined as an increase in concentration of a contaminant in animal tissues in successive members of a food chain. Determining whether a contaminant is biomagnified from one trophic level to the next is not simple (for review, see Reinfelder et al. 1998). In a comprehensive review of the literature, Suedel et al. (1994) found relatively few metals truly biomagnify in aquatic food webs. It occurs only where the contaminant is slowly eliminated from tissues of prey animals but is highly assimilated by the predator during digestion. MeHg has been found to have a long half-life and an exceptionally high assimilation efficiency (AE); its been reported to have a half-life as long as 1030 days in fish (for review, see Trudel and Rasmussen 1997) and an AE as high as 70–80% (Rodgers and Beamish 1982). Consequently, MeHg biomagnifies with concentrations in the tissues of top predators millions of times higher than the sub-part-per-trillion concentrations that occur in the some waterbodies. Therefore, top predators can be affected to a greater

D. G. Rumbold (✉)
Florida Gulf Coast University, Fort Myers, FL, USA
e-mail: drumbold@fgcu.edu

© Springer Nature Switzerland AG 2019

D. G. Rumbold et al. (eds.), *Mercury and the Everglades. A Synthesis and Model for Complex Ecosystem Restoration*, https://doi.org/10.1007/978-3-030-32057-7_7

extent from MeHg than the rest of the food web (for review, see Chap. 10, this volume). However, before it is biomagnified it must first be bioconcentrated from the water and bioaccumulated at the base of the food chain.

7.2 MeHg Entry into the Food Chain

Surprisingly, the mechanism and kinetics of MeHg uptake at the base of the food chain is poorly understood despite the fact that it is the largest bioconcentration step affecting MeHg accumulation (i.e., 10^5 to 10^6 times). An early study that included octanol-water partitioning tests and accumulation by a marine phytoplankton concluded that MeHg accumulation was through passive adsorption (Mason et al. 1996). They found that inorganic Hg was principally bound in phytoplankton membranes, whereas MeHg accumulated in the cell cytoplasm (Mason et al. 1996). Watras et al. (1998, p. 192) stated "passive uptake of MeHgCl by phytoplankton is indisputable"; however, based on results of their field study, they concluded that uptake into phytoplankton and bacterioplankton was dominated by active transport. Some of the earliest experimental work on uptake by freshwater phytoplankton was done in Florida by Moye et al. (2002). They tested four different species of phytoplankton and found the highest rate of MeHg uptake in the largest species studied (Moye et al. 2002). This was important because if uptake was passive, surface area to volume ratios would be critical in determining uptake with the greatest MeHg uptake expected in the smaller phytoplankton with greater surface area to volume ratio. More telling, Moye et al. (2002) also found MeHg uptake inhibited by uncouplers of phosphorylation, e.g., when cells were exposed to γ-radiation (at doses known to be lethal) and when cells were kept in the dark for prolonged periods. Based on these lines of evidence and based on the bimodal character of the uptake rate, they concluded that MeHg uptake was an active process. More recently, Pickhardt and Fisher (2007) found MeHg accumulation in freshwater plankton was an order of magnitude greater in live cells than dead cells, suggesting uptake was metabolically controlled. They further reported that inorganic Hg and MeHg were entirely bound to cell surface in dead cells (i.e., passive adsorption), whereas 59–64% of the MeHg and only 9–16% of the inorganic Hg was in the cytoplasm in living cells. Luengen et al. (2012) also found greater MeHg uptake by live marine diatoms versus dead cells. Yet, the debate continues as to whether the principal mechanism by which phytoplankton accumulate MeHg is through passive or active uptake. In a recent study using ^{203}Hg uptake by six marine phytoplankton species, Lee and Fisher (2016) found passive transport to be the major pathway by which the plankton concentrated MeHg out of the water (by as much as 6.4×10^6); volume concentration factors were also directly related to surface-to-volume ratios. Yet, they found one species, a dinoflagellate, where accumulation did not appear to be influenced in the same way by surface-to-volume ratio and that greater uptake occurred at warmer temperatures; all suggesting the possibility that uptake was active in that species (Lee and Fisher 2016). They speculated that the mixtrophic species took in MeHg bound to organic matter through phagocytosis (Lee and Fisher 2016) and cited earlier work by Tranvik et al. (1993) that metals bound to colloidal organic matter might be an important source for heterotrophic flagellates.

We have long known that different Hg species complex with colloids (Stordal et al. 1996). Babiarz et al. (2001), for example, determined the concentrations of inorganic Hg and MeHg in samples from 15 freshwater locations and reported that the colloidal phase comprised up to 72% of the unfiltered concentration. Samples from Everglades' sites had, on average, 20% MeHg in the colloidal phase (Babiarz et al. 2001). Importantly in terms of estuaries, they also report that the colloidal phase decreased with increasing specific conductance possibly suggesting ligand exchange (Babiarz et al. 2001).

Most studies report complexation with dissolved organic matter (DOM) reduces MeHg uptake by phytoplankton (Moye et al. 2002; Gorski et al. 2008; Luengen et al. 2012; for further review, see Chap. 4, this volume). It is thought that DOM can directly affect bioavailability of MeHg by forming large complexes that reduce transport across cell membranes. Yet, Pickhardt and Fisher (2007) reported higher volume concentration factors (2–2.6 times greater) in eukaryotic phytoplankton in water with high DOM. After reviewing these conflicting studies, Luengen et al. (2012) speculated that the differences may be attributable to variations in the composition of the DOM and other water quality variables (e.g., pH and Cl^-); in their research they found a significant difference in hydrophobic and transphilic DOM fractions. This is clearly an important phenomenon considering the high concentration of DOM in Florida waters and has been invoked previously to explain differences in bioaccumulation observed in different watersheds (Rumbold et al. 2011, 2018a). As discussed in Chap. 4 of this volume, while DOM may reduce MeHg bioavailability under certain conditions, it must be stressed that, depending on its quantity and quality, DOM often promotes MeHg production (by increasing bioavailability of inorganic Hg and stimulating bacteria) and transport (by keeping it in solution increasing water column concentration). The fate of the MeHg following degradation of the DOM complex (photochemcially or biotically) is unknown.

7.3 Biomagnification

The transfer of MeHg to zooplankton is thought to be primarily through dietary uptake rather than directly from water (Watras and Bloom 1992; Pickhardt et al. 2002; Lee and Fisher 2017). In reference to the previous discussion of inorganic Hg versus MeHg, Pickhardt et al. (2002) found zooplankton preferentially accumulated MeHg concentrated within algal cytoplasm as opposed to inorganic Hg on the cell surface. Lee and Fisher (2017) found that over 90% of a copepod's (i.e., a zooplankton) MeHg body burden was derived from their diet when fed an algal diet whose MeHg bioconcentration factor (BCF, ratio of the concentration of a substance in an organism to the concentration of the substance in the surrounding water) was $\geq 10^6$. A recently developed model that successfully predicted concentrations observed both in phytoplankton and zooplankton in the Atlantic Ocean incorporated a BCF for phytoplankton that ranged from $10^{2.4}$ to $10^{5.9}$ and bioaccumulation factors (BAF, defined as the ratio of the concentration of a substance in an organism to the concentration of the substance in the surrounding medium) for zooplankton that ranged from $10^{4.6}$ to $10^{6.2}$. The variability in the BCF and BAF values was a result of difference in ranges in DOC (40–500 μM) and productivities from ultraoligotrophic to hypereutrophic (Schartup et al. 2018). On the other hand, MeHg concentrations in zooplankton in lakes of Vancouver Island, Canada were better predicted using lipid biomarkers for a bacterial diet rather than an algal diet (Kainz and Mazumder 2005). In another study using dietary fatty acid biomarkers, ingestion of bacteria was also associated with increased MeHg uptake in primary consumers (e.g., insect nymphs and larvae) in humus-rich streams as compared to an algae diet (de wit et al. 2012). This is noteworthy in regards to the freshwater Everglades where periphyton mats (composed mainly of cyanobacteria, diatoms, and green algae) and, possibly more importantly, bacteria (that rework detritus, including dead or sloughed off material from the periphyton mats) are the base of the food web (Browder et al. 1994; Belicka et al. 2012; also see Chap. 8, this volume).

Food is also the dominant pathway of MeHg bioaccumulation in fish at water-column MeHg concentrations typically encountered (i.e., as opposed to industrially contaminated sites where water-column concentrations can be unusually high; Hall et al. 1997; Pickhardt et al. 2006). BAFs differ from BCFs in that exposure from the diet is implicitly considered and therefore includes potential biomagnification. They have been calculated as the ratio of total mercury (THg, all forms measured) or MeHg concentration in fish muscle (ng/kg wet weight) to the concentration of that Hg species in water (ng/L; with resulting units kg/L), so caution is warranted when making comparisons. As a measure of biomagnification, BAFs vary widely due to numerous factors. Southworth et al. (2004), for example, report BAFs ranging from 10^3 to 10^4 for THg and from 10^5 to 10^6 for MeHg (mean $= 1.4 \times 10^6$ or log BAF $= 6.14$) in adult redbreast sunfish (*Lepomis auritus*) from Poplar Creek, Oak Ridge, Tennessee. Pollman and Axelrad (2014) report mean BAFs (based on MeHg) for small mosquitofish (*Gambusia* spp.; for review of life history, see Chap. 8, this volume) from different areas of the Everglades to range from 1.7×10^5 to 8.8×10^5 (log BAF 5.2–5.9). Largemouth bass (*Micropterus salmoides*; normalized to age 3) from different areas of the Everglades exhibit BAFs (based on MeHg) ranging from 1.9×10^6 to 7.6×10^6 (log BAF $=$ 6.3–6.9; T. Lange, FFWCC, pers. comm.). By comparison, Brumbaugh et al. (2001) report average BAFs (based on MeHg) to range from 1.5×10^6 to 7×10^6 (log BAF $=$ 6.17–6.8) for black bass (*Micropterus* spp.; not normalized to age 3) from 20 watersheds across the United States, with the lowest BAFs in fish from the Everglades where MeHg concentrations in water were highest. They suggested that the apparent inverse relationship between the BAFs and water-column concentration suggested lower concentrations of MeHg were more efficiently biotransfered than higher concentrations (Brumbaugh et al. 2001). Several studies have reported a similar inverse relationship between BAFs for other metals, particularly at highly contaminated sites (for review, see DeForest et al. 2007); however, Pollman and Axelrad (2014) warn this inverse relationship is driven by spurious correlations. While DeForest et al. (2007) did not find a clear inverse relationship between BAFs and aqueous exposure concentrations of MeHg, they did find trophic transfer factors (TTF) inversely related to dietary exposure concentration, which they attributed to saturable uptake kinetics. A specialized form of BAF termed TTF or biomagnification factor (BMF) is calculated as the ratio of the concentration in predator to the concentration in its prey. Additional studies have also reported this inverse relationship between BMF and dietary exposure concentration (Jardine et al. 2013; Painter et al. 2016).

As discussed in other chapters of this volume, biomagnification and resulting concentrations in top predators are driven by Hg loading, net methylation rates and MeHg bioavailability to the base of the food web. Yet biomagnification through the food web is also a complex process that is influenced by many factors including, but not limited to: food source (Gorski et al. 2003), food chain length and community structure (Cabana and Rasmussen 1994; Stemberger and Chen 1998), productivity (Pickhardt et al. 2002) and water chemistry (Lange et al. 1993). Gorski et al. (2003), for example, found BAFs and BMFs to be higher in a lake with a pelagic-based food web as compared to a lake with a benthic-based food web. Stemberger and Chen (1998) found biomagnification was positively associated with food chain length; however, they also found concentrations in top predators were negatively correlated with the number of feeding links between species. They suggested that structurally complex food webs, comprising many lateral links, may attenuate the degree of biomagnification due to pathways that do not lead to top predators (Stemberger and Chen 1998). Pickhardt et al. (2002) found biomagnification to be reduced in algal-rich eutrophic systems as a result of biodilution, i.e., MeHg concentrations reduced per algal cell, which subsequently reduced concentrations in zooplankton two to three-fold. Lange et al. (1993) surveyed Hg in harvestable-sized, Largemouth bass from 53 Florida lakes and found that 45% of the variation in their tissue-Hg was explained by differences in alkalinity and chlorophyll *a* as a measure of productivity (Hg levels were significantly lower in lakes with higher chlorophyll). They also found that lake pH (which varied from 3.6 to 9.1) alone accounted for 41% of the variation. They could not, however, differentiate the potential direct influence of pH in MeHg production or bioavailability on biomagnification from its potential indirect effect on fish metabolism, diet or prey species composition (Lange et al. 1993).

Interestingly, two eutrophic lakes in Florida, Lake Okeechobee and Lake Trafford, consistently have fish with low Hg (T. Lange, FWC, pers. Comm.). When the state began efforts to reduce phosphorus loading in agricultural stormwater and resulting eutrophy in Everglades' marshes, some warned it might reduce biodilution and, thereby, exacerbate the Hg problem (PTI 1995; Exponent 1998). However, phosphorus enrichment has not been found to be a statistically significant factor in mitigating Hg concentrations in Everglades' mosquitofish through biodilution (Pollman 2014; see also Chap. 6 this volume). This may be a function of the greater dependence on periphyton, and associated bacteria at the base of the food web as compared vascular plants in the Everglades (see above regarding vascular plant biomarkers, Belicka et al. 2012). Moreover, while periphyton productivity is high in the Everglades, consumer standing stocks are relatively low compared to other systems, possibly due to the structural integrity of the mats (Trexler et al. 2015).

Stable isotope analysis (SIA) of nitrogen has long been used to investigate trophic relationships, including defining BAF relationships. Organisms preferentially excrete the lighter ^{14}N isotope, resulting in enrichment in ^{15}N. The ratio of ^{15}N to ^{14}N ($\delta^{15}N$) thus represents a time-integrated, continuous measure of trophic position. More recently, SIA has been used as a tool to examine biomagnification integrated across entire food webs. The slope of the regression of log-transformed, tissue-Hg concentration on $\delta^{15}N$ values in various taxa from a given community is termed the "Trophic Magnification Slope" (TMS; Fig. 7.1) and is considered to be a measure of biomagnification efficiency for that ecosystem (Jardine et al. 2006). Because $\delta^{15}N$ signatures can vary due to anthropogenic nitrogen inputs (Chang et al. 2002), many studies have standardized $\delta^{15}N$ based on values in a local, primary consumer. Those studies also sometimes translate $\delta^{15}N$ to trophic level using a diet-tissue discrimination factor (or enrichment factor; Jardine et al. 2006). This allows for estimation of the amount of basal Hg entering the food web (i.e., based on the y-intercept) and a food web magnification factor (FWMF, i.e., from the slope of log-transformed, tissue-Hg concentration on trophic level). This approach allows us to disentangle and assess separately the effect that variations in basal MeHg (either due to variations in loading and bioavailability of the various Hg species or activity of the methylating bacteria) from other factors affecting biomagnification (e.g., food chain length, community structure, etc.). In their early review of the literature, Riget et al. (2007) noted the high degree of similarity among estimates of biomagnification efficiency from different ecosystems. This led some authors to suggest that spatial variation observed in bioaccumulated Hg may largely be a result of variability in the amount of MeHg entering the base of the food web (Riget et al. 2007; Chasar et al. 2009). More recent studies have had the power to discern statistically significant differences in TMS values (and FWMFs) due to various factors including: natural variation in trophic structure among communities (e.g., food chain length, linkage strength, etc.), productivity, and water chemistry, particularly as it affects growth rates of organisms (Chasar et al. 2009; Swanson and Kidd 2010; Clayden et al. 2013; for review, see Lavoie et al. 2013). Based on a world-wide, meta-analysis involving 69 different studies, Lavoie et al. (2013) found higher TMS values at polar and temperate sites when compared to tropical sites. They speculated that, among other factors such as temperature (possibly as it affects growth rate and other factors), low species diversity and simpler food webs at higher latitudes might lead to higher Hg biomagnification (Lavoie et al. 2013). A multivariate linear regression model found that 25% of the variation in MeHg TMS was accounted for by variation in pH, DOC and total phosphorus (Lavoie et al. 2013). Nutrients were also a factor in differences in TMS values among 11 Canadian lakes (Clayden et al. 2013). Using principal component analysis they found shallower TMS in lakes with higher nutrients (Ca, total organic carbon, total nitrogen and total phosphate) possibly due to biodilution, faster growth of fish or both. Nutrient enrichment was recently also posited to explain a shallower TMS in the Caloosahatchee River Estuary in southwest Florida, which has frequent algal blooms, relative to other nearby estuaries (Rumbold et al.

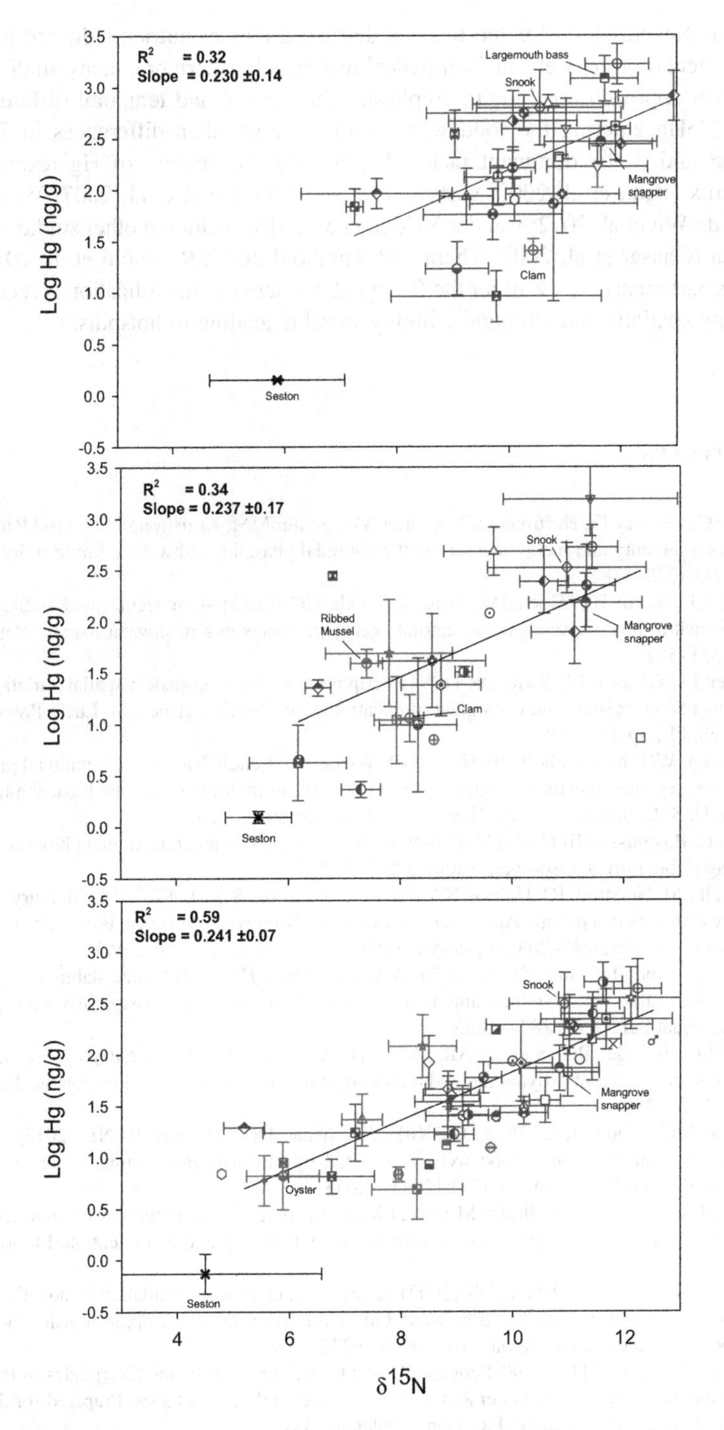

Fig. 7.1 Trophic magnification slopes (TMS) from regression of log Hg concentration on $\delta^{15}N$ (species means ±1 SD or range where n < 3; coefficient of determination and TMS ±95% CI) for different communities along Shark River Estuary, Florida (adapted from Rumbold et al. 2018a)

2018b). Nevertheless, for the reasons discussed above, nutrients do not have the same effect on food webs in Everglades' marshes. Furthermore, many studies using the TMS approach continue to emphasize that spatial and temporal differences in basal MeHg entering the food web are often larger than differences in TMS or FWMF and is the dominant factor determining the amount of Hg reaching top predators (Kidd et al. 2003; Campbell et al. 2005; Riget et al. 2007; Ward et al. 2010; de Wit et al. 2012; van der Velden et al. 2013) including other studies done in Florida (Chasar et al. 2009; Thera and Rumbold 2014; Rumbold et al. 2018a, b). This is particularly relevant for the Everglades where biomagnification is recognized as being spatially and temporally highly variable leading to hotspots.

References

Babiarz CL, Hurley JP, Hoffmann SR, Andren AW, Shafer MM, Armstrong DE (2001) Partitioning of total mercury and methylmercury to the colloidal phase in freshwaters. Environ Sci Technol 35(24):4773–4782

Belicka LL, Sokol ER, Hoch JM, Jaffé R, Trexler JC (2012) A molecular and stable isotopic approach to investigate algal and detrital energy pathways in a freshwater marsh. Wetlands 32 (3):531–542

Browder JA, Gleason PJ, Swift DR (1994) Periphyton in the Everglades: spatial variation, environmental correlates, and ecological implications. In: The Everglades. St. Lucie Press, Delray Beach, FL, pp 379–419

Brumbaugh WG, Krabbenhoft DP, Helsel DR, Wiener JG, Echols KR (2001) A national pilot study of mercury contamination of aquatic ecosystems along multiple gradients: bioaccumulation in fish. U. S. Geological Survey, [Reston, VA], United States (USA)

Cabana G, Rasmussen JB (1994) Modelling food chain structure and contaminant bioaccumulation using stable nitrogen isotopes. Nature 372:255–257

Campbell LM, Norstrom RJ, Hobson KA, Muir DCG, Backus S, Fisk AT (2005) Mercury and other trace elements in a pelagic Arctic marine food web (Northwater Polynya, Baffin Bay). Sci Total Environ 351–352:247–263. https://doi.org/10.1016/j.scitotenv.2005.02.043

Chang CC, Kendall C, Silva SR, Battaglin WA, Campbell DH (2002) Nitrate stable isotopes: tools for determining nitrate sources among different land uses in the Mississippi River Basin. Can J Fish Aquat Sci 59(12):1874–1885

Chasar LC, Scudder BC, Stewart AR, Bell AH, Aiken GR (2009) Mercury cycling in stream ecosystems. 3. Trophic dynamics and methylmercury bioaccumulation. Environ Sci Technol 43 (8):2733–2739

Clayden MG, Kidd KA, Wyn B, Kirk JL, Muir DC, O'Driscoll NJ (2013) Mercury biomagnification through food webs is affected by physical and chemical characteristics of lakes. Environ Sci Technol 47(21):12047–12053

de Wit HA, Kainz MJ, Lindholm M (2012) Methylmercury bioaccumulation in invertebrates of boreal streams in Norway: effects of aqueous methylmercury and diet retention. Environ Pollut 164:235–241

DeForest DK, Brix KV, Adams WJ (2007) Assessing metal bioaccumulation in aquatic environments: the inverse relationship between bioaccumulation factors, trophic transfer factors and exposure concentration. Aquat Toxicol 84(2):236–246

Exponent (formerly PTI) (1998) Ecological risks to wading birds of the Everglades in relation to phosphorus reductions in water and mercury bioaccumulation in fishes. Prepared for the Sugar Cane Growers Cooperative. Exponent, Bellevue, WA

Gorski PR, Cleckner LB, Hurley JP, Sierszen ME, Armstrong DE (2003) Factors affecting enhanced mercury bioaccumulation in inland lakes of Isle Royale National Park, USA. Sci Total Environ 304(1):327–348. https://doi.org/10.1016/S0048-9697(02)00579-X

Gorski P, Armstrong D, Hurley J, Krabbenhoft D (2008) Influence of natural dissolved organic carbon on the bioavailability of mercury to a freshwater alga. Environ Pollut 154(1):116–123

Hall B, Bodaly R, Fudge R, Rudd J, Rosenberg D (1997) Food as the dominant pathway of methylmercury uptake by fish. Water Air Soil Pollut 100(1):13–24

Jardine TD, Kidd KA, Fisk AT (2006) Applications, considerations, and sources of uncertainty when using stable isotope analysis in ecotoxicology. Environ Sci Technol 40(24):7501–7511

Jardine TD, Kidd KA, O'Driscoll N (2013) Food web analysis reveals effects of pH on mercury bioaccumulation at multiple trophic levels in streams. Aquat Toxicol 132:46–52

Kainz M, Mazumder A (2005) Effect of algal and bacterial diet on methyl mercury concentrations in zooplankton. Environ Sci Technol 39(6):1666–1672

Kidd KA, Bootsma HA, Hesslein RH, Lyle Lockhart W, Hecky RE (2003) Mercury concentrations in the food web of Lake Malawi, East Africa. J Great Lakes Res 29(Suppl 2):258–266. https://doi.org/10.1016/s0380-1330(03)70553-x

Lange TR, Royals HE, Connor LL (1993) Influence of water chemistry on mercury concentration in largemouth bass from Florida lakes. Trans Am Fish Soc 122(1):74–84

Lavoie RA, Jardine TD, Chumchal MM, Kidd KA, Campbell LM (2013) Biomagnification of mercury in aquatic food webs: a worldwide meta-analysis. Environ Sci Technol 47 (23):13385–13394. https://doi.org/10.1021/es403103t

Lee CS, Fisher NS (2016) Methylmercury uptake by diverse marine phytoplankton. Limnol Oceanogr 61(5):1626–1639

Lee CS, Fisher NS (2017) Bioaccumulation of methylmercury in a marine copepod. Environ Toxicol Chem 36(5):1287–1293

Luengen AC, Fisher NS, Bergamaschi BA (2012) Dissolved organic matter reduces algal accumulation of methylmercury. Environ Toxicol Chem 31(8):1712–1719

Mason RP, Reinfelder JR, Morel FM (1996) Uptake, toxicity, and trophic transfer of mercury in a coastal diatom. Environ Sci Technol 30(6):1835–1845

Moye HA, Miles CJ, Phlips EJ, Sargent B, Merritt KK (2002) Kinetics and uptake mechanisms for monomethylmercury between freshwater algae and water. Environ Sci Technol 36 (16):3550–3555. https://doi.org/10.1021/es011421z

Painter K, Janz D, Jardine T (2016) Bioaccumulation of mercury in invertebrate food webs of Canadian Rocky Mountain streams. Freshwater Science 35(4):1248–1262

Pickhardt PC, Fisher NS (2007) Accumulation of inorganic and methylmercury by freshwater phytoplankton in two contrasting water bodies. Environ Sci Technol 41(1):125–131

Pickhardt PC, Folt CL, Chen CY, Klaue B, Blum JD (2002) Algal blooms reduce the uptake of toxic methylmercury in freshwater food webs. Proc Natl Acad Sci 99(7):4419–4423. https://doi.org/10.1073/pnas.072531099

Pickhardt PC, Stepanova M, Fisher NS (2006) Contrasting uptake routes and tissue distributions of inorganic and methylmercury in mosquitofish (*Gambusia affinis*) and redear sunfish (*Lepomis microlophus*). Environ Toxicol Chem 25(8):2132–2142

Pollman CD (2014) Mercury cycling in aquatic ecosystems and trophic state-related variables—implications from structural equation modeling. Sci Total Environ 499:62–73

Pollman CD, Axelrad DM (2014) Mercury bioaccumulation factors and spurious correlations. Sci Total Environ 496:vi–xii

PTI (later to become Exponent) (1995) Ecological risks to wading birds of the Everglades in relation to phosphorus reductions in water and mercury bioaccumulation in fishes. Prepared for Sugar Cane Growers Cooperative. PTI Environmental Services, Bellevue, WA

Reinfelder JR, Fisher NS, Luoma SN, Nichols JW, Wang W-X (1998) Trace element trophic transfer in aquatic organisms: a critique of the kinetic model approach. Sci Total Environ 219 (2-3):117–135

Riget F, Møller P, Dietz R, Nielsen T, Asmund G, Strand J et al (2007) Transfer of mercury in the marine food web of West Greenland. J Environ Monit 9(8):877–883

Rodgers D, Beamish F (1982) Dynamics of dietary methylmercury in rainbow trout, Salmo gairdneri. Aquat Toxicol 2(5–6):271–290
Rumbold DG, Evans DW, Niemczyk S, Fink LE, Laine KA, Howard N et al (2011) Source identification of Florida bay's methylmercury problem: mainland runoff versus atmospheric deposition and in situ production. Estuar Coasts 34(3):494–513. https://doi.org/10.1007/s12237-010-9290-5
Rumbold DG, Lange TR, Richard D, DelPizzo G, Hass N (2018a) Mercury biomagnification through food webs along a salinity gradient down-estuary from a biological hotspot. Estuar Coast Shelf Sci 200:116–125
Rumbold DG, Lange TR, Richard D, DelPizzo G, Haas N (2018b) Mercury concentrations and ratios of stable isotopes of nitrogen and carbon in food webs of the Caloosahatchee estuary, Florida. Florida Sci 81(4):105–125
Schartup AT, Qureshi A, Dassuncao C, Thackray CP, Harding G, Sunderland EM (2018) A model for methylmercury uptake and trophic transfer by marine plankton. Environ Sci Technol 52 (2):654–662. https://doi.org/10.1021/acs.est.7b03821
Southworth GR, Peterson MJ, Bogle MA (2004) Bioaccumulation factors for mercury in stream fish. Environ Pract 6(2):135–143
Stemberger RS, Chen CY (1998) Fish tissue metals and zooplankton assemblages of northeastern U.S. lakes. Can J Fish Aquat Sci 55(2):339–352
Stordal M, Gill G, Wen LS, Santschi P (1996) Mercury phase speciation in the surface waters of three Texas estuaries: importance of colloidal forms. Limnol Oceanogr 41:52–61
Suedel BC, Boraczek JA, Peddicord RK, Clifford PA, Dillon TM (1994) Trophic transfer and biomagnification potential of contaminants in aquatic ecosystems. Rev Environ Contam Toxicol 136:21–89
Swanson HK, Kidd KA (2010) Mercury concentrations in Arctic food fishes reflect the presence of anadromous Arctic charr (*Salvelinus alpinus*), species, and life history. Environ Sci Technol 44 (9):3286–3292
Thera JC, Rumbold DG (2014) Biomagnification of mercury through a subtropical coastal food web off Southwest Florida. Environ Toxicol Chem 33(1):65–73
Tranvik L, Sherr EB, Sherr BF (1993) Uptake and utilization of 'colloidal DOM' by heterotrophic flagellates in seawater. Mar Ecol Prog Ser 92:301–309
Trexler JC, Gaiser EE, Kominoski JS, Sanchez J (2015) The role of periphyton mats in consumer community structure and function in calcareous wetlands: lessons from the Everglades. In: Microbiology of the Everglades ecosystem. Science Publications. CRC, Boca Raton, pp 155–179
Trudel M, Rasmussen JB (1997) Modeling the elimination of mercury by fish. Environ Sci Technol 31(6):1716–1722
van der Velden S, Dempson JB, Evans MS, Muir DCG, Power M (2013) Basal mercury concentrations and biomagnification rates in freshwater and marine food webs: effects on Arctic charr (*Salvelinus alpinus*) from eastern Canada. Sci Total Environ 444(0):531–542. https://doi.org/10.1016/j.scitotenv.2012.11.099
Ward DM, Nislow KH, Folt CL (2010) Bioaccumulation syndrome: identifying factors that make some stream food webs prone to elevated mercury bioaccumulation. Ann N Y Acad Sci 1195 (1):62–83
Watras CJ, Bloom NS (1992) Mercury and methylmercury in individual zooplankton: implications for bioaccumulation. Limnol Oceanogr 37(6):1313–1318
Watras CJW, Back RC, Halvorsena S, Hudson RJM, Morrison KA, Wente SP (1998) Bioaccumulation of mercury in pelagic freshwater food webs. Sci Total Environ 219:183–208
Westöö G (1966) Determination of methylmercury compounds in foodstuffs I. Methylmercury compounds in fish, identification and determination. Acta Chem Scand 20(8):2131–2137
Wiener JG, Krabbenhoft DP, Heinz GH, Scheuhammer AM (2003) Ecotoxicology of mercury. Handb Ecotoxicol 2:409–463

Chapter 8
Food Web Structures of Biotically Important Species

Peter C. Frederick, William F. Loftus, Ted Lange, and Mark Cunningham

Abstract Since mercury (Hg) exposure in wild vertebrates is primarily through food consumption, and since methylmercury is highly bioaccumulative, food web analysis can be especially important to understanding exposure in wild fauna. Here, we summarize extensive and intensive studies of food habits and Hg exposure for four well-researched groups of vertebrates that have outsized effects on community structure and function in the Everglades—mosquitofish, Florida Bass, long legged wading birds, and Florida Panthers. Generally these studies show that a high degree of variation in tissue Hg is attributable to geographic location within south Florida, usually because of known differences in either prey consumed or Hg concentrations in prey. Prey identity may be strongly shaped by local community structure—for example bass in canals have access to a more diverse group of prey and larger prey than those foraging in marshes. Similarly, shifts in panther food habits and consequently Hg exposure have been strongly affected by changes in available prey, driven by hydrology or ungulate management practices. While trophic position can be important in predicting Hg exposure (panthers, mosquitofish), it is interesting that in some cases (bass) geographic location may have an even stronger effect. Each of these species or species-groups has shown value as an indicator of risk, with large variation seen in Hg concentrations over both time and space. The ability to attribute that variation to location, trophic, and contaminant exposure effects has been a major contribution of these long term studies.

P. C. Frederick (✉)
University of Florida, Gainesville, FL, USA
e-mail: pfred@ufl.edu

W. F. Loftus
Aquatic Research & Communication, LLC, Vero Beach, FL, USA
e-mail: arc_wfl@bellsouth.net

T. Lange
Florida Fish and Wildlife Conservation Commission, Eustis, FL, USA
e-mail: ted.lange@myfwc.com

M. Cunningham
Florida Fish and Wildlife Conservation Commission, Gainesville, FL, USA
e-mail: Mark.Cunningham@MyFWC.com

© Springer Nature Switzerland AG 2019

D. G. Rumbold et al. (eds.), *Mercury and the Everglades. A Synthesis and Model for Complex Ecosystem Restoration*, https://doi.org/10.1007/978-3-030-32057-7_8

Keywords Food webs · Mosquitofish · Panther · Wading bird · Large-mouthed bass · Bioaccumulation

8.1 Introduction

As outlined in Chap. 7, aquatic food webs are often complex, have many trophic levels, and show high bioaccumulation potential. The Everglades is certainly in this category—as an example, the concentrations of Hg in adult Largemouth bass in the Everglades may be 1–3 million times the concentrations in water. Trophic position therefore can have a huge impact on Hg bioaccumulation, and risk to organisms. These effects may be unexpected—spiders eating emergent predatory aquatic insects may have higher Hg exposure than many omnivorous fishes. Similarly, highly predatory mammals like the Florida Panther (*Felis concolor coryi*) that typically consume terrestrial herbivores may experience an order of magnitude higher risk from Hg if consuming aquatic omnivores like raccoons (*Procyon lotor*).

Thus the details of food webs in aquatic systems like the Everglades can have a large impact on understanding and predicting risk in fish and wildlife. As in Chap. 7 (this volume), food webs can shift dramatically as a result of externalities like nutrients, hydrology, and land cover; in the Everglades these externalities are strongly affected by water management, climate, sea level, and land use patterns. The Comprehensive Everglades Restoration Plan aims specifically to change many of those drivers, potentially causing large scale geographic shifts in both Hg availability, and in food webs. Thus understanding Hg bioaccumulation and exposure in relation to food web dynamics has immediate value for predicting the effects of large scale restoration and management of this system.

In this chapter, we present information on four animal groups for which the details of food habits, dietary Hg concentrations, and Hg exposure are well known. These case histories serve several purposes. First, they stand as well-documented examples of bioaccumulation in wetland species in a complex aquatic food web that can be of use in understanding and predicting effects in other wetland systems worldwide. Second, each of these species has been studied over a considerable time, allowing rare insight into long term fluctuations in diet and exposure—in some cases over a period of declining Hg concentrations. Finally, these examples have allowed a comprehensive picture of contamination of indicator species across an entire food web.

These specific taxa have been the focus of intensive Hg research for several reasons. Mosquitofish (*Gambusia holbrooki*) are perhaps the most ubiquitous marsh fish in the Everglades, are important food items for most of the carnivorous consumers in the marsh, and can be found virtually anywhere in the system that has surface water. This means the species can function as a sentinel over an extremely wide geographic area, and because they do not move far during their lifetimes, are indicators of Hg concentration at a fine geographic grain. Largemouth Bass (*Micropterus salmoides*) are a piscivorous predator at or near the top of the

Everglades aquatic food web. This species is also an important target for sport and recreational fishing, and a possible route for human exposure. Bass can be found in most deeper and longer-hydroperiod habitats throughout the system, and Hg concentrations in musculature can be standardized by year-class through the use of otoliths. Wading birds (ibises, egrets, storks, spoonbills and herons) are also at or near the top of the food web, and their mobility (ca. 35 km from central roost and breeding sites daily) allows them to serve as integrative sampling systems for huge portions of the wetted Everglades during short time periods. Within some size limits, wading bird food habits are generally reflective of available prey—wading birds therefore serve as mirrors of shifting prey community composition in marshes and coastal areas. Sampling can be completely nondestructive in birds, with blood and feather tissue being reliable indicators. Florida Panthers (*Felis concolor coryi*) have been studied intensively for decades because of their highly endangered status, and both food habits' information and health information can be linked to Hg exposure in this species. Depending on diet, panthers may be feeding at low (deer) or very high (raccoon) trophic positions within the Everglades ecosystem. Together, studies of these taxa give an unusually comprehensive view of dietary variability and Hg exposure across trophic levels, habitats, and time in the Everglades ecosystem.

8.2 Eastern Mosquitofish

The Eastern mosquitofish (*Gambusia holbrooki*) is a small species (maximum length ~6.5 cm) that is the most abundant and widely distributed fish in the Everglades ecosystem (Loftus and Kushlan 1987). It ranges from Lake Okeechobee south through the Everglades and Big Cypress Swamp regions (Zokan et al. 2015) where it inhabits all aquatic habitats including alligator ponds, marshes, swamps and canal margins. It is also abundant in estuarine and shallow coastal habitats (Loftus and Kushlan 1987; Rehage and Loftus 2007). Its aggressive nature, wide diet, small size, and ability to survive in poor water-quality habitats (Lewis 1970; Loftus 2000) combine with a long reproductive season to make this the most successful fish species in southern Florida freshwaters. *G. holbrooki* are prey for many top-level predators in the Everglades system, particularly during dry periods (Gunderson and Loftus 1993).

8.2.1 Food Habits and Dietary Makeup

Within the water column, *G. holbrooki* is primarily a surface dweller, with small groups moving about constantly while foraging. Groups quickly converge on any disturbance in search of food. They feed selectively by increasing their consumption of diatoms when periphyton mats are disrupted, selecting palatable species and avoiding unpalatable, filamentous cyanobacteria abundant in the mats (Geddes and

Trexler 2003). As a generalist omnivore, *G. holbrooki* consume algae and both plant and animal prey from aquatic and terrestrial sources. Reversals in predator-prey roles with growth occur between *G. holbrooki* and centrarchid sunfishes (Pagan 2000) in which *Gambusia* raid nests to feed on larval sunfishes. However, as they grow, sunfishes prey on *Gambusia*, possibly contributing to Hg burden in sunfishes. Cannibalism and predation by *G. holbrooki* on other small fishes also occur (Taylor et al. 2001).

Everglades fishes feed most heavily during the seasonal transition period between high and low water when prey are concentrated (Loftus and Kushlan 1987; Loftus 2000). This may be an important period in the annual hydrological cycle for Hg uptake by fishes in the Everglades aquatic food web. Cleckner et al. (1999) and Stober et al. (2001) reported strong seasonal changes in diets of Everglades *G. holbrooki*. In Everglades National Park (ENP), gut-content data showed that small *G. holbrooki* fed mainly on invertebrate prey at high-water seasons in alligator ponds, with algae more important during low water (Loftus 2000). Larger individuals in ponds fed primarily on algae and detritus during both seasons. In sawgrass habitat, both size groups took proportionately more animal prey, probably because shading by sawgrass limits algal production (Loftus 2000). In spikerush marshes, both small and large *G. holbrooki* consumed chironomid midges at high water, switching to periphyton and a variety of invertebrates at low water. Those seasonal changes in diet in spikerush habitat were accompanied by a decrease in trophic position as water levels decreased, resulting from a shift to lower trophic-level food items. Williams and Trexler (2006) also found that the trophic position increased with time following a dry-down. Likewise, a regression model developed by Sargeant et al. (2010) indicated that dry-downs had a negative effect on the trophic diversity of individual *G. holbrooki* at a site, as estimated by $\delta^{15}N$ range, because that range was positively correlated with time since flooding. A model of Hg assimilation efficiency confirmed that dietary pathways dominated methylmercury (MeHg) accumulation in *G. affinis* from California, with aqueous uptake comprising only 0.6–1.6% of the total predicted MeHg concentration in fish (Pickhardt et al. 2006), consistent with earlier studies (Watras et al. 1998).

From gut data, Loftus (2000) calculated trophic scores to assign trophic classes (1 = herbivore/detritivore to 5 = carnivore) to small and large *G. holbrooki*. Small individuals ranged from trophic class 1 in ponds at low water to trophic class 3 in ponds, spikerush and sawgrass at high water. Large individuals ranged from trophic class 1 in high-water ponds to trophic class 3 in spikerush and sawgrass at high water. To independently examine the dietary classification of the fishes into trophic classes, Loftus (2000) used nonmetric multidimensional scaling (MDS) to ordinate the fishes by volumetric diet proportions. Those ordinations corroborated the trophic patterns from the gut analyses and trophic-class designations.

In a broader spatial sampling of *G. holbrooki* from north to south across the Everglades, Trexler et al. (Appendix D in Stober et al. 2001) reported that periphyton comprised 36% of the diet by biomass in gut contents, with insect, crustacean, arachnid, and piscine prey accounting for the remaining 64%. Adult chironomids accounted for 33.5% of the diet biomass while larvae composed an additional 9.6%.

Heterandria formosa (least killifish), spiders, ants, and beetles comprised another 14.8% of the diet biomass. About 50% of the individual fish had plant matter in their guts, and about 45% had adult midges. Similar to the results of Loftus (2000), Trexler et al. (Appendix D in Stober et al. 2001) reported very few empty stomachs in their samples. They also concluded that piscivory by *G. holbrooki* was rare.

Loftus (2000) sampled biota at various levels in the Everglades aquatic food web for total Hg. Total Hg levels in the food items of *G. holbrooki* varied greatly and increased as prey trophic level increased. Primary producers ranged from 41.9 ± 16.9 ng/g ww (mean ± 1 sd) to 90.4 ± 49.6 ng/g. These values were representative of the late 1990s when the samples were collected and may not be representative of later times. Total Hg levels in aquatic invertebrates typically varied within and among taxa as would be expected by their varied diets, sizes and life histories. There was a significant positive relationship among the trophic classes of the invertebrates and their mean Hg values ($r_s = 0.76$, $p < 0.001$). The lowest trophic class, comprised mainly of herbivores, had the lowest Hg levels, typically less than 40 ng/g ww. Intermediate trophic species included the majority of omnivorous invertebrates, such as copepods, cladocerans, and chironomid midges that were the major animal prey for *G. holbrooki*. The $\delta^{15}N$ values for invertebrates corroborated the positive relationship between trophic position and Hg levels ($r_s = 0.75$, $p = 0.02$). The $\delta^{15}N$ values of Everglades fishes also correlated significantly with mean total Hg levels ($r_s = 0.681$, $p = 0.002$).

Loftus (2000) used stable isotopes of carbon and nitrogen in another independent examination of the trophic positions of *G. holbrooki*, supporting findings from the gut and MDS analyses. Independently collected data sets from guts and trophic scores were positively correlated ($r_s = 0.72$, $p < 0.005$). Based on the $\delta^{13}C$ values, potential basal carbon sources for *G. holbrooki* appeared to be *Utricularia* spp. (bladderwort) and/or green algae (mainly *Spirogyra*), later confirmed by Belicka et al. (2012). The mean (±1 se) $\delta^{15}N$ value for *G. holbrooki* was 9.8 (0.2), just slightly lower than top-level predatory fishes with $\delta^{15}N$ values from 11 to 12. The trophic scores for all Everglades fish species analyzed, including *G. holbrooki*, were significantly correlated with the $\delta^{15}N$ values ($r_s = 0.72$, $p < 0.005$), showing the positive correspondence of these two independently derived measurements in depicting the trophic placement of the fishes (Loftus 2000). Williams and Trexler (2006) examined $\delta^{15}N$ values for *G. holbrooki* from 20 Everglades sites and found a similar value of 10.0 ± 1.0 sd, while Sargeant et al. (2010) measured a $\delta^{15}N$ value of about 9. Abbey-Lee et al. (2013) measured $\delta^{15}N$ values in adult male and female and juvenile *G. holbrooki* from 21 Everglades sites; their values did not differ significantly by sex/age groups (males: mean = 9.3; females: mean = 9.1; and juveniles: mean = 9.3). Belicka et al. (2012) measured $\delta^{15}N$ in two groups of *G. holbrooki* near and far from a canal in Taylor Slough (TS). Their values were much higher than in other Everglades studies ($\delta^{15}N$ means of 16 near canal and about 15 far from canal) and probably reflect nutrient enrichment by phosphorus from the canal. They used fatty-acid composition of food items and fishes to estimate the relative contributions of autotrophic bacteria and algae (using markers of cyanobacteria and green algae) and detritus (using markers for heterotrophic bacteria) inputs to lower trophic-level consumers. This study indicated a very small contribution of vascular plants to *G. holbrooki* diets (<1%). Fatty-acid composition was similar in *G. holbrooki* and other small marsh fishes, consistent with an invertebrate diet for these omnivorous fish. Their analyses also suggested that heterotrophic bacteria metabolizing detritus in flocculent sediments are an important food source for chironomids and amphipods. The similarity of microbial markers between chironomids and floc indicated that heterotrophic bacteria likely form the link between the floc and consumers.

The diet of *G. holbrooki* showed some seasonal differences by hydroperiod (Loftus 2000). In summer samples from ENP, individuals from short-hydroperiod (<8 months of flooding) marshes consumed a slightly higher volume of food than fish at long-hydroperiod sites. The short-hydroperiod fish took larger volumes of chironomids, while palaemonid shrimp comprised over 20% of the prey volume for long-hydroperiod fish. In winter samples, short-hydroperiod fish fed increasingly on copepods, cladocerans, emerging chironomid pupae, and miscellaneous items including bryozoans; long-hydroperiod fish consumed a much greater volume of food in winter, including increased amounts of periphyton and small shrimp. Niche breadth (B) measures were lowest for short-hydroperiod fish ($B = 4.29$ and 3.08 in winter and summer, respectively) and higher for long-hydroperiod fish ($B = 7.37$ and 3.85 in winter and summer, respectively).

Seasonal food availability in the marshes may also play a role in Hg accumulation by *G. holbrooki*. Loftus (2000) used caged, captive-reared fish in field experiments to compare with wild-fish data at six ENP locations in paired short- and long-hydroperiod marshes. Fish at two of the three long-hydroperiod locations in winter had lower Hg levels than the short-hydroperiod fish, and had eaten greater volumes of algae and fewer chironomids. That diet appeared to result in less Hg uptake. Summer samples showed the same pattern by location, and although the volume of algae consumed was similar between fish from different hydroperiods, more individual long-hydroperiod fish ate algae. The diet data also showed that short-hydroperiod fish with higher Hg levels received more food volume from chironomids emerging at the water surface. It is likely that chironomids, which spend their larval lives exposed to MeHg in sediments and periphyton, are important vectors of MeHg into the water column where they are eaten by fishes.

Cleckner (personal communication) speculated that nocturnal migration by zooplankton facilitates Hg transport from the benthos to *G. holbrooki* in the water column. However, Obaza et al. (2011) demonstrated that *G. holbrooki* are relatively inactive at night, with nighttime minnow trap captures generally the lowest of those during 24-h sampling; no *G. holbrooki* were collected on moonless nights. The abundance of zooplankton in the water column of the Everglades is relatively low, as evidenced by the high water clarity (Turner and Trexler 1997; Trexler and Loftus 2016). Thus, this mechanism appears inadequate to explain patterns of *G. holbrooki* Hg, and it is unlikely that zooplankton migration is an important vector of Hg to diurnally active *G. holbrooki*.

Differences in seasonal water temperatures and in substrate organic matter may be important in understanding Hg tissue burden of *G. holbrooki* in the Everglades. Hg methylation is positively correlated with higher water temperatures (Wiener and

Spry 1996). Mason and Lawrence (1999) found that estuarine amphipods living in substrates with high organic content accumulated less Hg than those in more inorganic substrates. The benthic prey of mosquitofish, particularly chironomid larvae, living in the organic-poor sediments of short-hydroperiod marshes, may have been exposed to more available MeHg, which was passed on to the fish. This may explain the results of caged fish and wild fish analyses in Loftus (2000) in which the highest Hg levels usually occurred in short-hydroperiod marshes, particularly during cooler-season samples. The highest growth rates for *G. holbrooki* in the Loftus (2000) study also occurred in the short-hydroperiod marshes, likely the result of the higher daily-maximum temperatures and the resulting high metabolic rates. Higher temperatures result in higher metabolic rates, greater food consumption, and faster growth. Uptake of Hg by fish has been shown to relate both to respiration and feeding rates. Small fish, like *Gambusia*, have higher size-specific respiration and consumption rates than larger fishes, thereby accumulating Hg at faster rates (Post et al. 1996).

8.2.2 *Hg Concentrations in* Gambusia*: Geographical and Temporal Variation*

There is wide geographic variation in the Hg concentrations in *G. holbrooki* in the Everglades, with Hg 'hotspots' located at sites throughout the ecosystem (Stober et al. 2001). In addition, the concentrations reported here were typical of the time that samples were collected and may be most valuable for comparisons across sites and taxa in a relative sense. Concentrations in the whole-fish samples from the Everglades north of Alligator Alley in the 1990s trended lower than farther south in the central Water Conservations Areas and northern ENP (Stober et al. 2001). Total Hg concentrations in ENP ranged from 60 to >400 ng g^{-1} ww (Loftus 2000), with fish from northern Shark River Slough (SRS) having the highest Hg levels, consistent with other studies (Stober et al. 2001). In the EPA's spatially extensive Regional Environmental Monitoring and Assessment Program (R-EMAP) program during the 1990s (Stober et al. 2001), fish from canal habitats were usually much lower in Hg than fish from marsh habitats. Stober et al. (2001) found that in more than half the area in the marsh (62% or more than 400,000 ha), *G. holbrooki* exceeded 100 ng g^{-1} in total Hg. However, at only 17% of the canal kilometers sampled (about 210 km total) did *G. holbrooki* have Hg concentrations in excess of 100 ng g^{-1}. Stober et al. (2001) concluded that the marshes were the primary areas of Hg contamination in the Everglades.

Hg levels in Everglades *G. holbrooki* (Stober et al. 2001) revealed several hotspots in Water Conservation Area 3A between Alligator Alley and Tamiami Trail and northern SRS. However, subsequent Hg levels in *G. holbrooki* in the R-EMAP area declined significantly throughout the marsh between 1995/1996 and 1999 samples. Stober et al. (2001) attributed the decline to corresponding declines in atmospheric wet deposition. More recent work indicated that the trend towards lower levels of Hg continues in some geographical compartments (Rumbold et al. 2008). Hg appears more readily available in some regions of ENP than others, as indicated by *G. holbrooki* Hg levels (Stober et al. 2001).

Geographical differences in Hg levels in fish may be the result of differential food-web complexity affecting bioaccumulation or from spatially distinct biogeochemical processes that affect methylation and bioavailability, or from a combination of these processes (Stober et al. 2001). Because only a low proportion of Hg moves passively from the environment into the fish (e.g. via gills), therefore most Hg in fish tissues must be derived from their diet (Pickhardt et al. 2006). In the Loftus (2000) study, results of a caged-fish study showed the same patterns as wild–fish samples from the same locations. Fish from northern SRS and TS usually had higher Hg levels in the short- versus long-hydroperiod marshes, while the converse was found in middle SRS, similar to Stober et al.'s (2001) findings of lower levels in that region. Concentrations varied temporally and spatially both within and among sites. Though hydroperiod was a significant factor in explaining Hg levels in *G. holbrooki*, site-specific patterns in Hg uptake appear to reflect local factors affecting Hg methylation processes, Hg input, or biomagnification. Abbey-Lee et al. (2013) summarized two competing hypotheses for the origins of patchy levels of Hg contamination and hotspots in Everglades marshes: either they reflect local areas of higher Hg methylation resulting from biogeochemical processes (e.g., Bates et al. 2002), or they reflect local areas with greater food-web complexity, leading to greater biomagnification. The results from both Loftus (2000) and Abbey-Lee et al. (2013) suggested that local food availability and niche partitioning are present in *G. holbrooki* during the wet season. Thus, the conditions for spatial variation in food-chain length and local heterogeneity of biomagnification are present. Additional research is needed to untangle those hypotheses and it is likely both play a role in forming local Hg hotspots.

8.2.3 Conclusions

G. holbrooki have been used in ecotoxicological and bioaccumulation studies of contaminants because they are hardy, ubiquitous, livebearing fish with a life span of only a few months (Haake and Dean 1983). Because of their short life span and small size, contaminant levels must result from recent accumulation in contrast with long-term accumulation by larger, multi-year, more mobile fishes.

Because *G. holbrooki* food uptake may vary as a function of hydrologic alteration, nutrient input or other factors, this could affect Hg dietary bioaccumulation. Restoration of hydroperiod and deeper waters in Everglades may lead to lower accumulation of Hg during winter's cooler water temperatures (Loftus 2000). Because methylation of Hg is affected by agricultural inputs of sulfate into Everglades water via canals (Compeau and Bartha 1985; Bates et al. 2002; Chap. 2, this volume), restoration of marsh sheet flow in place of canal deliveries could reduce sulfate input and may reduce bioaccumulation of Hg by *G. holbrooki*.

8.3 Largemouth Bass

The Florida Largemouth Bass (*Micropterus salmoides floridanus*), described by Bailey and Hubbs (1949) is the native sub-species of the Largemouth Bass (*M. salmoides*) occurring south of the Suwannee River (Barthell et al. 2010) and throughout the Everglades ecosystem. Florida bass are renowned for attaining trophy size (Chew 1975; Crawford et al. 2002) and fishing effort in the Everglades for these fishes has ranged between 28 and 65% of total fishing effort over the past decade in Water Conservation Area 3 (FWC, unpublished data). Bass are an appropriate indicator for human health risk assessment (Wiener et al. 2007) and have proven useful as a sentinel of temporal and spatial trends in Hg bioaccumulation (Atkeson et al. 2003; DEP 2013; Loftus and Bass 1992; Lange et al. 1993; Wiener et al. 2007). Their utility comes from their widespread distribution, popularity as a sport fish, role as an apex predator and their ability to integrate Hg over large temporal and spatial scales. Florida Bass are found throughout the Everglades ecosystem, commonly occurring in both urban and water conservation area canals as well as within shallow vegetated sloughs and marshes of the water conservation areas (Fury et al. 1998; Chick et al. 2004). To the south, within Everglades National Park, surface waters are shallower, and suitable habitat is limited to long-hydroperiod wetlands and estuarine tidal creeks (Loftus 2000; Rehage and Loftus 2007). Bass are highly mobile, and the extensive system of canals, levees and water control structures designed primarily for flood control has allowed bass populations to persist throughout the shallow wetlands of the Everglades by utilizing these deep water refugia (Loftus and Kushlan 1987; Light and Dineen 1994; Trexler et al. 2000).

Bass and other large-bodied fish move in response to hydropattern (Batzer 1998; Turner et al. 1999) and during periods of dry-down, bass typically abandon marsh slough habitat and seek refuge in alligator holes (Kushlan 1974; Carlson and Duever 1977), coastal rivers (Rehage and Loftus 2007; Rehage and Boucek 2013), canals (Fury et al. 1998; Gandy and Rehage 2017) and isolated deep water marsh sloughs (primarily *Eleocharis* sloughs) located in close proximity to canals (Fury et al. 1995; Trexler et al. 2001). Correspondingly, during periods of high rainfall, vast expanses of marshes become seasonally inundated allowing re-colonization by prey (fish and crustaceans) and large bodied fish (i.e. bass, gar, bowfin) as well as other apex predators such as alligators. Particularly in the WCAs where most canals have direct contact to wetlands (Rehage and Trexler 2006) canals and wetlands undergo strong seasonal cycles of fish abundance based on hydroperiod (Fury et al. 1995). As bass respond to hydroperiod their opportunistic feeding nature can result in marked changes in composition of diet across habitats.

8.3.1 *Food Habits and Dietary Makeup*

Florida bass are a top predator in the Everglades aquatic food web and occupy a similar trophic position as other piscivorous species including Florida gar (*Lepisosteus platyrhyncus*, Hunt 1953) and warmouth (*Lepomis gulosus*, Loftus 2000). Typical of all Black Bass species, Florida Bass are opportunistic feeders, utilizing a wide variety of available prey as adults (Fig. 8.1, also see, Summers 1981; Maceina and Murphy 1989). Bass collected across seasons for 2-years from canal and marsh habitats within the WCAs (Northern and Central Everglades), generally fit the mold of a piscivorous predator with fish occurring in 74% of stomachs and contributing 91% of the total weight of all food consumed by all size classes (Lange et al. 1998, 1999). The most frequently encountered groups of fish in bass stomachs were livebearers (family Poeciliidae; e.g. mosquitofish) at 15% of all bass stomachs, sunfish (family Centrarchidae) at 13%, and killifish (family Cyprinodontidae) at 12%. These families are the most numerous fishes in the Everglades marshes and ponds (Gunderson and Loftus 1993) explaining why they were most commonly observed in bass stomachs. Lange et al. (1998, 1999) reported that unidentified fish remains (digested beyond identification) occurred in 47% of all bass stomachs. Two species of sunfish, warmouth (*Lepomis gulosus*) and spotted sunfish (*L. punctatus*) were the most important prey items by weight in bass stomachs across all habitats.

Decapod grass shrimp (*Palaemonetes paludosus*) and crayfishes (*Procambarus fallax* and *P. alleni*) were important numerically but contributed only 2.3% of the

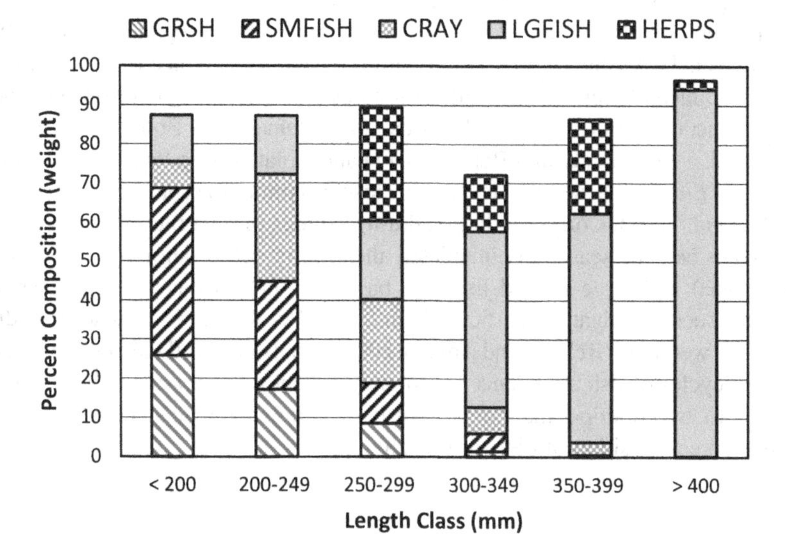

Fig. 8.1 Major prey groups consumed by 2651 Florida bass of different sizes during monthly assessments conducted 1996–1998 in Water Conservation Areas 2 and 3. Prey groups include grass shrimp (GRSH), small prey fish <30 mm total length (TL) (SMFISH), crayfish (CRAY), frogs and snakes (HERPS), and large prey fish >30 mm TL (LGFISH). From Lange et al. (1998, 1999)

total weight of food consumed. Grass shrimp frequency of occurrence ranged from 31 to 62% in bass stomachs and occurred more commonly in marsh habitats than in deep-water canals (Lange et al. 1998). The remainder of the prey consumed by bass consisted of insects, amphibians, reptiles, and plant material with a total of 78 taxa identified as prey items of bass in the WCAs.

Diet of bass in the Everglades is partly determined by habitat. Loftus (2000) reported few bass encountered in long-hydroperiod marshes in northern SRS in Everglades National Park (Southern Everglades) but, for those encountered, decapod crustaceans represented a higher biomass (56%) in diet than fish (43%). Bass in SRS were not common due to frequent dry downs (Chick et al. 2004) and were only found in alligator ponds during periods of high water where high fish species richness but low overall numbers of prey were observed. The most commonly consumed decapods in SRS were grass shrimp and crayfish (Loftus 2000). Within estuarine creeks at the southern terminus of SRS (Southern Everglades), Rehage and Loftus (2007) reported the greatest abundance of Largemouth bass during the dry season as marsh water levels receded at the end of the wet season. Bass transitioned from a diet rich in freshwater invertebrates to prey subsidized from the drying marsh (Boucek and Rehage 2013) which consisted of cyprinodontoid fish (40% by number), invertebrates (42%) and lepomid sunfish (16%), while co-occurring marine prey species contributed little to the diet. Presumably, the bass sampled by Boucek and Rehage (2013) migrated from the long-hydroperiod marsh habitats as water levels receded. In contrast, diet of bass in coastal rivers on the east side of the Everglades where extensive wetlands have been eliminated included a high percentage of marine derived prey species as well as fresh water non-native species (Dutka-Gianelli et al. 2011).

Food webs supporting bass also varied among habitat types. Slough and wet prairie habitat in the WCA marshes offer the most suitable depths for bass and are characterized by thick growth of submerged aquatic vegetation (SAV). Within marsh habitats, large fish (>30 mm TL) contributed little to biomass of prey consumed by intermediate size bass (<300 mm), (Fig. 8.2) and instead small prey fish (<30 mm TL), decapods, and herpetofauna were more readily consumed. However, large prey fish (>30 mm) including juvenile Lepomids and non-native species were present at marsh sites but presumably hunting efficiency by bass was reduced by SAV. These observations are consistent with those of Cailteux et al. (1996) who reported delays in ontogenetic shifts to piscivory in bass with increased habitat complexity. Similarly, Savino and Stein (1982) suggested that prey-capture efficiency decreases with increasing submerged aquatic vegetation (SAV) which can lead to decreased piscivory within complex vegetated marsh habitats.

Age and size can also have a large influence on diet of Everglades bass. Bass undergo a series of ontogenetic diet shifts as they age, transitioning from a diet rich in zooplankton and macroinvertebrates as juveniles and fish as adults (Wicker and Johnson 1987; Cailteux et al. 1996). Because Lange et al. (1998) sampled few bass <100 mm TL from WCA sites, zooplankton prey were not assessed but a consistent and expected trend in prey selection with increasing size of bass was observed

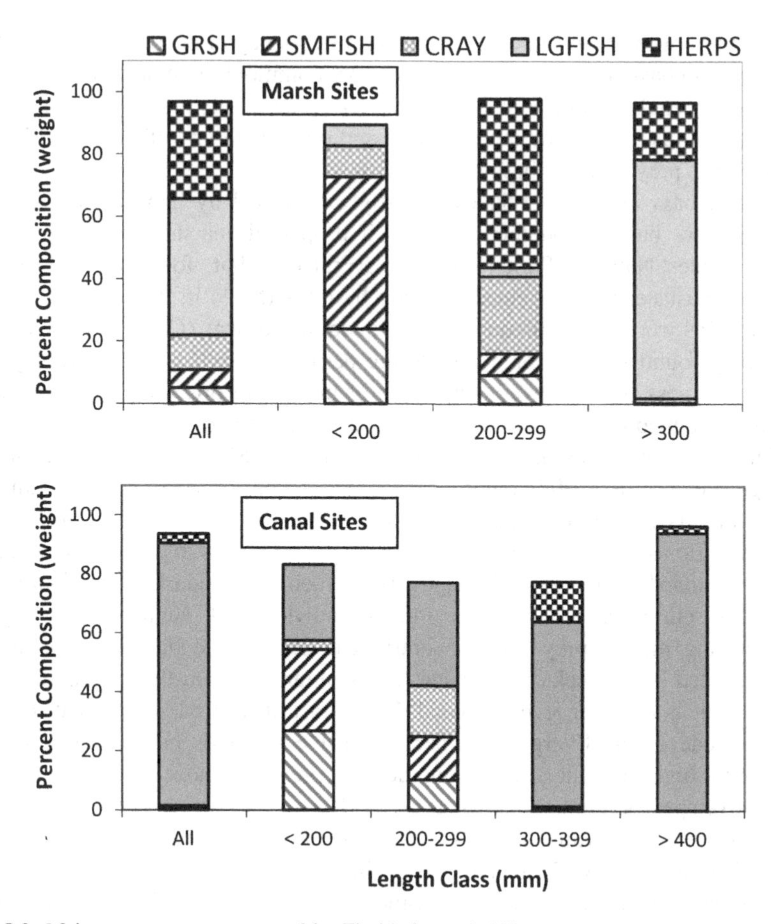

Fig. 8.2 Major prey groups consumed by Florida bass of different sizes during monthly assessments conducted 1996–1998 at (top) Marsh Sites in Water Conservation Area (WCA) 2 and 3 (n = 194) and (bottom) Canal Sites WCA2 and 3 (n = 1870). Composition of diet is presented for bass of all sizes combined (ALL) as well as for specific size classes. Prey groups include grass shrimp (GRSH), small prey fish <30 mm TL (SMFISH), crayfish (CRAY), frogs and snakes (HERPS), and large prey fish >30 mm TL (LGFISH). From Lange et al. (1998, 1999)

(Fig. 8.2) where groupings include grass shrimp, crayfish, small prey fish, large prey fish, and frogs and snakes.

Bass of all sizes were piscivorous, but species composition and size of prey changed with size, as did their proportion of biomass consumed (Fig. 8.1). Bass from the smallest size range (<200 mm total length (TL)) commonly consumed both grass shrimp and small prey fish, which occurred in 47% and 48%, respectively, of stomachs (Lange et al. 1999). By contrast, these items were of negligible importance for bass >300 mm TL. For small bass (<200 mm TL) small prey fish contributed 43% of the biomass of all food consumed and consisted primarily of Eastern mosquitofish (*M. holbrooki*), Least Killifish (*Heterandria formosa*), and Bluefin killifish (*Lucania goodei*). Large prey fish made up 12% of the biomass (Fig. 8.1) and consisted of juveniles of sunfish (*Lepomis spp.),* Mayan cichlid (*Cichlasoma uropthalmus*) and spotted tilapia (*Tilapia mariae*).

For intermediate size bass (200–399 mm), crayfishes were an important prey item contributing 28% of overall mass of diet (Fig. 8.1). These same sized bass consumed low numbers of herpetofauna including Pig Frogs (*Rana grylio*), Southern Leopard Frogs (*Lithobates sphenocephalus*), water snakes (*Nerodia spp.*), and swamp snakes (*Seminatrix sp.*). In spite of low rates of predation, herpetofauna contributed between 14 and 24% by weight to the diet of intermediate size bass (Fig. 8.1).

A shift to a diet dominated by large prey fish occurred around 300 mm TL in the WCAs, at age class 2–4 depending on individual bass growth rate (Lange et al. 1998). The contribution of large prey fish to the biomass of diet increased from 45% in bass 300–399 mm to 94% in bass in excess of 400 mm TL (Fig. 8.1). These larger bass fed primarily on Lepomid sunfish (warmouth, spotted sunfish, bluegill (L. *macrochirus*) and redear sunfish (L. *microlophus*)) as well as other native and non-native species of fish.

There is also a gradual shift away from herpetofauna as bass grow. Amphibians and reptiles contributed 54% of the biomass eaten by intermediate sized bass (200–299 mm) at marsh sites (Fig. 8.2) while occurring in only 4% of bass stomachs (Lange et al. 1998, 1999). Herpetofauna also continued to constitute a significant portion of the biomass consumed by bass <400 mm TL at marsh sites. Herpetofauna were not consumed frequently in any habitat; however, owing to their large mass relative to other prey, their contribution to biomass was very high for some individual bass at marsh sites. This selective feeding was due perhaps to a lower prey diversity in the marsh driving a higher utilization of suitably sized prey, including reptiles and amphibians (Jordan and Arrington 2001).

Food webs varied most between canal and marsh sites probably due to habitat structure affecting prey availability. The transition by bass to piscivory, particularly of large fish prey (i.e. sunfish), came at a substantially smaller size in canal bass. At marsh sites, large prey fish (>30 mm TL) contributed little to biomass consumed by intermediate size bass (Fig. 8.2) and instead small prey fish, decapods, and herpetofauna were more readily consumed. Juvenile lepomids and non-native species were present at marsh sites but presumably hunting efficiency by bass was reduced in the heavily vegetated slough habitat where bass reside. Instead, amphibians and reptiles contributed 54% of the biomass and crayfish 25% of the biomass to intermediate size bass at marsh sites (Fig. 8.2). Conversely, at canal sites large prey fish (>30 mm TL) contributed 25% of biomass in the diet of bass <200 mm, increasing to 63% of biomass for bass 300–399 mm.

Another obvious difference between canal and wetland food webs included the more dominant role of exotic species in bass diet in canal habitats. Spotted tilapia (*Tilapia mariae*) made up 33% of the weight of prey consumed by intermediate bass (200–399 mm TL) in the L67A Canal and effectively replaced crayfish as a food item in this transitional size (Lange et al. 1999). Crayfish were apparently abundant in canals, based on warmouth stomachs, which contained 75% by weight crayfish. Bass simply selected other, possibly more easily captured, prey which in this case was spotted tilapia. Given that the Everglades has low standing stocks of native fishes relative to other wetlands (Turner et al. 1999), non-native species can be an important prey item for bass, particularly in WCA canals where non-native species can be very abundant (up to 20% of canal fish biomass (Trexler et al. 2000).

8.3.2 Hg Concentrations in Bass and Their Prey

Hg concentrations of important prey items of bass collected from the WCAs are compiled in Table 8.1 (USGS ACME Project unpubl. data—project description by Krabbenhoft et al. 1996; Lange et al. 1998, 1999). Prey items represent species and sizes observed in bass stomachs by Lange et al. (1998, 1999) that were subsequently collected for Hg analyses. Collections include prey from both canal and wetland habitats in WCAs 2 and 3 and are grouped by major functional groups as well as important prey types that occurred frequently in bass stomachs (Lange et al. 1998). Gradients in Hg levels of prey (reported as MeHg or total Hg [THg]) increased from crayfish to insects to grass shrimp and fish and were generally higher in marsh than canal habitats.

Patterns of Hg concentration in whole bass were consistent with their major prey items at the four WCA study sites where the bass were captured (Table 8.1). These Hg concentrations were consistent with spatial distribution observed in similarly timed assessments of mosquitofish (Stober et al. 2001), prey fish (Cleckner et al. 1999), and great egrets and alligators (Rumbold et al. 2002; Axelrad et al. 2007). Findings of increasing Hg in bass from WCA2 to WCA3 (Table 8.1) were also consistent with the decades long pattern of north to south increases in Hg bioaccumulation that has been evident across the Everglades in prey fish (Scheidt and Kalla 2007) and large bodied fish, including bass (Axelrad et al. 2007; Julian et al. 2017).

We evaluated the relationship between bass trophic position from stomach contents, and Hg concentration among sites. We calculated trophic scores as described by Adams et al. (1983) and utilized by Winemiller (1990) and Loftus et al. (1998) where trophic level of a predator was based on the sums of the trophic scores of its prey multiplied by the proportion of the diet comprised by each prey type. Where possible, small fish and invertebrate trophic scores were derived from food habit data from researchers working at co-located sample sites (Lange et al. 1998, 1999; USGS ACME Program data, Loftus 2000) or from literature review. In Fig. 8.3, bass trophic scores ranged from 2.9 to 3.34 at canal and marsh sites in WCA2 and 3 (Lange et al. 2003) and were consistent with the score for bass in long-hydroperiod SRS in ENP of 3.18 reported by Loftus (2000). Because marsh bass had a more omnivorous diet, trophic scores were lower than the corresponding canal sites within the same WCA (Fig. 8.3) suggesting that among site differences in food webs were less critical to bass Hg concentrations than the MeHg available at the base of the food web. This is consistent with Bemis et al. (2003) who utilized stable isotope analysis to assess bass Hg concentrations at 12 marsh and canal locations in the

Table 8.1 Mean Hg concentration in whole-body organisms that constitute major prey groups consumed by bass from marsh and canal sites within the Everglades Water Conservation Areas (WCA) 2 and 3

		CRAY	Insect	GRSH	SM FISH	Gambusia	Frog	LG FISH	Spotted tilapia	Bass	
Location	Type	MeHg	MeHg	MeHg	THg	THg	THg	THg	THg	N	THg
WCA2	Canal		12.9	29.2	38.9	45.5		119.8	100	91	460
	Marsh	15.8	35.7	59.7	74.1	96.3	44.0	91.8		83	390
WCA3	Canal		8.7	15.6	19.8	22.2		126.4	50	73	620
	Marsh	19.9	63.5	165.1	83.1	110.0	176.0	288.3		71	790

Samples were collected between March 1995 and November 1999 as part of the USGS ACME (Aquatic Cycling of Mercury in the Everglades) project (description by Krabbenhoft et al. 1996) and reported as THg or MeHg in fresh weight whole-body homogenates as ng/g wet weight. Hg in whole bass are reported for collections made between September 1997 and October 1999. Prey groups include insects, grass shrimp (GRSH), crayfish (CRAY), small fish <30 mm total length (SMFISH), large fish 30–130 mm (LGFISH), and Spotted Tilapia. All results are unpublished USGS data except LGFISH, Spotted Tilapia and bass samples, which were reported by Lange et al. (1998, 1999). Frog data are from Lange (2006) reported as THg in leg muscle tissue collected in 2006

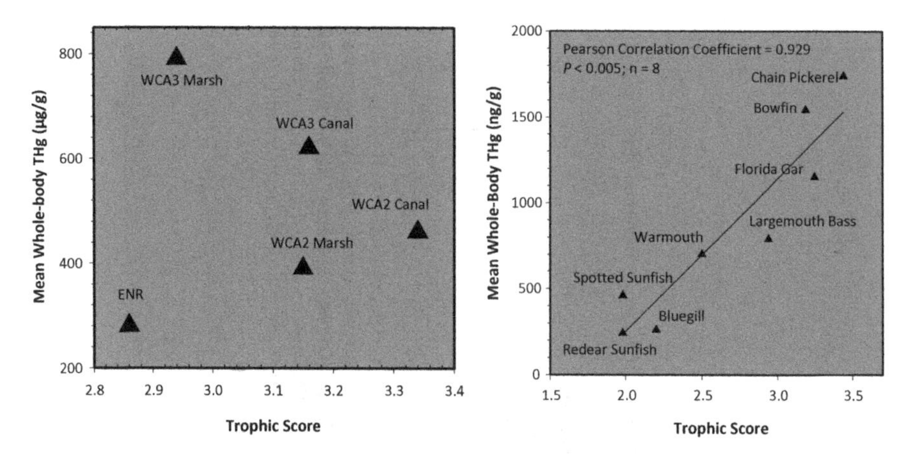

Fig. 8.3 Trophic scores of fish based on weight proportions of prey items assessed monthly during 1996–1998 plotted versus THg in fish tissue for (left) study sites in WCAs 2 and 3 (from Table 8.1) and the Everglades Nutrient Removal Project (ENR) and (right) fish collected at WCA3 site 3A15. From Lange et al. (2003)

WCAs. They found that δ15N (estimate of food web length) explained little (5%) of the variance in bass Hg across sites and rather spatial differences (i.e. latitude) and variations in fish size (i.e. length) explained the majority of variance in bass Hg concentrations, 38% and 15%, respectively.

We also evaluated relationships between trophic position and food web characteristics at individual sites utilizing trophic scores calculated in the same manner as above for large fish at each study site (Lange et al. 2003). Results presented in (Fig. 8.3 for WCA3 marsh site) suggest that trophic position is a strong local predictor of fish Hg concentration. Interestingly, based on food habits data bass are less piscivorous than three other common predators including Florida Gar (*Lepisosteus platyrhincus*), Bowfin (*Amia calva*), and Chain Pickerel (*Esox niger*) at the WCA3 marsh site. These findings are consistent with Bemis and Kendall (2004) who found that δ15N values were good predictors of Hg in taxa from most, but not all, canal and marsh sites within the WCAs. It was surmised that where poor relationships between δ15N derived estimates of trophic position and Hg of local taxa were found, that several factors could contribute including localized biogeochemical and hydrologic conditions influencing food webs at a micro scale (Wankel et al. 2003), different food webs for different species and large-scale fish migrations in and out of localized sites (Kendall et al. 2003).

8.3.3 Conclusions

As a sport fish that routinely reaches trophy size, Florida bass are an important part of Everglades sport fisheries and represent a worst-case scenario for Hg bioaccumulation in an apex predator. Levels of Hg in bass often have reached levels that warrant advisories on consumption which are currently the only viable method to manage risks to human consumers. Fishing effort directed at bass is predominantly focused on canals and their near vicinity during the dry season when bass and other aquatic animals migrate from shallow marsh habitat to deeper water. Food webs vary between habitats and bass occupying canals have an earlier onset of piscivory, which should result in higher concentrations of Hg in bass in these habitats.

It has been demonstrated that the variability in Hg concentrations in bass and other fish can be accounted for by food web interactions at the local site level; however, food web interactions do not accurately reflect Hg concentrations in bass across the water conservation areas. Clearly, site-specific characteristics of the food web along with ontogenetic shifts in diet control the bioaccumulation of Hg in bass. The lack of strong relationships between trophic position and Hg concentration in bass across multiple sites and habitats suggests that localized site characteristics are important.

Bass feed opportunistically and generally consume the most available prey regardless of habitat. As restoration efforts affect the timing and duration of flooding, as well as the quality of the water, food web interactions and fish movement will certainly change. For example, increased duration and extent of hydroperiod should be accompanied by increased biomass of small prey fish for extended periods of time (DeAngelis et al. 1997) and therefore, increased piscivory and Hg bioaccumulation rates in bass occupying marsh habitat. Restored conditions may also force changes in the availability of MeHg at the base of the food web, and species composition and distribution of non-native fish species.

The high rate of piscivory in bass and other top-level predators support a high rate of Hg bioaccumulation in bass across the Everglades, regardless of location. In spite of declines in Hg concentrations in bass (Chap. 2, Vol. III), Hg concentrations continue to exceed the criterion for the protection of human health across vast areas of the Everglades.

8.4 Wading Birds

8.4.1 Food Habits and Dietary Makeup

Long legged wading birds use a variety of methods to capture prey (Kelly et al. 2003), but all are limited to foraging in shallow water (5–30 cm). Stabbing, sight-foraging, sediment probing, active chase, tactile foraging, and combinations of these are used depending on bird species, habitat, prey density, depth and prey type (Kelly et al. 2003). Birds are homeothermic and among aquatic organisms have high metabolic and prey intake rates relative to body mass, making them good indicators of Hg in the food web (Cristol et al. 2008). Wading birds are considered generalist feeders and will take a wide variety of prey types and sizes, in a range of wetland habitats. Fishes, insects, snails, shrimps, crayfishes, and other invertebrates are typical in their diets, though there are preferences by avian taxa. With the exception of Ogden et al. (1976), all food habits information summarized here are from collections of food regurgitated by nestlings. At least in the White Ibis, no difference has been found between types of food fed to nestlings and food consumed by breeding adults (Kushlan 1974). The advantage of using nestlings as a sampling tool is that their food must come from a known foraging radius around the colony (20–30 km), and Hg concentrations from nestlings are therefore not skewed by Hg sources encountered in the much larger annual range of the adults.

In the Everglades, White Ibises (*Eudocimus albus*) tend to specialize on preying on crayfishes and shrimps, probably because bill shape and size facilitates capture by tactile probing, and gut morphology is specialized for grinding shelled organisms (Kushlan and Kushlan 1975; Heath et al. 2009). Crayfish make up a substantial portion of the diet across all studies (55–85% by mass, Kushlan and Kushlan 1975; Boyle et al. 2012, 2014), with high proportions of crayfish particularly during strong reproductive years (Boyle et al. 2014).

Sight-foraging herons and egrets tend to consume predominantly fish in their diet. In the Everglades, fish made up >95% of the biomass consumed by Great Egret (*Ardea albus*) chicks (Frederick et al. 1999), with centrarchid fishes making up between 40 and 80% of the biomass depending on year. Snowy Egrets (*Egretta thula*) and Tricolored Herons (*Egretta tricolor*) are also fish eaters (92–95% of biomass) across a wide variety of habitat and water conditions (Strong et al. 1997; Boyle et al. 2012; Klassen et al. 2016). Though Little Blue Herons (*Egretta caerulea*) consume large proportions of fish in their diet in south Florida (76%, Klassen et al. 2016), they have been found to consume more grass shrimp, insects, and amphibians in their diet than the other two *Egretta* species (Smith 1997; Klassen et al. 2016).

Wood Storks (*Mycteria americana*) tend to specialize on large-bodied fish species, and large individual prey items. Their diet consists mostly of fishes, and the majority of biomass consumed (56% and 84%, Ogden et al. 1976; Klassen and Gawlik 2014 respectively) has been comprised of sunfishes (*Lepomis spp.*) and bullhead (*Ameiurus spp.*). However, their diet can also include smaller species like flagfish (*Jordanella floridae*), sailfin mollies (*Poecilia lattipinna*) and marsh killifish (*Fundulus confluentis*), which comprise the highest frequency of prey items taken (Ogden et al. 1976).

In all cases, herons, egrets, ibises and storks tend to select for larger species and individuals than are available in the marsh. Fishes greater than 1.9 cm were preferentially consumed by *Egretta* herons (Klassen et al. 2016; Boyle et al. 2012), and Great Egrets consumed fishes averaging over 9 cm in length (Frederick et al. 1999). White Ibises consumed crayfishes with average total length of 3.3 cm total length, considerably larger than crayfishes available in the surrounding marshes (Dorn et al. 2011). Wood Storks take considerably larger prey than are available in the marsh (Ogden et al. 1976), and average size has ranged from 4.5 to 5.5 cm (Klassen et al. 2016).

8.4.2 Hg Concentrations in Food

Hg concentrations in fish and invertebrates that make up the bulk of wading bird diets are highly dependent upon location, prey animal size, and prey species. Further, Hg concentrations have changed markedly over time, and it should be remembered that the concentrations reported are typical of specimens collected during that time period. Loftus (2000) comprehensively measured Hg content in most fish and macroinvertebrate species in the Everglades (see Table 8.2). Frederick et al. (1999) found total Hg concentrations in whole food items taken from Great Egret chick boluses averaged 0.22–3.55 mg/kg ww. These authors used both biomass and prey size information to derive an estimation of Hg exposure for Great Egret chicks during the late 1990s, reporting a 3-year average of 0.41 mg/kg ww in diet, (range of annual means 0.37–0.47 mg/kg ww). Similar studies of diet have not been done for other bird species, but the Great Egret example is probably a reasonable proxy for exposure in Wood Storks, which eat largely the same species and sizes of fishes. For *Egretta* herons, exposure can be reasonably inferred for the same time period because diet is constrained to a small set of marsh fishes (Table 8.2, 0.3–0.7 mg/kg ww). Similarly White Ibis exposure may inferred from crayfish Hg concentrations because of the predominance of crayfish in the diet (0.6 ppm during the late 1990s).

Within wading birds, virtually 100% of Hg in feather tissue is in the MeHg form (99.8%, Frederick et al. 2002, 93% Spalding et al. 2000). In liver tissue however, the proportion of Hg methylated increased from 56% in low exposure animals to 73% in high exposure, suggesting this organ could be a site of demethylation. Given the high proportion of MeHg in the diet of these birds, it seems likely that physiological exposure in these birds is largely through the methylated form of Hg.

Hg exposure in Everglades aquatic birds is likely to increase with increasing trophic level of prey, and with increasing size of prey items. Typically, both size and trophic level of predatory fish increases with increasing hydroperiod in the Everglades suggesting that water management can be an important variable to consider in Hg exposure. However, total Hg exposure also increases with volume of prey

Table 8.2 Hg concentrations (mg/kg ww) in fish and macroinvertebrates that form a high proportion of food items in wading bird diets in the Everglades

Species	Mean	s.d.
Fundulus chrysotus	0.684	0.143
Fundulus confluentus	0.394	0.158
Jordanella floridae	0.084	0.023
Gambusia holbrooki	0.321	0.063
Poecilia latipinna	0.157	0.056
Lepomis macrochirus	0.478	0.274
Lepomis punctatus	0.416	0.092
Micropterus salmoides	0.967	0.216
Cichlasoma uropthalmus	0.322	0.121
Procambarus fallax	0.064	0.016
Paleomenetes spp.	0.186	0.078

From Loftus (2000)

eaten—and in years when food availability is high, exposure may also increase. While effects of Hg toxicity clearly increase with exposure in food, it is unclear how this relationship extends to effects on reproduction. When birds are in good condition they may be less susceptible to Hg effects, even though they are receiving higher total loads of Hg. This potential tradeoff represents a new frontier for research in this area.

Location of nesting is an important source of variation in nestling feather Hg concentrations (Frederick et al. 1999). While concentrations in the freshwater Everglades north of Tamiami Trail declined markedly during the period 1990–2005 (Chap. 2, Vol. III), there is less evidence that the same has happened for either birds or fish south of Tamiami Trail. This may be because there are higher Hg concentrations in freshwater discharge from canals feeding the southern marshes or it may have to do with sulfate dynamics in the coastal zone. In any case there is the potential for higher exposure in ENP than elsewhere, and this is of considerable concern because one goal of CERP is to restore fish populations and wading bird nesting to the coastal zone of ENP.

8.5 Florida Panther

The Florida Panther (*Puma concolor coryi*) is an endangered subspecies of puma (also called mountain lion and cougar) that was once contiguous with other puma populations and subspecies across North America (Young and Goldman 1946). With the advent of European colonization the puma in eastern North America was eventually extirpated across almost all of its range. By the early 1990s only an isolated and inbred population numbering as few as 30 individuals remained in the Big Cypress and Everglades ecosystems of south Florida (Nowak and McBride 1974). Some protection was afforded when the panther was classified as a game species in 1950, and complete state protection followed in 1958 (Onorato et al. 2010); the panther became federally protected in 1967 with the passage of the endangered species act (Federal Register 1967). A heroic multi-agency recovery program began in 1981 and included research and management actions such as land acquisition, habitat management, highway underpasses, law enforcement protection, prey management, and genetic rescue. The latter involved the release of eight pumas from Texas into south Florida to give the population a genetic boost (Seal 1994). As a result of these actions, the Florida panther has grown to between 120 and 230 individuals. Despite this reversal from near extirpation, the panther still faces a number of threats including poaching, infectious diseases, vehicular trauma, invasive species, climate change, environmental contaminants, and poor quality habitat in much of their range (Roelke et al. 1991; Roelke et al. 1993; Cunningham et al. 2008; Onorato et al. 2010; Cunningham 2012).

8.5.1 Food Habits and Dietary Makeup

In general, pumas are adaptable and can survive in a myriad of habitat types. They have one of the most extensive geographic ranges of any mammal with populations virtually contiguous from Alaska to Patagonia (Young and Goldman 1946). They feed on a range of prey items from mice and birds to guanacos and elk (Guggisberg 1975). Typically, pumas favor ungulates across their range (Guggisberg 1975), and these hoofed animals—specifically white-tailed deer (*Odocoileus virginiana*) and feral swine (*Sus scrofa*)—are the primary prey for Florida panthers (Maehr et al. 1990; Dalrymple and Bass 1996). The Florida panther, however, is somewhat unique in that the consumption of midsize prey, such as raccoon (*Procyon lotor*)—the third most common prey item across their range—and nine-banded armadillo (*Dasypus novemcinctus*) is more common than that seen in western pumas (Maehr et al. 1990; Caudill et al. 2019). Other prey species include rabbits (*Slyvilagus* spp.), rodents, birds, alligators (McBride and McBride 2010), domestic cats (Cunningham et al. 2008), otter (Dalrymple and Bass 1996), and livestock (Caudill et al. 2019).

Not surprisingly, food habits of panthers vary across their range within south Florida. North of I-75, ungulates, especially feral swine, comprise a larger proportion of their diet (Maehr et al. 1990). South of I-75 raccoon and deer were the most important prey species, comprising equal proportions of panther diets (Caudill et al. 2019). Panther diet also has varied temporally. Significant declines in deer and feral swine in some areas south of I-75 have been observed (Garrison et al. 2011; FWC 2016). The cause is unknown but may be due to high water levels and increased predation (Garrison et al. 2011; Caudill et al. 2019). These declines likely are the cause for a shift in panther prey selection from ungulates to raccoons over the past two decades (Caudill et al. 2019).

8.5.2 Hg Concentrations in Food

Variation in panther diet will, of course, lead to variation in Hg concentrations in panthers. Prey that are part of the aquatic food chain will naturally be expected to have higher Hg burdens. Hg concentrations in white-tailed deer in the Everglades are relatively low (<0.5 ppm) in hair (Ware et al. 1990; Forrester 1992) and are comparable to concentrations in deer in other areas (Kleinert and Degurse 1972; Cumbie and Jenkins 1974). Hg concentrations in feral swine in south Florida are unknown but are likely higher compared to deer due to their omnivorous feeding habits (Wren 1986; Dobrowolska and Melosik 2002). Dobrowolski and Melosik (2002) found that 93% of hunter-killed wild boar in Poland had kidney Hg concentrations exceeding 0.05 ppm while red deer collected from the same area had comparatively low values. Nevertheless, feral swine are not considered an important source of Hg to panthers in south Florida.

As a top prey species south of I-75, raccoons probably present the highest risk for Hg exposure in panthers. All raccoons sampled in the Everglades had detectable hair Hg concentrations (averaging 7.6 ppm; Porcella et al. 2004); however, raccoons from some areas—especially south of Tamiami Trail (US 41) had concentrations up to 72.5 ppm (Roelke et al. 1991). Raccoon hair Hg concentrations in Long Pine Key, an area commonly frequented by Florida panthers, averaged 9.4 ppm and Shark Valley Slough south of US 41 averaged 26.7 ppm (Porcella et al. 2004). Alligators comprise a relatively small proportion of panther diet; however, their consumption may be more common in aquatic habitats. Hg concentrations in alligator liver and muscle in the Everglades ecosystem averaged 4.9 and 0.64 ppm respectively; however, samples from Everglades National Park (ENP) were twofold higher (Rumbold et al. 2002). Alligators sampled from Shark Valley Slough had mean liver Hg concentrations of 50 ppm with highest levels in larger alligators (Heaton-Jones et al. 1997). Panthers have been shown to kill and consume relatively large alligators (up to 2.7 m) (McBride and McBride). Otters are relatively unimportant as a prey species for panthers; however, otter consumption by panthers does occur in ENP (Dalrymple and Bass 1996), and this piscivorous species may be an additional source of Hg. Hg concentrations in otters, have been reported at up to 37 ppm (FWC unpubl. data).

8.5.3 Hg Concentrations in Florida Panthers

Hg is detectable in hair from Florida panther museum specimens dating to the nineteenth century (Newman et al. 2004). All panthers sampled since the late-1970s have had detectable levels of Hg in hair with almost 4% having concentrations ≥30 mg/kg (Table 8.3; also see Brandon 2011; FWC unpub. data; for review of toxicity reference value, see Chap. 10, this volume). This variation in Hg distribution appears to be geographically based. Although individuals with hair concentrations ≥30 ppm were found in all regions, panthers sampled in ENP had the largest proportion exceeding 30 ppm, ranging up to 100 ppm (Brandon 2011).

Elevated Hg burdens in panthers from ENP are likely due to increased availability of MeHg in the local environment combined with greater dependence on aquatic

Table 8.3 Mean, standard deviation, and range by region of total Hg concentrations in hair Florida panthers >4 months of age sampled in south Florida 1978–2017

Region	Sample size	Mean hair Hg concentration (ppm)	STD (ppm)	Range (ppm)	
				Min	Max
Southwest Florida north of I-75	260	3.49	4.67	0.09	49
Southwest Florida south of I-75	126	6.92	7.48	0.16	67
Southeast Florida (Everglades ecosystem)	21	24.59	26.67	0.94	100

prey (e.g. raccoons, alligators, otters). Across most of their range, hair Hg concentrations in panthers declined from the levels seen in 1978–1991 (Brandon 2011), and this mirrors observations in other species (Lange et al. 2000; Rumbold et al. 2001, 2002; Frederick et al. 2002; Lange et al. 2005). Concentrations, however, have been on the rise in some areas. Mean hair Hg concentrations in panthers from the Everglades ecosystem rose slightly from a mean of 23.2 ppm in 1986–2000 to 28.7 ppm in 2001–2014 (FWC unpubl data). A small increase was also observed in mean hair-Hg samples collected from the southern portion of the Big Cypress Swamp ecosystem 1991–2010 compared to 2011–2017 (6.7 mg/kg and 7.4 mg/kg, respectively). This increase may be due to a shift from ungulates to raccoons and other small prey. Feral swine and deer numbers have decreased dramatically in some areas of this ecosystem (Garrison et al. 2011; FWC 2017) while raccoon predation by panthers has increased over the past two decades (Caudill et al. 2019).

Hg concentrations varied by gender in panthers with males having higher levels in hair than females (Brandon 2011). Differences in food habits, however, between males and females were not observed (Caudill et al. 2019). Brandon (2011) suggested that excretion of Hg through avenues not available to males (e.g. parturition and lactation) could explain the difference. Indeed, concentrations in hair up to 40 ppm Hg have been observed in some neonatal panther kittens. Although significant differences in Hg concentration were not generally observed among age groups (possibly due to small sample size) (Brandon 2011), subadult panthers predominately consumed raccoons and thus would be expected to have higher Hg burdens (Caudill et al. 2019).

Despite the overall decline in Hg seen in wildlife in the Everglades ecosystem, individual panthers continue to have high Hg burdens—especially in ENP and the Conservation areas to the north. Although the sample size for this region is small, concentrations up to 100 ppm were detected in panthers from ENP in 2004–2014. Elevated levels—up to 67 ppm in hair—also were detected in southern Big Cypress National Preserve in this time period (FWC, unpubl. data).

8.5.4 Conclusions

White-tailed deer are the preferred prey for Florida panthers. A decrease in feral swine and deer in some areas of panther range may be causing a shift in diet to prey more closely associated with the aquatic food chain. Further, higher water levels and longer hydroperiods associated with the restoration of historic water flow in the Everglades may further decrease ungulate populations and contribute to a greater dependence on the aquatic food chain. This shift away from ungulates will serve not only to increase the risk of Hg toxicity but also results in exposure to other contaminants (polychlorinated biphenyls and organochlorines [aquatic food chain], and rodenticides [rodents, other carnivores]). Increased predation on raccoons also increases the risk of rabies infection since this species is the primary reservoir for this virus in the southeastern U.S.. Management to increase deer numbers apparently reduced Hg exposure in the Fakahatchee Strand State Preserve (Roelke et al. 1991). An increase in deer numbers would not only benefit panther nutrition (Schortemeyer et al. 1991) but would also reduce exposure to environmental contaminants and infectious diseases. Without these management changes, Hg will continue, or indeed may become a larger factor, as a stressor on panther health and reproduction.

8.6 Discussion

We have here summarized studies of food habits and Hg exposure for four groups of vertebrates that are important regarding community structure and function in the Everglades—mosquitofish, Florida Bass, long legged wading birds, and Florida Panthers.

Hg accumulation in each of these species is strongly affected by diet. Dietary characteristics can in turn be driven strongly by community composition, which can be affected in this aquatic system by antecedent hydrology (hydroperiod, depth, drying patterns). Hg content of the same prey items can also vary over space, and can be driven by geographic differences in Hg deposition and methylation, the latter of which is also affected by hydroperiod and water chemistry. Each of the ultimate drivers of Hg availability for these species (hydrology, deposition, water chemistry) are factors that can be affected by water management and contamination rulemaking.

While a solid basis of understanding exists for the connection between food habits, exposure and accumulation in these species, there is a lack of understanding about how exposure affects top predators, the prey animals upon which they depend, and their interactions. The ability to predict population-level effects in wild species is a current frontier of ecotoxicology in general, but the Everglades is one of the few places in which studies like these can link land and water management to risk. The next step is clearly to better understand cumulative effects in free-ranging populations.

References

Abbey-Lee RN, Gaiser EE, Trexler JC (2013) Relative roles of dispersal dynamics and competition in determining the isotopic niche breadth of a wetland fish. Freshw Biol 58:780–792

Adams SM, Kimmel BL, Plosky GR (1983) Sources of organic matter for reservoir fish production: a trophic-dynamics analysis. Can J Fish Aquat Sci 40:1480–1495

Atkeson T, Axelrad D, Pollman C, Keeler G (2003) Integrating atmospheric Hg deposition and aquatic cycling in the Florida Everglades, an approach for conducting a Total Maximum Daily Load analysis for an atmospherically derived pollutant. Final Report, Florida Department of Environmental Protection, Tallahassee, FL

Axelrad D, Atkeson T, Lange T, Pollman C, Gilmour C, Orem W, Mendelssohn I, Frederick PF, Krabbenhoft D, Aiken G, Rumbold D, Scheidt D, Kalla P (2007) Chapter 3B: Hg monitoring, research and environmental assessment in South Florida. In: South Florida Environment Report. South Florida Water Management District, West Palm Beach, FL. http://my.sfwmd.gov/portal/

page/portal/pg_grp_sfwmd_sfer/portlet_prevreport/volume1/chapters/v1_ch_3b.pdf. Accessed 9 Jan 2018

Bailey RM, Hubbs CL (1949) The Black Basses (Micropterus) of Florida, with description of a new species. Occasional papers of the Museum of Zoology, No. 516, University of Michigan, Ann Arbor, 42 pp

Barthell BL, Lutz-Carrillo DJ, Norberg KE, Porak WF, Tringali MD, Kassler TW, Johnson WE, Readel AW, Krause RA, Philipp DP (2010) Genetic relationships among populations of Florida Bass. Trans Am Fish Soc 139:1651–1641

Bates AL, Orem WH, Harvey JW, Spiker EC (2002) Tracing sources of sulfur in the Florida Everglades. J Environ Qual 31:287–299

Batzer DP (1998) Trophic interactions among detritus, benthic midges, and predatory fish in a freshwater marsh. Ecology 79:1688–1698

Belicka LL, Sokol ER, Hoch JM, Jaffé R, Trexler JC (2012) A molecular and stable isotopic approach to investigate algal and detrital energy pathways in a freshwater marsh. Wetlands 32:531–542

Bemis BE, Kendall C (2004) Isotopic views of food web structure in the Florida Everglades. US Geological Survey Fact Sheet FS 2004-3138. 4 pp

Bemis BE, Kendall C, Wankel SD, Lange T, Krabbenhoft DP (2003) Using nitrogen and carbon isotopes to explain mercury variability in largemouth bass. Greater Everglades Science Program: 2002 Biennial Report, US Geological Survey, Palm Harbor. FL Open-File Report 03-54

Boucek RE, Rehage JS (2013) No free lunch: displaced marsh consumers regulate a prey subsidy to an estuarine consumer. Oikos 122:1453–1464. https://doi.org/10.1111/j.1600-0706.2013.20994.x

Boyle RA, Dorn NJ, Cook MI (2012) Nestling diet of three sympatrically nesting wading bird species in the Florida Everglades. Waterbirds 35:154–159

Boyle RA, Dorn NJ, Cook MI (2014) Importance of crayfish prey to nesting White Ibis (*Eudocimus albus*). Waterbirds 37:19–29

Brandon AL (2011) Spatial and temporal trends in mercury concentrations in the blood and hair of Florida Panthers (*Puma concolor coryi*). Unpublished Master's Thesis, Florida Gulf Coast University

Cailteux RL, Porak WF, Crawford S, Connor LL (1996) Differences in largemouth bass food habits and growth in vegetated and unvegetated north-central Florida lakes. Proc Annu Conf SEAFWA 50(1996):201–211

Carlson JE, Duever MJ (1977) Seasonal fish population fluctuations in a south Florida swamp. Proc Annu Conf SEAFWA 31:603–611

Caudill G, Onorato DP, Cunningham MW, Caudill D, Leone EH, Smith LM, Jansen D (2019) Temporal trends in Florida panther food habits. Hum-Wildl Interact 13:87–97

Chew RL (1975) The Florida largemouth bass. In: Stroud RH, Clepper H (eds) Black bass biology and management. Washington, DC, Sport Fishing Institute, pp 450–458

Chick JH, Ruetz CR III, Trexler JC (2004) Spatial scale and abundance patterns of arge fish communities in freshwater marshes of the Florida Everglades. Wetlands 24(3):652–664

Cleckner LB, Garrison PJ, Hurley JP, Olson ML, Krabbenhoft DB (1999) Trophic transfer of methyl mercury in the northern Florida Everglades. Biogeochemistry 40:347–361

Compeau GC, Bartha R (1985) Sulfate-reducing bacteria: principal methylators of mercury in anoxic estuarine sediment. Appl Microbiol 50:498–502

Crawford S, Porak WF, Renfro DJ, Cailteux RL (2002) Characteristics of trophy Largemouth Bass populations in Florida. In: Philipp DP, Ridgway MS (eds) Black bass: ecology, conservation, and management. American Fisheries Society, Symposium 31, Bethesda, MD, pp 567–582

Cristol DA, Brasso RL, Condon AM, Fovargue RE, Friedman SL, Hallinger KK, Monroe AP, White AE (2008) The movement of aquatic mercury through terrestrial food webs. Science 320:335

Cumbie PM, Jenkins JH (1974) Mercury accumulation in native mammals of the Southeast. Proceedings of the Annual Conference of the Southeastern Association of Game and Fish Commissioners

Cunningham MW (2012) Geographic distribution of environmental contaminants in the Florida panther. Gainesville, FL, Florida Fish and Wildlife Conservation Commission: 42

Cunningham MW, Brown MA et al (2008) Epizootiology and management of feline leukemia virus in the Florida panther. J Wildl Dis 44(3):537–552

Dalrymple GH, Bass OL (1996) The diet of the Florida panther in Everglades National Park, Florida. Bull Fla Mus Nat Hist 39:251–254

DeAngelis DL, Loftus WF, Trexler JC, Ulanowicz RE (1997) Modeling fish dynamics and effects of stress in a hydrologically pulsed ecosystem. J Aquat Ecosyst Stress Recover 6:1–13

DEP (Florida Department of Environmental Protection) (2013) Mercury TMDL for the State of Florida. Final Report, Tallahassee, FL, 104 pp

Dobrowolska A, Melosik M (2002) Mercury contents in liver and kidneys of wild boar (*Sus scrofa*) and red deer (*Cervus elaphus*). Z Jagdwiss 48(1):156–160

Dorn NJ, Cook MI, Herring G, Boyle RA, Nelson J, Gawlik DE (2011) Aquatic prey switching and urban foraging by the White Ibis *Eudocimus albus* are determined by wetland hydrological conditions. Ibis 153:323–335

Dutka-Gianelli J, Taylor R, Nagid E, Whittington J, Johnson K, Strong W, Tuten T, Trotter A, Marsh S, Young J, Berry A, Yeiser B, Nault K (2011) Habitat utilization and resource partitioning of apex predators in coastal Rivers of southeast Florida. Florida Fish and Wildlife Conservation Commission, Final Report, St Petersburg

Federal Register (1967) Native fish and wildlife: endangered species. Washington, DC, Federal Register-Department of the Interior-Fish and Wildlife Service: 4001

Forrester DJ (1992) Parasites and diseases of wild mammals in Florida. University Press of Florida, Gainesville, FL

Frederick PC, Spalding MG, Sepulveda MS, Williams GE Jr, Nico L, Robbins R (1999) Exposure of Great Egret nestlings to mercury through diet in the Everglades of Florida. Environ Toxicol Chem 18:1940–1947

Frederick PC, Spalding MG et al (2002) Wading birds as bioindicators of mercury contamination in Florida, USA: annual and geographic variation. Environ Toxicol Chem 21(1):163–167

Fury JR, Wilkert JD, Cimbaro J, Morello F (1995) Everglades fisheries investigations 1992–1995. Florida Game and Freshwater Fish Commission. Federal Aid in Fish Restoration Project, Project F-56, Completion Report, Tallahassee, FL

Fury JR, Cimbaro JS, Morello F (1998) Everglades fisheries investigations 1995–1998. Florida Game and Freshwater Fish Commission, Federal Aid in Fish Restoration Project, Project F-56, Completion Report, Tallahassee, FL

FWC (2016) FWC wildlife management area harvest reports. https://myfwc.com/hunting/harvest-reports/. Accessed 13 May 2016

FWC (2017) Wildlife management area harvest reports. Create WMA harvest report by region across multiple years. http://myfwc.com/hunting/harvest-reports/region/. Accessed 9 Apr 2017

Gandy DA, Rehage JS (2017) Examining gradients in ecosystem novelty: fish assemblage structure in an invaded Everglades Canal system. Ecosphere 8(1):e01634. https://doi.org/10.1002/ecs2.1634

Garrison E, Leone EH et al (2011) Analysis of hydrological impacts on white-tailed deer in the stairsteps unit, big Cypress national preserve. Tallahassee, Florida, Florida fish and wildlife conservation commission: 21

Geddes P, Trexler JC (2003) Uncoupling of omnivore-mediated positive and negative effects on periphyton mats. Oecologia 136:585–595

Guggisberg CAW (1975) Wild cats of the world. Taplinger Publishing Company, New York, NY

Gunderson LH, Loftus WF (1993) The Everglades. In: Martin WH, Boyce SG, Echternacht AC (eds) Biodiversity of the Southeastern United States. Wiley, New York, pp 199–255

Haake PW, Dean JM (1983) Age and growth of four Everglades fishes using otolith techniques. Technical Report SFRC-83/03. Everglades National Park, Homestead, FL

Heath JA, Frederick PC, Kushlan JA, Bildstein KL (2009) White Ibis (*Eudocimus albus*). In: Poole A (ed). The birds of North America Online. Cornell Lab of Ornithology, Ithaca, NY. Retrieved from the Birds of North America Online: http://bna.birds.cornell.edu/bna/species/009

Heaton-Jones T, Homer B, Heaton-Jones D, Sundlof S (1997) Mercury distribution in American Alligators (Alligator mississippiensis) in Florida. J Zoo Wildl Med 28(1):62–70

Hunt BP (1953) Food relationships between Florida spotted gar and other organisms in the Tamiami Canal, Dade County, Florida. Trans Am Fish Soc 82(1):13–33

Jordan F, Arrington DA (2001) Weak trophic interactions between large predatory fishes and herpetofauna in the channelized Kissimmee River, Florida, USA. Wetlands 21(1):155–159

Julian P II, Gu B, Weaver K (2017) Chapter 3B: Mercury and sulfur environmental assessment for the Everglades. In: South Florida Environmental Report, South Florida Water Management District, West Palm Beach, FL. http://apps.sfwmd.gov/sfwmd/SFER/2017_sfer_final/v1/sfer_toc_v1.pdf. Accessed 1 Feb 2018

Kelly JF, Gawlik DE, Kieckbusch DK (2003) An updated account of wading bird foraging behavior. Wilson Bull 115:105–107

Kendall C, Bemis BE, Trexler J, Lange T, Stober JQ (2003) Is food web structure a main control on mercury concentrations in fish in the Everglades? Greater Everglades ecosystem restoration (GEER) meeting, April 2003, Palm Harbor, FL. Program and Abstracts

Klassen JA, Gawlik DE (2014) Wood stork prey composition at a coastal and interior colony in Everglades National Park. In: Cook MI, Kobza M (eds) South Florida Wading Bird Nesting Report, vol 20. South Florida Water Management District, West Palm Beach, Florida, pp 40–41

Klassen JA, Gawlik DE, Frederick PC (2016) Linking wading bird prey selection to number of nests in a food-limited population. J Wildl Manag 80:1450–1460

Kleinert SJ, Degurse PE (1972) Mercury levels in Wisconsin fish and wildlife. Wisconsin Department of Natural Resources Technical Bulletin 52

Krabbenhoft DP, Hurley JP, Aiken G, Gilmour C, Marvin-DiPasquale M, Orem WH, Harris R (1996) Mercury cycling in the Florida Everglades: a mechanistic field study. Verh Intenat Verein Limnol 27:1–4

Kushlan JA (1974) Ecology of the White Ibis in southern Florida: a regional study. PhD Dissertation, University of Miami, Miami, FL, 130 pp

Kushlan JA, Kushlan MG (1975) Food of the white Ibis in Southern Florida. Florida Field Nat 3:31–38

Lange TR (2006) Final report: everglades pig frog mercury study. Report to Florida department of environmental protection, Contract SP377. Prepared by Florida fish and wildlife conservation commission, Tallahassee, FL

Lange TR, Royals HE, Connor LL (1993) Influence of water chemistry on mercury concentration in Largemouth Bass from Florida lakes. Trans Am Fish Soc 122:74–84

Lange TR, Richard DA, Royals HE (1998) Trophic relationships of mercury bioaccumulation in fish from the Florida Everglades. Annual report to the Florida Department of Environmental Protection, Tallahassee, FL

Lange TR, Richard DA, Royals HE (1999) Trophic relationships of mercury bioaccumulation in fish from the Florida Everglades. Annual report to the Florida Department of Environmental Protection, Tallahassee, FL

Lange T, Richard D et al (2000) Long-term trends of mercury bioaccumulation in Florida's largemouth bass. Proceedings of the Annual Meeting South Florida Mercury Science Program, Tarpon Springs, FL

Lange TR Richard DA, Sargent BE (2003) Interactions of trophic position and habitat with mercury bioaccumulation in Florida Everglades Largemouth Bass. American Society of Limnology and Oceanography Aquatic Sciences Meeting, Salt Lake City, Utah, 8–14 February

Lange T, Richard D et al (2005) Annual fish mercury monitoring report, August 2005. In: Long-term monitoring of mercury in Largemouth Bass from the Everglades and Peninsular Florida. Florida Fish and Wildlife Conservation Commission, Eustis, FL

Lewis WM Jr (1970) Morphological adaptations of cyprinodontoids for inhabiting oxygen deficient waters. Copeia 1970:319–326
Light SS, Dineen JW (1994) Water control in the Everglades: a historical perspective. In: Davis SM, Ogden JC (eds) Everglades: the ecosystem and its restoration. St. Lucie Press, Delray Beach, FL, pp 47–84
Loftus WF (2000) Accumulation and fate of mercury in an Everglades aquatic food web. PhD dissertation, Florida International University, Miami, FL
Loftus WF, Bass OS Jr (1992) Mercury threatens wildlife resources and human health in Everglades NP. Park Sci 12:18–21
Loftus WF, Kushlan JA (1987) Freshwater fishes of southern Florida. Bull Florida State Mus Biol Sci 31(4):147–344
Loftus WF, Trexler JC, Jones RD (1998) Mercury transfer through an aquatic food web. Report submitted to Florida Department of Environmental Protection, Tallahassee, FL
Maceina MJ, Murphy BR (1989) Florida, northern, and hybrid Largemouth Bass feeding characteristics in Aquila Lake, TX. Proc Annu Conf SEAFWA 42(1988):112–119
Maehr DS, Belden RC et al (1990) Food-habits of panthers in Southwest Florida. J Wildl Manag 54 (3):420–423
Mason RP, Lawrence AL (1999) Concentration, distribution, and bioavailability of mercury and methylmercury in sediments of Baltimore Harbor and Chesapeake Bay, Maryland, USA. Environ Toxicol Chem 18:2438–2447
McBride R, McBride C (2010) Predation of a large alligator by a Florida panther. Southeast Nat 9 (4):854–856
Newman J, Zilloux E, Rich E, Liang L, Newman C (2004) Historical and other patterns of monomethyl and inorganic mercury in the Florida panther (*Puma concolor coryi*). Arch Environ Contam Toxicol 48(1):75–80
Nowak RM, McBride RT (1974) Status survey of the Florida panther. World Wildlife Fund Yearbook 1973–74, pp 237–242
Obaza A, DeAngelis DL, Trexler JC (2011) Using data from an encounter sampler to model fish dispersal. J Fish Biol 78:495–513
Ogden JC, Kushlan JA, Tilmant JT (1976) Prey selectivity by the Wood Stork. Condor 78:324–330
Onorato D, Belden C et al (2010) Long-term research on the Florida panther (*Puma concolor coryi*): historical findings and future obstacles to population persistence. In: Macdonald D, Loveridge A (eds) Biology and conservation of wild felids. Oxford University Press, Oxford, pp 453–469
Pagan XO (2000) Effects of water level and predation on survival of spotted sunfish larvae in the Florida Everglades. Unpublished MS thesis, Florida International University, Miami, 60 pp
Pickhardt PC, Stepanova M, Fisher NS (2006) Contrasting uptake routes and tissue distributions of inorganic and methylmercury in mosquitofish (*Gambusia affinis*) and Redear Sunfish (*Lepomis microlophus*). Environ Toxicol Chem 25:2132–2142
Porcella DB, Zillioux EJ et al (2004) Retrospective study of mercury in raccoons (*Procyon lotor*) in South Florida. Ecotoxicology 13(3):207–221
Post JR, Vanderbos R, McQueen DJ (1996) Uptake rates of food-chain and waterborne mercury by fish: field measurements, a mechanistic model, and an assessment of uncertainties. Can J Fish Aquat Sci 53:395–407
Rehage JS, Boucek RE (2013) Local and regional factors influencing fish communities within a sub-tropical estuary. Mar Sci 820:625–645
Rehage JS, Loftus WF (2007) Seasonal fish community variation in the upper stretches of mangrove creeks in the southwestern Everglades: the role of creeks as dry-down refuges. Bull Mar Sci 80:625–645
Rehage JS, Trexler JC (2006) Assessing the net effect of anthropogenic disturbance on aquatic communities in wetlands: community structure relative to distance from canals. Hydrobiologia 569(1):359–373

Roelke ME, Schultz DP et al (1991) Mercury contamination in Florida panthers. A report of the Florida Panther Technical Subcommittee to the Florida Panther Interagency Committee. Tallahassee, FL, Florida Game and Freshwater Fish Commission: 26

Roelke ME, Martenson JS et al (1993) The consequences of demographic reduction and genetic depletion in the endangered Florida panther. Curr Biol 3:340–349

Rumbold D, Niemczyk S et al (2001) Mercury in eggs and feathers of great egrets (Ardea albus) from the Florida Everglades. Arch Environ Contam Toxicol 41(4):501–507

Rumbold DG, Fink LE, Laine KA, Niemczyk SL, Chandrasekhar T, Wankel SD, Kendall C (2002) Levels of mercury in alligators (*Alligator mississippiensis*) collected along a transect through the Florida Everglades. Sci Total Environ 297:239–252

Rumbold DG, Lange TR, Axelrad TM, Atkeson TD (2008) Ecological risk of methylmercury in Everglades National Park, Florida, USA. Ecotoxicology 17:632–641

Sargeant BL, Gaiser EE, Trexler JC (2010) Biotic and abiotic determinants of intermediate-consumer trophic diversity in the Florida everglades. Mar Freshw Res 61:11–22

Savino JF, Stein RA (1982) Predator-prey interactions between largemouth bass and bluegills as influenced by simulated, submersed vegetation. Trans Am Fish Soc 111:255–266

Scheidt DJ, Kalla PI (2007) Everglades ecosystem assessment: water management and quality, eutrophication, mercury contamination, soils and habitat: monitoring for adaptive management: a R-EMAP status report. USEPA Region 4, Athens, GA. EPA 904-R-07-001. 98 pp

Schortemeyer JL, Maehr DS et al (1991) Prey management for the Florida panther: a unique role for wildlife managers. Trans North Am Wildl Nat Resour Conf 56:512–526

Seal US (1994) A plan for genetic restoration and management of the Florida panther (*Felis concolor coryi*), USA. Report to the Florida Game and Freshwater Fish Commission. Conservation Breeding Specialist Group, Apple Valley, MN, p 22

Smith JP (1997) Nesting season food habits of 4 species of herons and egrets at Lake Okeechobee, Florida. Waterbirds 20:198–220

Spalding MG, Frederick PC, McGill HC, Bouton SN, McDowell LR (2000) Methylmercury accumulation in tissues and its effects on growth and appetite in captive Great Egrets. J Wildl Dis 36:411–422

Stober QJ, Thornton K, Jones R, Richards J, Ivey C, Welch R, Madden M, Trexler J, Gaiser E, Scheidt D, Rathbun S (2001) South Florida ecosystem assessment: Phase I/II – Everglades stressor interactions: hydropatterns, eutrophication, habitat alteration, and mercury contamination. In: Monitoring for adaptive management: implications for ecosystem restoration. EPA 904-R-01-002, Athens, GA, 63 pp

Strong AM, Bancroft GT, Jewell SD (1997) Hydrological constraints on Tricolored Heron and Snowy Egret resource use. Condor 99:894–905

Summers GL (1981) Food of adult Largemouth Bass in a small impoundment with a dense aquatic vegetation. Proc Annu Conf SEAFWA 34(1980):130–136

Taylor RC, Trexler JC, Loftus WF (2001) Separating the effects of intra- and interspecific age-structured interactions in an experimental fish assemblage. Oecologia 127:143–152

Trexler JC, Loftus WF (2016) Invertebrates of the Florida everglades. In: Batzer D, Boix D (eds) Invertebrates in freshwater wetlands. Springer, Cham

Trexler JC, Loftus WF, Jordan F, Lorenz JJ, Chick JH, Kobza RM (2000) Empirical assessment of fish introductions in a subtropical wetland: an evaluation of contrasting views. Biol Invasions 2:265–277

Trexler JC, Loftus WF, Jordan CF, Chick J, Kandl KL, McElroy TC, Bass OL (2001) Ecological scale and its implications for freshwater fishes in the Florida Everglades. In: Porter JW, Porter KG (eds) The Everglades, Florida Bay, and coral reefs of the Florida Keys: an ecosystem sourcebook. CRC Press, Boca Raton, pp 153–181

Turner A, Trexler JC (1997) Sampling invertebrates from the Florida Everglades: a comparison of alternative methods. J N Am Benthol Soc 16:694–709

Turner AM, Trexler JC, Jordan CF, Slack SJ, Geddes P, Chick JH, Loftus WF (1999) Targeting ecosystem features for conservation: standing crops in the Florida Everglades. Conserv Biol 13:898–911

Wankel SD, Kendall C, McCormick P, Shuford R (2003) Effects of microhabitats on stable isotopic composition of biota in the Florida Everglades. Greater Everglades Science Program: 2002 Biennial Report, US Geological Survey, Palm Harbor, FL. Open-File Report 03-54

Ware FJ, Royals H et al (1990) Mercury contamination in Florida largemouth bass. Proc Annu Conf SEAFWA 44:5–12

Watras CJ, Back RC, Halvorsen S, Hudson RJM, Morrison KA, Wente SP (1998) Bioaccumulation of mercury in pelagic freshwater food webs. Sci Total Environ 219:183–208

Wicker AM, Johnson WE (1987) Relationships among fat content, condition factor, and first-year survival of Florida Largemouth Bass. Trans Am Fish Soc 116:264–271

Wiener JG, Spry DJ (1996) Toxicological significance of mercury in freshwater fish. In: Beyer WN, Heinz GH, Redmon-Norwood AW (eds) Environmental contaminants in wildlife: interpreting tissue concentrations. Special Publication Society for Environmental Toxicology and Chemistry, Boca Raton, FL, pp 297–339

Wiener JG, Bodaly RA, Brown SS, Lucotte M, Newman MC, Porcella DB, Reash RJ, Swain EP (2007) Monitoring and evaluating trends in methylmercury accumulation in aquatic biota. In: Harris R, Krabbenhoft DP, Mason R, Murray WM, Reash R, Saltman T (eds) Ecosystem responses to mercury contamination, indicators of change. Society for Environmental Toxicology and Chemistry, Pensacola, FL

Williams AJ, Trexler JC (2006) A preliminary analysis of the correlation of food-web characteristics with hydrology and nutrient gradients in the southern Everglades. Hydrobiologia 569:493–504

Winemiller KO (1990) Spatial and temporal variation in tropical fish trophic networks. Ecol Monogr 60:331–367

Wren CD (1986) A review of metal accumulation and toxicity in wild mammals. Environ Res 40 (1):210–244

Young SP, Goldman EA (1946) The puma, mysterious American cat. Part I. History, life habits, economic status, and control. The American Wildlife Institute, Washington, DC

Zokan M, Ellis G, Liston S, Lorenz J, Loftus WF (2015) The ichthyofauna of the Big Cypress National Preserve, Florida. Southeast Nat 14:517–550

Chapter 9
Comparison of Everglades Fish Tissue Mercury Concentrations to Those for Other Fresh Waters

Ted Lange

Abstract The purpose of this chapter is to provide information to resource managers about the distribution of Hg in fish within the Everglades ecosystem. To understand the causes of the variability and often elevated fish mercury levels in the Everglades and south Florida, the myriad environmental factors that influence Hg methylation and its bioavailability must be considered. The unique characteristics of the Everglades across both fresh and marine waters result in strong spatial gradients in fish Hg concentrations that vary by species that occupy different habitats and trophic regimes within the ecosystem. This chapter provides descriptions of the spatial and species distributions of Hg in fish and provides a comparison of Hg in Everglades predatory fish to level in fish from other areas of Florida and the United States.

Keywords Largemouth bass · Sport fish · Non-native fish · Mercury · Fish consumption · Spatial trends

9.1 Introduction

Fish stocks worldwide are impacted by mercury (Hg), which is one of the few metals that accumulate in food webs, reaching concentrations that are much higher in upper trophic level organisms than in primary producer and consumer organisms. Worldwide, fish Hg concentrations vary widely, and mean values differ as much as 100-fold (USEPA 1997). In Florida, about one-third of all freshwater fish tested for Hg exceed the 0.3 mg/kg USEPA (2001) methylmercury (MeHg) criterion for the protection of human health, leaving recreational and subsistence anglers at risk. Currently, in Florida, human exposure to Hg through consumption of fish is managed nearly exclusively through consumption advisories covering 31 freshwater and 60 marine species as well as other game species within the Everglades (FLDOH

T. Lange (✉)
Florida Fish and Wildlife Conservation Commission, Eustis, FL, USA
e-mail: ted.lange@myfwc.com

© Springer Nature Switzerland AG 2019

D. G. Rumbold et al. (eds.), *Mercury and the Everglades. A Synthesis and Model for Complex Ecosystem Restoration*, https://doi.org/10.1007/978-3-030-32057-7_9

2019). Because of variations in Hg among species of fish and locations, consumption advisories may not be fully protective for recreational and subsistence anglers who follow consumption recommendations (see Chap. 11, this volume). The state of Florida completed a statewide Total Maximum Daily Load (TMDL) assessment for mercury as a pollutant in 2012, under Section 303(d)(1)(C) of the US Clean Water Act, estimating that an 86% reduction of atmospheric Hg deposition was needed to reduce fish Hg levels to safe levels for human consumption (FDEP 2013).

In Florida, atmospheric deposition is the source of nearly all Hg in the environment, with most originating from sources outside the state (see Chap. 5, Volume 1). Although industrial point sources of Hg are known to contaminate fish, atmospheric deposition and subsequent bioaccumulation of Hg is largely responsible for the widespread distribution of Hg in fish in the Everglades and elsewhere (Krabbenhoft and Sunderland 2013; see also, Chap. 4, Volume 1). Accumulation of MeHg in upper trophic level fish is controlled by the rate of its formation in aquatic environments as well the physical and chemical characteristics of the waterbody and by the characteristics of food webs (for review of biomagnification, see Chap. 7, this volume). Consequently, fish Hg levels vary by species and, within species, among locations, with high concentrations commonly reported from aquatic systems with no known anthropogenic point source or geologic sources (Wiener et al. 2006). In the Everglades and to the south in Florida Bay, atmospheric deposition of Hg, at rates higher than the rest of the country (see Chap. 4, Volume 1), continue to sustain high rates of MeHg bioaccumulation in Everglades fish and wildlife.

Elevated Hg in fish and wildlife is commonly associated with dark-water and low alkalinity lakes and streams (Grieb et al. 1990; Spry and Wiener 1991; Lange et al. 1993), newly flooded reservoirs (for review see, Bodaly et al. 1997), and wetlands or wetland-influenced ecosystems such as the Everglades (Snodgrass et al. 1999; Naimo et al. 2000; Cleckner et al. 1998; Ware et al. 1990). A listing of unique characteristics of the Everglades that are reviewed elsewhere in this volume (i.e. shallow slow-moving wetlands, sub-tropical rainfall patterns, high water recession rates and a warm environment), result in predatory fish with some of the highest concentrations of Hg in the country and most exceed the USEPA criterion for the protection of human health (Lange et al. 1998). Surveys show elevated Hg in Everglades wading birds (Spalding and Forrester 1991), pig frogs (Lange 2006) alligators (Hord et al. 1990; Heaton-Jones et al. 1997) raccoons, and panthers (Roelke et al. 1991). Much like aquatic systems throughout the US, strong spatial gradients in bioaccumulation are evident (Stober et al. 2001). Elevated Hg concentrations in Everglades fish have been documented in both the freshwater reaches of the Everglades (Loftus 2000; Gabriel et al. 2010; Lange et al. 1999), and in downstream estuaries (Kannan et al. 1998; Evans and Crumley 2005; Rumbold et al. 2018) and the Gulf of Mexico (Adams et al. 2018).

Several significant synoptic or long-term monitoring programs have advanced our understanding of the spatial and species distributions of Hg in Everglades fish. These decade-long monitoring programs have served to assess both ecological and human health risks and to establish baseline conditions from which to assess Hg risks in a post-restoration Everglades. The Florida Fish and Wildlife Conservation Commission (FWC) conducted synoptic surveys of Hg levels in native and non-native sport fish commonly harvested by anglers to assess human health risks and temporal and spatial trends in fish Hg within the freshwater Everglades (as described in Lange et al. 1998, 1999). The South Florida Water Management District (SFWMD) conducted broad-based fish sampling mandated by the Everglades Forever Act (Florida Statutes Chapter 373.4592), which focused on the ecosystem, to assess fish Hg concentrations within the Everglades Protection Area (EvPA) and other waters associated with the Everglades Construction Project (South Florida Environmental Reports, 1999–2019, www.sfwmd.gov). Large-bodied fish collected by the SFWMD and FWC biomonitoring programs provide an integrated assessment of species distributions of THg in fish commonly harvested for consumption from both canal and wetland habits. These species represent a gradient in trophic levels and life history characteristics which result in varying Hg concentrations among species and habitats. The most commonly sampled species, Florida Largemouth Bass (bass) (for detailed discussion on bass, see Chap. 8, this volume), is a highly mobile top-level predator that has proven to be an ideal indicator of temporal and spatial trends in Hg bioaccumulation (Wiener et al. 2007).

Here we report on portions of the SFWMD and FWC programs intended to monitor Hg concentrations in axial muscle tissue (i.e. fillet) in fish species commonly consumed by recreational and sustenance anglers and their families. Bass data were accessed from the SFWMD data portal DBHYDRO (www.sfwmd.gov) and from the FWC biomonitoring program in support of fish consumption advisories (FDOH 2019). All other data is from FWC fish advisory monitoring reported by the Florida Department of Health (FDOH). We used data collected between 2000 and 2017. We report on Hg in axial muscle tissue for 5393 freshwater fish representing 9 native and 6 non-native species commonly harvested for consumption in the Everglades. Using these data, species differences and the areal extent of biomagnification of Hg in common native and non-native sport fish are contrasted. Bass were the most common (3428 fish) and widely collected species (49 locations) and are contrasted with results for bass from lakes and rivers in Florida and elsewhere in the United States.

9.2 Spatial Gradients in Hg Bioaccumulation

In a statewide synoptic survey of Hg in fish and wildlife species, Ware et al. (1990) found the highest concentrations of Hg in bass in south Florida, including the Water Conservation Areas (WCAs) and Everglades National Park (ENP) which together comprise the Everglades Protection Area (EvPA). Ogden et al. (1974) had previously noted higher and more widespread Hg levels in fish from freshwater than estuarine habitats within ENP. These findings of high levels of Hg in fish and wildlife in Florida, and the Everglades in particular, resulted in more focused Hg surveys in freshwater as well as estuarine habitats. Subsequent surveys in the Everglades identified a north to south increase in Hg bioaccumulation in fish across the Water Conservation Areas (WCAs) and south into the Shark River Slough (SRS) in the ENP (Lange et al. 1998; Loftus 2000; Julian et al. 2018). In part, this north to south gradient (Fig. 9.1) was driven by Hg hotspots—areas with high rates of MeHg bioaccumulation in fish—that were identified by Stober et al. (2001) in southwest WCA3 and further south in SRS in ENP. Stober found that Hg in small prey fish (mosquitofish) (see also, Chap. 8, this volume), was almost twofold higher in these areas than the Everglades basin-wide average (Stober et al. 1998). Similarly, Hg levels in top-predators, including bass, from these areas exceeded levels found in other areas of the Everglades (Gu et al. 2012; Julian et al. 2014; Rumbold et al. 2018). Subsequent surveys of Hg in piscivorous wildlife from the SRS revealed high levels in wading birds, racoons, alligators, Florida panthers and Burmese pythons (for review see Rumbold et al. 2008; Axelrad et al. 2011) as well as in fish (Fig. 9.1). The status of SRS as a bioaccumulation hotspot has persisted, with consistently high levels of Hg in bass and other freshwater fish (Gabriel et al. 2011; Gu et al. 2012; see Fig. 9.2, bottom). Downstream estuarine waters of the SRS were also identified as a MeHg hotspot with Hg concentrations in Grey Snapper 1.7–2.5 times higher than

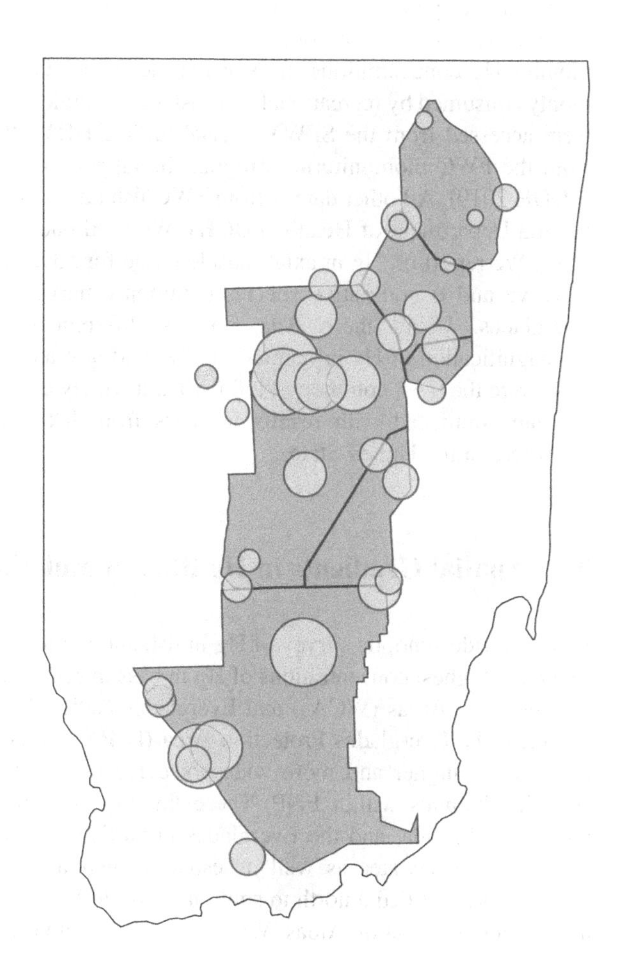

Fig. 9.1 Bubble plot of Largemouth Bass (bass) mercury (THg) concentrations from the Everglades that were collected between 2000 and 2017. THg concentrations were normalized to a 12-in. bass at each site to represent a size common for harvest. Bubble size is proportional to concentration of THg in bass at each site which ranged from 0.256 to 1.016 mg/kg. Of 49 available sites, 35 were included that met the minimum requirement of 12 bass

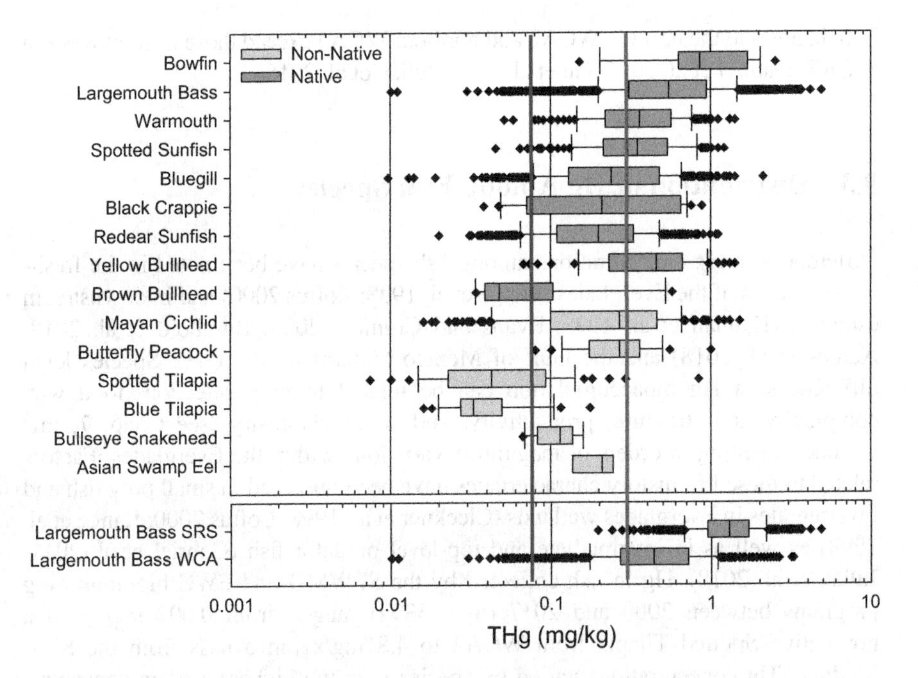

Fig. 9.2 Fish tissue mercury (THg) concentrations (mg/kg) for species collected from the Everglades between 2000 and 2017. THg is reported for individual fish axial muscle tissue except mosquitofish which were composites of 100 whole fish. Top Panel: fish collected from the Everglades Protection Area (EvPA). Bottom Panel: Largemouth bass from the Shark River Slough (SRS) in the Everglades National Park (ENP) and from the Water Conservation Areas (WCA). Box plots represent the median, 25th and 75th percentiles, error bars the 10th and 90th percentiles and points are outliers. Red line is the USEPA MeHg criterion for human health (USEPA 2001) and blue line is the USEPA MeHg predator protection criterion as regards trophic level 3 fish (USEPA 1997)

other south Florida estuaries and offshore waters (Adams et al. 2018), yet Rumbold et al. (2018) noted down-estuary declines in Hg concentrations in predatory fish (Common Snook, Grey Snapper) along a salinity gradient in SRS. They attributed these declines to relatively lower concentrations of Hg in down estuary primary consumers (i.e. less availability of MeHg for bioaccumulation), but changes in predator trophic position down-estuary may also cause the declines in downstream predator Hg levels (Farmer et al. 2010).

To the north in the WCAs, where R-EMAP surveys by USEPA in 1995 and 1996 also identified a bioaccumulation hotspot in southwestern WCA3 (Stober et al. 2001), more current findings suggest a more dynamic situation where MeHg hotspots have transitioned over time. Stober identified this hotspot based on mosquitofish Hg levels twofold higher than surrounding areas and similar spatial trends were observed in sunfish and bass (Lange et al. 1998; Gabriel et al. 2011). However, over time, the spatial patterns of fish Hg across all trophic levels has changed within the WCAs and now fish

THg in areas to the north in WCA2 and northern WCA3 exceed those in southwestern WCA3 (Gabriel et al. 2011; Gu et al. 2012; Julian et al. 2018).

9.3 Distribution of Hg Among Fish Species

Differences in Hg concentrations among fish species have been noted in the freshwater reaches of the Everglades (Lange et al. 1998; Loftus 2000) and in downstream estuaries (Kannan et al. 1998; Evans and Crumley 2005; Rumbold et al. 2018; Adams et al. 2018) and the Gulf of Mexico (Adams et al. 2003). Species level differences in Hg bioaccumulation can be related to prey selection, food web complexity and structure, productivity, and water chemistry (see Chap. 7, this volume) resulting in orders of magnitude variations within the Everglades. Factors related to these life-history characteristics have been observed in small prey fish and invertebrates in Everglades wetlands (Cleckner et al. 1998; Loftus 2000; Lange et al. 1998) as well as in intermediate and top-level predator fish (Gabriel et al. 2011; Julian et al. 2018). Hg in fish collected by the SFWMD and FWC biomonitoring programs between 2000 and 2017 ($n = 5394$) ranged from 0.007 mg/kg in a non-native Spotted Tilapia from WCA2 to 4.8 mg/kg in a bass from the SRS. Median THg concentrations varied by species with the highest median concentrations in longer lived piscivorous species (Bowfin, bass, Yellow Bullhead) (Fig. 9.2 top; for more information on bass food webs, see Chap. 8, this volume). Of nine native species, five had median THg concentrations in excess of the 0.3 mg/kg USEPA (2001) MeHg human health criterion. In contrast, the lowest THg concentrations were found in non-native Blue Tilapia and Spotted Tilapia; they are primarily herbivores, feeding on plankton, macrophytes and aquatic invertebrates.

Similarly, Lange et al. (1999) found that Brown Bullhead in the WCAs fed predominately on crustaceans and insects, resulting in a low trophic position and low concentration (0.05 mg/kg) of Hg (Fig. 9.2, top). In contrast, the closely related Yellow Bullhead fed predominantly on fish and crustaceans resulting in a median Hg concentration of 0.48 mg/kg. Concentrations of Hg in sunfish increased with the degree of piscivory by species (Lange et al. 1999), including Redear Sunfish (0.22 mg/kg), Bluegill (0.27 mg/kg), Warmouth (0.37 mg/kg) and Spotted Sunfish (0.40 mg/kg) (Fig. 9.2, top). Julian et al. (2018) reported similar species relationships in whole sunfish THg concentrations in the Everglades Protection Area (EvPA) for the period 1998–2017 and attributed these differences to trophic position. Sunfish are an important part of the diet in bass (see Chap. 8, this volume) and 87% of all sunfish exceeded the 0.077 mg/kg USEPA (1997) MeHg criterion for the protection of fish-eating wildlife (Fig. 9.2, top).

There were ten species of fish commonly harvested by anglers with median THg concentrations less than the 0.3 mg/kg USEPA (2001) MeHg criterion for the protection of human health including six non-native and four native species (Fig. 9.2, top). These species generally feed at a lower trophic level than those with higher Hg concentrations. In contrast, species with a more highly piscivorous diet are Bowfin, bass, Warmouth, and Yellow Bullhead catfish (Loftus 2000; Lange et al. 1998) which results in the highest overall Hg concentrations in Everglades fish (Fig. 9.2, top). As a long-lived top-level predator, bass exhibit high THg concentrations across the Everglades. Although differences in bass THg were observed between regions, 79% of WCA bass and nearly 99% of SRS bass exceeded the 0.3 mg/kg USEPA (2001) human health criterion (Fig. 9.2, bottom). Because of their widespread distribution in the Everglades and the United States, popularity as a sport-fish, and role as a top predator, bass are a useful tool to contrast Hg bioaccumulation in the Everglades with other areas. Because of their long-lifespan and piscivorous feeding habits, bass increase in Hg concentration with age and size (Lange et al. 1994).

9.4 Hg Distributions in Everglades Fish vs. US Fish

To contrast bass THg concentrations among Florida waters and regions of the US, we selected harvestable-size bass greater than 10″ (254 mm) in length. Bass THg concentrations were contrasted with similar-sized bass collected from Florida lakes and rivers as part of the Florida Mercury Total Maximum Daily Load (TMDL) for Hg (FDEP 2013) and Florida Department of Health (FDOH 2019) human health risk assessment program (FWC unpublished data). Florida data were then contrasted with fish Hg data from the National Study of Chemical Residues in Lake Fish Tissue (USEPA 2009) where THg in composite samples of Black Bass, *Micropterus* spp., were assessed from 301 lakes between 2000 and 2003. Results were also contrasted with Hg concentrations for Black Bass (*Micropterus* spp.) collected between 1998 and 2005 from 149 streams and rivers across the conterminous United States and reported as both composited concentrations and as individual means (Scudder et al. 2009).

Cumulative frequency distributions of Hg in bass (Fig. 9.3) were constructed individually for each of the four surveys listed above and our Everglades data set using site average or composite values (US rivers only). We further report the 50th and 90th percentiles of Hg concentration, maximum concentrations, and the number of sites with means higher than the 0.3 mg/kg USEPA (2001) human health criterion (Table 9.1). We tested for distributional differences individually between the Everglades data set and each of the other data sets using non-parametric two-way Kolmogorov-Smirnov (KS) tests (alpha (p) $\leq$ 0.05) in R v3.3.3 (R Development Team 2017). All comparisons used data for Black Bass (*Micropterus* spp.) including predominantly *M. salmoides* but also *M. coosae, M. dolomieui, M. notius*, and *M. punctatus* as no one species of bass could be collected representatively across the US (Table 9.1). All Florida locations were removed from the national databases prior to assessment.

The median (50%) Hg concentrations for each database ranged from 0.27 in US rivers to 0.67 mg/kg in the EvPA (Table 9.1). The highest individual site concentration was observed in the US lakes dataset (USEPA 2009), which is consistent with

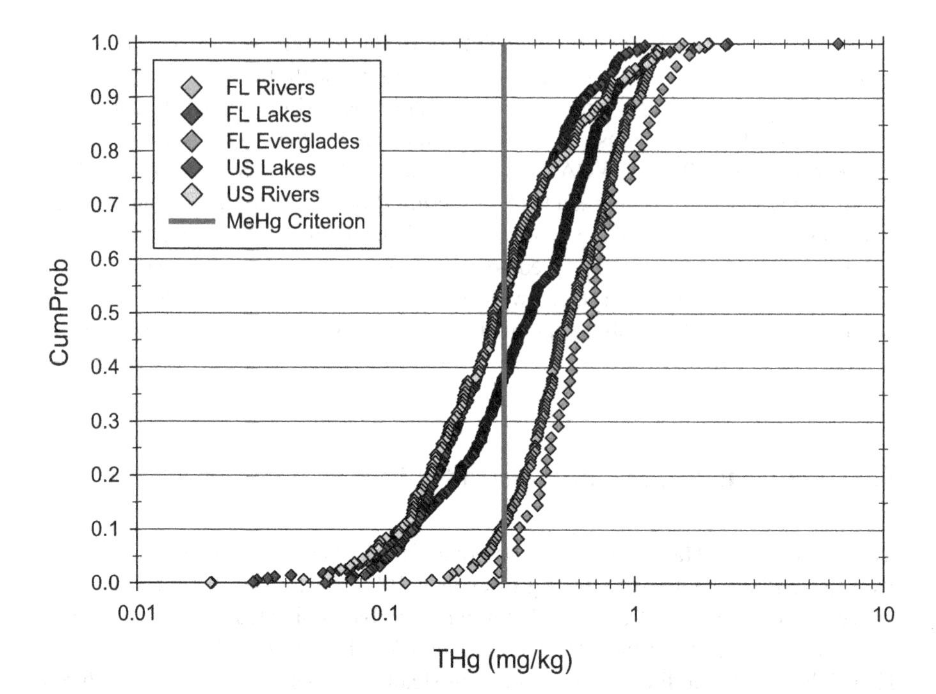

Fig. 9.3 Cumulative probability distributions of bass mean THg concentrations from Florida waters between 2000 and 2017 and national sampling programs (see Table 9.1 for data descriptions)

lakes having a wider range of biomagnification potential based on variations in trophic states across the US as compared to rivers and the EvPA. The EvPA, which includes both the compartmentalized Water Conservation Areas and Everglades National Park had the highest overall distribution of mean bass THg concentrations with 96% of all sites higher than the 0.3 mg/kg human health criterion. Within the EvPA, bass from SRS had higher overall THg concentrations than those from the WCAs (Fig. 9.2, bottom). There was no significant difference between the distribution of means in the EvPA and FL rivers (Fig. 9.3) where 89% of all riverine sites exceeded the 0.3 mg/kg criterion (KS: D $=$ 0.217, $p = 0.085$; D represents the difference in cumulative distribution value corresponding to where the maximum difference in distribution occurs for any given concentration). However, both the EvPA and FL rivers were significantly higher than FL lakes (KS; D $=$ 0.444, $p < 0.001$ and D $=$ 0.294, $p < 0.001$, respectively). Assessments of bass THg concentrations across US waters include both lakes (n $=$ 301) and rivers (n $=$ 149) where nearly 50% of all sites exceeded the 0.3 mg/kg human health criterion (Table 9.1). Although the upper limit of bass concentrations observed in US lakes exceed that in all Florida waters included in this analysis, the overall distribution of mean concentrations was nonetheless clearly and significantly higher in the EvPA than US lakes (KS; D $=$ 0.56, $p < 0.001$). Conversely, the upper limit on the distribution of bass concentrations in US Rivers was substantially lower than each

Table 9.1 Summary descriptors of cumulative frequency distributions of Black Bass THg concentrations from United States rivers and lakes, Florida rivers and lakes, and the Everglades Protection Area (EvPA)

Area	Species	Sample type	Sites	THg (mg/kg) 50th percentile	90th percentile	Sites > 0.3	Maximum	Study period	Source
US Lakes	*Micropterus spp*	C	301	0.29	0.63	49%	6.61	2000–2003	USEPA (2009)
US Rivers	*Micropterus spp*	C and M	149	0.27	0.77	47%	1.95	1998–2005	Scudder et al. (2009)
FL Lakes	*Micropterus spp*	M	341	0.38	0.80	62%	5.20	2000–2016	FWC/FLDEP (2013)
FL Rivers	*Micropterus spp*	M	163	0.54	0.96	89%	2.50	2000–2016	FWC/FLDEP (2013)
FL Everglades	*M. salmoides floridanus*	M	49	0.67	1.25	96%	4.80	2000–2017	FWC/DBHYDRO

C composite samples, *M* mean of individuals

of the three Florida ecosystems. Not surprisingly, the KS test indicates that bass concentrations in the EvPA were significantly higher compared to US rivers ($D = 0.61, p < 0.001$; Fig. 9.3) as well.

9.5 Conclusions

Using Hg results for commonly encountered sport and non-native fish found in the Everglades, this assessment provided an overview of both the spatial and species gradients in Hg bioaccumulation. Because of the complexity of these relationships, efforts to communicate risks to those who consume Everglades fish (i.e. recreational and subsistence anglers) is challenging.

Risks posed by consumption of Everglades fish is similar to that of eating fish from Florida rivers but elevated over fish from many Florida lakes and from lakes and rivers of the rest of the United States. Although Hg concentrations decline in predators down-estuary as salinity increases, risks to consumers remain, with 60 species of marine fish—many estuarine dependent—under advisory. Species selection by anglers can profoundly influence risk, as generally more desirable predatory species typically have more Hg. The charismatic and popular Largemouth Bass represents a worst-case scenario for those who consume Everglades fish yet there are some areas in the Everglades where consumption of bass poses low risk. Non-native species tend to have lower Hg concentrations than native species; however, non-natives generally have more limited ranges and are more commonly associated with canal habitats where they are more accessible to anglers. Localized areas of high Hg bioaccumulation (i.e. hotspots) pose the highest risks to consumers as do larger, long-lived predator fish.

Hydrologic restoration of the Everglades will result in changes in the environmental factors related to Hg bioaccumulation and fish distributions. Decompartmentalization of WCAs and restoration of hydroperiods in highly drained areas (e.g. northern WCA3) could have the most significant impact by changing the size and distribution of existing Hg hotspots. For example, changes in hydrology likely will result in changes in the dynamics of dissolved organic carbon, which in turn are expected to influence Hg methylation dynamics (Chaps. 4 and 6, this volume). Changes in the areal extent of existing hot spots could profoundly impact bioaccumulation in fish and wildlife. Risks to those who consume fish from the Everglades remain high and communication of risk is complicated by a high degree of spatial and species level variations in Hg.

References

Adams DH, McMichael RH Jr, Henderson GE (2003) Mercury levels in marine and estuarine fishes of Florida 1989–2001. Florida Marine Research Institute Technical Report TR-9, 2nd edn. St Petersburg, FL, p 57

Adams DH, Tremain DM, Evans DW (2018) Large-scale assessment of mercury in sentinel estuarine fishes of the Florida Everglades and adjacent coastal ecosystems. Bull Mar Sci 94 (4):1413–1427

Axelrad DM, Lange T, Gabriel MC (2011) Chapter 3B: Mercury and sulfur monitoring, research and environmental assessment for the Florida Everglades. In: 2011 Everglades Consolidated Report. South Florida Water Management District, West Palm Beach, FL, p 53. http://www.sfwmd.gov/sfer/. Accessed 19 Jul 2012

Bodaly RA, St Louis VL, Paterson MJ, Fudge RJP, Hall BD, Rosenberg DM, Rudd JWM (1997) Bioaccumulation of mercury in the aquatic food chain in newly flooded areas. In: Sigel A, Sigel H (eds) Mercury and its effects on environment biology, Metal ions in biological systems, vol 34. Marcel Dekker, New York, pp 259–287

Cleckner LB, Garrison PJ, Hurley JP, Olson ML, Krabbenhoft DP (1998) Trophic transfer of methylmercury in the northern Florida Everglades. Biogeochemistry 40:347–361

Evans DE, Crumley PH (2005) Mercury in Florida Bay fish: spatial distribution of elevated concentrations and possible linkages to Everglades restoration. Bull Mar Sci 77(3):321–345

Farmer TM, Wright RA, DeVries DR (2010) Mercury concentration in two estuarine fish populations across a seasonal salinity gradient. Trans Am Fish Soc 139:1896–1912

FDEP (Florida Department of Environmental Protection) (2013) Mercury TMDL for the state of Florida. Final Report. Watershed evaluation and TMDL section. Tallahassee, FL, p 104. http://www.sfwmd.gov/sfer/. Accessed 17 Apr 2014

FDOH (Florida Department of Health) (2019) your guide to eating fish caught in Florida. http://www.floridahealth.gov/. Accessed 12 Feb 2019

Gabriel MC, Axelrad DM, Lange T, Dirk, L (2010) Chapter 3B: Mercury and sulfur monitoring, research and environmental assessment in South Florida. In: 2010 South Florida Environmental Report. South Florida Water Management District, West Palm Beach, FL, p 48. http://www.sfwmd.gov/sfer/. Accessed 4 Jun 2010

Gabriel MC, Howard N, Atkins S (2011) Appendix 3B-1: Annual permit compliance monitoring report for mercury in downstream receiving waters of the Everglades protection area. In: 2011 South Florida Environmental Report. South Florida Water Management District, West Palm Beach, FL, p 34. http://www.sfwmd.gov/sfer/. Accessed 15 Apr 2012

Grieb TM, Driscoll CT, Gloss SP, Schofield CL, Bowie GL, Porcella DB (1990) Factors affecting mercury accumulation in fish in the upper Michigan peninsula. Environ Toxicol Chem 9:919–930

Gu B, Axelrad DM, Lange T (2012) Chapter 3B: Regional mercury and sulfur monitoring and environmental assessment. In: 2012 Everglades Consolidated Report. South Florida Water Management District, West Palm Beach, FL, p 42. http://www.sfwmd.gov/sfer/. Accessed 8 Feb 2012

Heaton-Jones TG, Homer BI, Heaton-Jones DL, Sundlof SF (1997) Mercury distributions in American alligators (Alligator mississippiensis) in Florida. J Zoo Wildlife Med 28:62–70

Hord LJ, Jennings M, Brunnell A (1990) Crocodiles, Proceedings of the 10th Working Meeting of the Crocodile Specialist Group of the Species Survival Commission of the World Conservation Union, Gainesville, FL, 23–27 April 1990, 1:229–240

Julian PJ, Gu B, Frydenborg R, Lange T, Wright AL, McCray JM (2014) Chapter 3B: Mercury and sulfur environmental assessment for the Everglades. In: 2014 Everglades Consolidated Report. South Florida Water Management District, West Palm Beach, FL, p 59. http://www.sfwmd.gov/sfer/. Accessed 31 Mar

Julian PJ, Gu B, Weaver K (2018) Chapter 3B: Regional mercury and sulfur environmental assessment for the Everglades. In: 2018 Everglades Consolidated Report. South Florida Water

Management District, West Palm Beach, FL, p 34. http://www.sfwmd.gov/sfer/. Accessed 30 Apr 2018

Kannan K, Smith RG Jr, Lee RF, Windom HL, Heitmuller PT, Macauley JM, Summers JK (1998) Distribution of total mercury and methylmercury in water, sediment, and fish from South Florida estuaries. Arch Environ Contam Toxicol 34:109–118

Krabbenhoft DP, Sunderland EM (2013) Global change in mercury. Science 341:1457–1458

Lange TR (2006) Final Report: Everglades pig frog mercury study. Report to Florida Department of Environmental Protection, Contract SP377. Prepared by Florida Fish and Wildlife Conservation Commission, Tallahassee, FL

Lange TR, Royals HE, Connor LL (1993) Influence of water chemistry on mercury concentration in largemouth bass from Florida lakes. Trans Am Fish Soc 122:74–81

Lange TR, Royals HE, Connor LL (1994) Mercury accumulation in largemouth bass (Micropterus salmoides) in a Florida lake. Arch Environ Conam Toxicol 27:466–471

Lange TR, Richard DA, Royals HE (1998) Trophic relationships of mercury bioaccumulation in fish from the Florida Everglades. Annual report to the Florida Department of Environmental Protection. Tallahassee, FL

Lange TR, Richard DA, Royals HE (1999). Trophic relationships of mercury bioaccumulation in fish from the Florida Everglades. Annual report to the Florida Department of Environmental Protection. Tallahassee, FL

Loftus WF (2000) Accumulation and fate of mercury in an Everglades aquatic food web. PhD dissertation, Florida International University, Miami, FL

Naimo TJ, Wiener JG, Cope WG, Bloom NS (2000) Bioavailability of sediment associated mercury to Hexagenia mayflies in a contaminated floodplain river. Can J Fish Aquat Sci 57:1092–1102

Ogden JC, Robertson WB Jr, Davis GE, Schmidt TW (1974) Pesticide, polychlorinated biphenyls and heavy metals in upper food chain levels, Everglades National Park and vicinity. National Park Service, Everglades National Park, Homestead, FL

R Development Team (2017) R: A language and environment for statistical computing. R Foundation for Statistical Computing, Vienna

Roelke M, Schultz D, Facemire C, Sundlof S, Royals H (1991) Mercury contamination in Florida panthers. Prepared by the Technical Subcommittee of the Florida Panther Interagency Committee

Rumbold DG, Lange TR, Axelrad DM, Atkeson TD (2008) Ecological risk of methylmercury in Everglades National Park, Florida, USA. Ecotoxicology 17:632–641

Rumbold DG, Lange TR, Richard D, DelPizzo G, Hass N (2018) Mercury biomagnification through food webs along a salinity gradient down-estuary from a biological hotspot. Estuar Coast Shelf Sci 200:116–125

Scudder BC, Chasar LC, Wentz DA, Bauch NJ, Brigham ME, Moran PW, Krabbenhoft DP (2009) Mercury in fish, bed sediment, and water from streams across the United States, 1998–2005: U.S. Geological Survey Scientific Investigations Report 2009–5109, p 74

Snodgrass JW, Jagoe CH, Bryan AL Jr, Brant HA, Burger J (1999) Effects of trophic status and wetland morphology, hydroperiod, and water chemistry on mercury concentrations in fish. Can J Fish Aquat Sci 57:171–180

Spalding MG, Forrester DJ (1991) Effects of parasitism and disease on the nesting success of colonial wading birds (Ciconiiformes) in southern Florida. Report to the Florida Game and Fresh Water Fish Commission. Report no. NG88-008. Tallahassee, FL

Spry DJ, Wiener JG (1991) Metal Bioavailability and toxicity to fish in low-alkalinity lakes: a critical review. Environ Pollut 71:243–304

Stober J, Scheidt D, Jones R, Thornton K, Gandy L, Stevens D, Trexler J, Rathbun S (1998) South Florida ecosystem assessment: monitoring for ecosystem restoration. Final Technical Report – Phase I. EPA 904-R-98-002. USEPA Region 4 Science and Ecosystem Support Division and Office of Research and Development, Athens, GA, p 285 plus appendices

Stober QJ, Thornton K, Jones R, Richards J, Ivey C, Welch R, Madden M, Trexler J, Gaiser E, Scheidt D, Rathbun S (2001). South Florida ecosystem assessment: Phase I/II – Everglades stressor interactions: hydropatterns, eutrophication, habitat alteration, and mercury contamination. In: Monitoring for adaptive management: implications for ecosystem restoration. EPA 904-R-01-002, Athens, GA, p 63

USEPA (1997) Mercury study report to Congress, vol VI: an ecological assessment for anthropogenic mercury emissions in the United States. U.S. Environmental Protection Agency, Washington, DC. EPA-452/R-97-008

USEPA (2001) Water quality criterion for the protection of human health: methylmercury. Washington, DC. EPA/823/R-01-001

USEPA (2009) The national study of chemical residues in lake fish tissue. EPA-823-R-09-006. U.S. Environmental Protection Agency, Office of Water, Washington, DC

Ware FJ, Royals H, Lange T (1990) Mercury contamination in Florida largemouth bass. Proc Annu Conf SEAFWA 44:5-12

Wiener JG, Knights BC, Sandheinrich MB, Jeremiason JD, Brigham ME, Engstrom DR et al (2006) Mercury in soils, lakes, and fish in Voyageurs National Park (Minnesota): importance of atmospheric deposition and ecosystem factors. Environ Sci Technol 40:6261–6268

Wiener JG, Bodaly RA, Brown SS, Lucotte M, Newman MC, Porcella DB et al (2007) Monitoring and evaluating trends in methylmercury accumulation in aquatic biota. In: Harris R, Krabbenhoft DB, Mason R, Murray MW, Reash R, Saltman T (eds) Ecosystem responses to mercury contamination: indicators of change. Society of Environmental Toxicology and Chemistry (SETAC) North America Workshop on Mercury Monitoring and Assessment, CRC Press, New York, pp 47–87

Chapter 10
Regional-Scale Ecological Risk Assessment of Mercury in the Everglades and South Florida

Darren G. Rumbold

Abstract The purpose of this chapter is to provide information to resource managers about current ecological risk from mercury exposure to fish and wildlife in the Everglades and the wider south Florida environment. The chapter begins with an overview of the history of previous ecological risk assessments in south Florida. Next, methods to assess toxicological effects and the difficulties in assessing exposure to fish and wildlife across this varied landscape are reviewed. Risks to south Florida wildlife are then characterized based on multiple lines of evidence for a variety of ecological receptors. The chapter closes with a discussion of potential future risk following the Comprehensive Everglades Restoration Plan.

Keywords Ecological risk · Toxicity · Fish · Wildlife · Mercury

10.1 Background

Ecological risk assessments (ERAs) are designed and conducted to provide information to risk managers about potential adverse effects (USEPA 1998). Numerous previous studies have assessed the ecological risk of mercury (Hg) in the Everglades since high levels of the toxic metal were first reported in fish and wildlife over 30 years ago (Ware et al. 1990; Table 10.1). Risk assessments are initiated for various reasons that can influence the process and the products. Previous risk assessments in the Everglades were initiated for one of two reasons: (1) exposure-initiated in response to the reports of high Hg levels or, (2) stressor-initiated to assess if various management actions were likely to worsen baseline risks.

One of the first ecological risk assessments focusing on Everglades biota was carried by the Technical Subcommittee working on the endangered Florida Panther, *Puma concolor coryi* (Roelke et al. 1991). The assessment was based on Hg concentrations ([Hg]) measured in hair and whole blood of free-ranging animals

D. G. Rumbold (✉)
Florida Gulf Coast University, Fort Myers, FL, USA
e-mail: drumbold@fgcu.edu

© Springer Nature Switzerland AG 2019

D. G. Rumbold et al. (eds.), *Mercury and the Everglades. A Synthesis and Model for Complex Ecosystem Restoration*, https://doi.org/10.1007/978-3-030-32057-7_10

Table 10.1 Select studies that have focused on Hg risks in the Everglades and south Florida

Source	Approach, ecological receptor, and findings
Roelke et al. (1991)	Based on examination of dead **panthers** concluded chronic exposure to Hg may be resulting in mortality and lower reproductive success
Jurczyk (1993)	Based on modeled daily dietary dose, estimated **great egrets** consumed 3.9× the reported LOAEL
PTI (1995)	Used model to predict Hg concentration in prey based on water-column phosphorus concentrations and estimated dietary dose to **wading birds**—predicted risks would increase substantially from efforts to reduce phosphorus
Sundlof et al. (1994)	Based on measured liver-[Hg] in 114 birds from 4 different **wading bird species**—found between 30% and 80% of the potential breeding-age birds exceed critical tissue TRV associated with reproductive impairment
Fleming et al. (1994)	Used a spatially explicit, individual based model to assess several hypotheses regarding declining **wood stork** populations including Hg impact; was not supported but was not a robust assessment of Hg risks
Frederick et al. (1997)	Based on dosing of captive **great egrets** and average concentration measured in food regurgitated from birds in the wild, great egrets at risk
Exponent (1998)	Updated PTI's model to predict Hg concentration in prey based on water-column phosphorus concentrations and estimated dietary dose to **wading birds**—predicted risks would increase substantially from efforts to reduce phosphorus
Bouton et al. (1999)	Based on dosing captive birds, with levels similar to that of exposure of **wild birds**, activity, food intake and willingness to hunt prey were reduced
Rumbold et al. (1999)	Based on modeled dietary dose, estimated HQs ranging from 2.4 to 5.2 for **wood stork, great blue heron, and great egrets** foraging at WCA-3A15
Duvall and Barron (2000)	Based on modeled daily dose, estimated probability of exceeding chronic risk thresholds ranged from 86% to 100% for **raccoons, alligators** and **great egrets** from south-central region
Spalding et al. (2000)	Based on dosing of captive birds, at levels similar to current exposure in Everglades, appetite and growth of **great egrets** negatively affected, putting them at risk
Rumbold (2000)	Based on a probabilistic model of dietary dose for **wood stork, great blue heron, and great egret**, as well as tissue residues measured in feathers and eggs, hazard quotients were, on average, less than 1 for birds feeding at the post-ECP reference site
Exponent (2000)	Modeled dietary dose, based on allometric regressions, for **great blue heron, wood stork** and **Everglades mink**, concluded that reductions in phosphorus concentrations in runoff in the northern Everglades, will result two- to fivefold increase in Hg risk
Rumbold et al. (2001)	Based on measured egg-[Hg], 27% of **great egret** eggs from WCA-3 had concentration > no effect level but < probable effect level; none of the measured feather-[Hg] exceeded LOAEL based on feather concentrations. Brood size and fledging success appeared normal
Barron et al. (2004)	Based on modeled daily dose, estimated **panthers** prior to 1992 had 17% probability of exceeding HQ of 1, with HQ ranging as high as 5.4; ASSUMING a 70–90% reduction in Hg across system by 2002, only a 4.6% probability of exceeding HQ of 1

(continued)

Table 10.1 (continued)

Source	Approach, ecological receptor, and findings
Heath and Frederick (2005)	Based on statistical relationships between Hg levels, hormone concentrations and nesting effort of wild caught **white ibis**—found nesting was negatively correlated with Hg exposure index
Rumbold (2005)	Based on modeled daily dose, estimated **great egrets** and **bald eagles** had 75% and 65% probability of exceeding LOAEL in WCA3 and ENP, respectively
Rumbold et al. (2008)	Based on modeled daily dose, estimated probability of exceeding LOAEL would range from 14% to 56% for **great egrets**, **wood storks** and **bald eagles** foraging in ENP
Herring et al. (2009)	Based on Feather-[Hg] and physiological measurements on wild **white ibises** and **great egrets** from WCA-2 and -3, suggest no negative effects in nestlings
Frederick and Jayasena (2010)	Based on captive study using chronic exposure levels experienced in the wild, **white ibis** exhibited altered reproductive behavior
Brandon (2011)	Based on measured Hg concentrations in blood and hair and benchmarks used previously by Roelke et al., **panthers** were still risk

and liver and other tissues from salvaged dead panthers. They compared observed tissue-[Hg] with eco-epidemiology data they had been collecting on panther health and reproductive success, and published toxicological data from tests done on domestic cats (*Felis spp.*). They concluded "chronic exposure of Florida panthers to Hg appears to be compromising their relative health and possibly productivity, especially in Everglades portions of their range; Hg must be considered a threat to the continuing existence of this already endangered taxon (Roelke et al. 1991, p. 20)." A subsequent risk assessment focused on several wildlife species and compared modeled dietary exposure to a lowest observable adverse effect level (LOAEL) from the published literature (Jurczk 1993). The author estimated that the greatest risk was to great egrets (*Ardea alba*) that were consuming Hg at a daily rate 3.9 times higher than the LOAEL observed in common loons (*Gavia immer*).

Because Hg methylation and the degree of biomagnification are known to be influenced by numerous environmental factors (Fig. 10.1), there was a concern that Hg risks would be unintentionally exacerbated by management actions to improve water quality in the Everglades (which included creation of six constructed wetlands termed stormwater treatment areas or STAs to remove phosphorus).

Early risk assessments therefore focused on impacts from the STAs and downstream waters of the Everglades Protection Area, which as defined in the Everglades Forever Act is comprised of Everglades National Park including Florida Bay and the Water Conservation Areas (WCA-1, WCA-2A, WCA-2B, WCA-3A, and WCA-3B) encompassing a huge area of about 9055 km^2 (e.g., PTI 1995; Exponent 1998, 2000; Rumbold et al. 1999; Rumbold 2000, 2005; Table 10.1). More recently the focus of the stressor-initiated assessments have evolved and now focus on possible increased Hg methylation, biomagnification and risk resulting from management activities

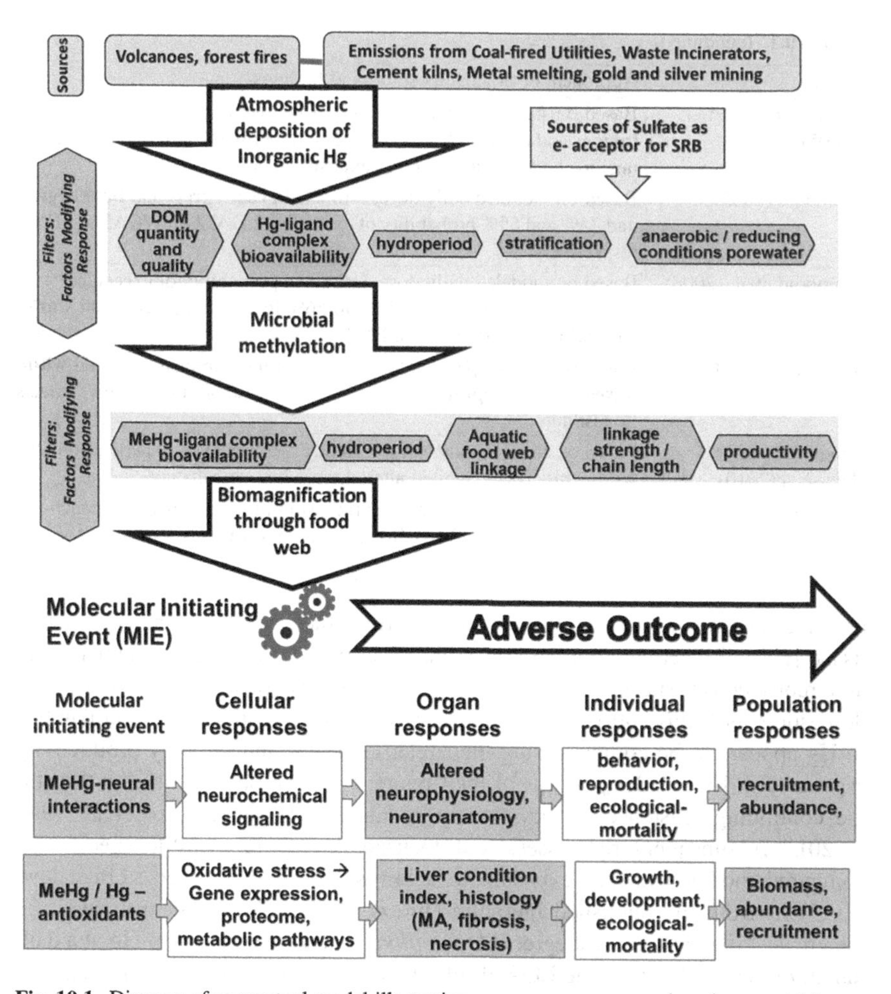

Fig. 10.1 Diagram of conceptual model illustrating sources, processes and environmental factors influencing mercury biomagnification and risk of different adverse outcomes (neurochemical-based AOP adapted from Basu 2015)

(particularly changes in hydrology and distribution of sulfate laden water) associated with the Comprehensive Everglades Restoration Plan (CERP) to restore the south Florida ecosystem (e.g., Brandon 2011; the present assessment). The CERP came out of "restudy" of the ecological impacts of construction of the Central and Southern Florida Project (authorized by US Congress in 1948) consisting of a regional network of canals, levees, storage areas and water control structures from Orlando to Florida Bay affecting an area of about 46,620 km^2; McLean and Bush 1999). Completing the numerous components of CERP is expected to take more than 30 years.

Not surprisingly, stakeholders interested in Everglades Hg risk have evolved over time but have included resource managers at various state and federal agencies, farmers operating within the Everglades Agricultural Area and their environmental consultants (e.g., Exponent [formerly PTI] 1998), academic researchers (notably researchers with University of Florida (UF) and UF's Institute of Food and Agricultural Sciences (e.g., Schueneman 2001), as well as non-governmental organizations, such as the Everglades Foundation (e.g., Corrales et al. 2011).

Depending on its scope and the reason for initiating a particular study, early risk assessments had differing management goals but they were often based on the state's "no toxics in toxic amounts" rule (i.e., FAC 62-302.530: Substances in concentrations which injure, are chronically toxic to, or produce adverse physiological or behavioral response in humans, plants, or animals—None shall be present) or the Federal Endangered Species Act (ESA; 16 U.S.C. § 1531 et seq.).

Given all of the above, it is not surprising that previous risk assessments also had differing assessment endpoints (i.e., explicit expression of what was to be protected; consisting of an ecological entity and its attribute; for examples, see Table 10.1). In many cases the choice of assessment endpoint was based on data availability and ease or economy of measurement rather than signal-to-noise ratio or statistical power. Furthermore, due to limited data and validated models, the receptor's attribute was more often at the organismal level (e.g., survival, recruitment) rather than at the population level (e.g., population growth rate). While risk to individuals of endangered species (such as the Florida panther, bald eagle, or wood stork) garnered concern by resource managers, attempts to characterize risk to non-listed species often relied on extrapolations to population-level effects, which were a source of considerable uncertainties due to multiple feedback loops.

10.2 Conceptual Model

A key finding from early biological surveys was that the degree of methylmercury (MeHg) biomagnification was spatially highly variable across the Everglades leading to hotspots (Spalding and Forrester 1991; Roelke et al. 1991; Stober et al. 1998; Lange et al. 1999). This was best illustrated by results of extensive surveys done across the Everglades in 1995 and 1996 by the USEPA (1998) that identified several hotspots where [Hg] in mosquitofish (*Gambusia holbrooki*) were much higher than surrounding areas. These hotspots for [Hg] in mosquitofish were also found to coincide with locations of wading bird nesting colonies where UF researchers had found much higher [Hg] in feathers of great egret chicks than surrounding colonies (Stober et al. 1998).

As discussed in detail in other chapters, there are many sources of inorganic Hg to the atmosphere and many environmental factors that influence its methylation rate and entry into a food web once it is deposited (Fig. 10.1). These environmental factors (or filters) are known to result in Hg hotspots, not just in the Everglades but other regions as well (Evers et al. 2007; Chen et al. 2012). This makes risk assessment much more difficult, particularly for wide-ranging animals, because risk must be defined within a multidimensional state space (Suter 2007).

A regional-scale risk assessment in the Everglades, for example, must consider an ecological entity moving across the landscape experiencing varying concentrations of Hg over differing exposure periods. As discussed in the ensuing sections, depending on an individual's susceptibility, which differs both within and among species, entities moving across this landscape will likely manifest an array of different effects. The result is that different proportions of a population may show some type of binary effect (e.g., live versus dead) while others may exhibit varying severities of a continuous response (e.g., degree of altered neurochemistry, endocrine disruption, altered behavior, motor coordination, response rate or histopathology, reduction in body mass, etc.).

10.3 Effect Characterization

Hg toxicity has been reviewed extensively elsewhere (Wiener et al. 2003; Scheuhammer et al. 2007, 2015; Depew et al. 2012a, b; Basu 2012) and so this chapter will only touch on issues and specific metrics relevant to this risk assessment. This discussion cannot proceed, however, without first acknowledging the difficulties inherent in toxicity testing on wildlife. Observing effects from field studies, either manipulative or eco-epidemiological wildlife surveys, often leaves a lot of uncertainty due to uncontrolled variables. Toxicity testing on wildlife in more controlled environments often suffer from small sample size, fewer treatment levels, fewer replicates or pseudoreplication due to the inherent difficulty in maintaining large numbers of wildlife. There is also the concern that captive animals have a different set of potential stressors than wild animals. Consequently, we are often forced to extrapolate from surrogate species and toxic endpoints at levels of biological organization (e.g., molecular, cellular, etc.) that are not particularly relevant for risk assessments.

10.3.1 Adverse Outcome Pathways (AOP)

Recently, a new conceptual framework termed adverse outcome pathways (AOP) has been proposed to synthesize and translate disparate sources of information on the mechanistic linkages among different levels of biological organization (Ankley et al. 2010). This should allow us to translate molecular or cellular events into endpoints meaningful for ecological risk assessments. AOPs are not chemical specific, instead they describe how a molecular initiating event (MIE) by a given chemical stressor may lead to an adverse effect; thus, multiple chemicals can initiate the same MIE and have same AOP (Fay et al. 2017). Different AOPs can also share key events in the sequence (allowing for the creation of networks) or even the same outcome. A number of AOPs have been proposed (Ankley et al. 2010; Basu 2015; Fay et al. 2017). Basu (2015), for example, proposed a conceptual neurochemical-based AOP resulting from exposure to environmental contaminants that have neurotoxic properties. It begins with toxicant-neural interactions leading to altered neurochemical signaling pathways and altered neurophysiology or anatomy which then may cause altered growth, development, reproduction or behavior ultimately potential leading to changes in population structure, recruitment and possible local extinction. While focusing on two well-known examples (domoic acid and dieldrin), he also reviews information on neurotoxicity biomarkers from MeHg exposure in various organisms (Basu 2015) that could easily serve as the MIE and early key events in a similar AOP (Fig. 10.1). With improved understanding of key events later in this sequence, such an AOP would help to explain reported behavioral toxicity of MeHg in fish and wildlife. Although species–dependent, these effects include, but are not limited to: lethargy, abnormal performance on learning tasks, poor maze performance, abnormal startle reflex, impaired righting response, impaired escape and avoidance behavior, altered resting position, altered spawning behavior, altered brooding behavior, altered time activity budgets and, finally, altered motor function in fish and wildlife (Wiener et al. 2003; Depew et al. 2012a, b; Basu 2015). Although complicated by numerous confounders, we should not forget that humans suffering from low-level MeHg exposure have given voice to similar neurotoxicity complaints (Hightower 2008; Karagas et al. 2012). Hightower (2008) diagnosed Hg toxicity in human patients complaining of various health issues and who were high-end consumers of fish (e.g., Sushi, swordfish, Ahi steaks). She reported that patients described it as having a constant hangover and dubbed it "fish fog". Imagine an egret or panther having to survive in the Everglades with a constant hangover.

As might be expected, a given chemical such as Hg can initiate different molecular events at different specific sites of action and, consequently, produce different adverse outcomes. For example, MeHg can both inhibit production of natural antioxidants in a body thus allowing reactive oxygen species (ROS) naturally produced to build up and can also generate ROS itself (i.e., two different MIE). This can lead to oxidative stress as an early key event in the damage of a variety of tissues (Ercal et al. 2001; Valavanidis et al. 2006; Mieiro et al. 2010; Hoffman et al. 2011; Henry et al. 2015; Kirkpatrick et al. 2015). Henry et al. (2015), for example, found indications of oxidative stress in livers of zebra finches (*Taeniopygia guttata*) with increased blood [Hg]. Similar studies done with the common loon (*Gavia immer*) and great egrets also found oxidative stress in the liver (Kenow et al. 2008; Hoffman et al. 2005). Liver damage has been linked to oxidative stress in variety animals exposed to Hg, both methylated and inorganic, including fish (Berntssen et al. 2003; for review, see Mela et al. 2007; Barst et al. 2016). As an example of shared key events in different AOPs, oxidative stress can cause lipid peroxidation, brain lesions and neurotoxicity in fish (Mieiro et al. 2010).

10.3.1.1 Ecological Mortality

As suggested above, effects from Hg exposure can be extremely varied (Wiener et al. 2003; Scheuhammer et al. 2007; Depew et al. 2012a, b; Scheuhammer et al. 2015).

In general, exposure levels that result in neurotoxicity and adverse behavioral effects are thought to be much lower (as much as two orders of magnitude lower) than exposure levels associated with overt toxicity (Hoffman 1995; Wiener et al. 2003; Depew et al. 2012a, b). It should be recognized, however, that many of these sublethal effects can result in reduced fitness leaving the organism at a competitive disadvantage (e.g., unable to find food causing malnutrition, increased susceptibility to disease, depressed reproduction) or at an increased risk of predation (Wiener et al. 2003). Ecological mortality, i.e., the concept that a toxicant–related reduction in fitness of an individual can be of a magnitude equivalent to somatic death (Newman 2015), may be the answer to the often-posed question "if the problem is so bad, where are all the dead animals?" Predators or scavengers have removed the less fit individuals. Additionally, as discussed below, where animal carcasses are recovered, their death may be attributed to the more proximate cause (e.g., disease, starvation, vehicular collision) rather than the underlying stress from Hg toxicity.

10.3.1.2 The Concept of Dose

Before proceeding, we must first review the concept of "dose" and dose metrics used in ERAs particularly with regard to their inherent uncertainties. The molecular events described above will be initiated only when a certain amount of the toxicant, in a bioactive form, remains at a site within the organism where it may cause toxic effects (i.e., the site of action) for a sufficient duration. This can only occur when the rate of exposure and uptake of the toxicant is greater than the combined rate at which it can be excreted or detoxified (McCarty et al. 2011). The latter includes biochemical and physiological mechanisms that fortuitously serve to control the active fraction of a metal forming the effective dose (e.g., metal is bound to internal ligands or sequestered to a site away from the site action). This biologically effective dose at the site of action usually remains unknown even during conventional toxicity testing (McCarty et al. 2011; Escher et al. 2011). There are a number of approaches to define dose-response relationships. Traditional exposure-based toxicity testing typically only measures administered dose (e.g., dietary exposure that focuses on amount of metal ingested per unit mass of the organism, etc.) or concentrations in exposure media (e.g., water, air, sediment/soil and, more recently, prey items). Thus, we are often left with surrogate dose measurements that introduce substantial uncertainty in the interpretation and use of toxicological data (McCarty et al. 2011).

Recently, use of the tissue residue-effects approach (TRA), where dose is defined by tissue or internal concentration, has expanded in both dose-response toxicity testing and ERAs (Meador et al. 2011; Sappington et al. 2011). This approach reduces uncertainty and variability in toxicity data (e.g., disaggregating species-specific differences in toxicokinetics and toxicodynamics, etc.; for review, see McCarty et al. 2011) and in assessing exposure in the field (e.g., variability in bioavailability, integration of exposure over time and space, etc.; for review, see Sappington et al. 2011). However, TRA is not a panacea (McCarty et al. 2011; Sappington et al. 2011). We must recognize its limitations and understand that it remains a proxy for biologically effective dose (Escher et al. 2011). First, one must consider the choice of tissue "how well does it reflect the effective dose?" Obviously, it is desirable to select a tissue that is closest to the site of action (Sappington et al. 2011); however, as discussed below, in an effort to use non-intrusive, non-lethal sampling (e.g., feathers, fur etc.) choice of tissue often reduces interpretability. Furthermore, as mentioned above, kinetics (i.e., time dependence) is important and, because a tissue concentration may not precisely reflect the rate of accumulation, more sophisticated time-resolved toxicokinetic–toxicodynamic models may be required (Escher et al. 2011).

10.3.2 *Toxicity Reference Values (TRVs)*

The different approaches to assess dose-response have resulted in a number of different metrics for use as toxicity reference values (TRVs; i.e., screening benchmarks that are believed to constitute the threshold of toxicity) in ERAs.

10.3.2.1 Daily-Dietary Dose TRVs

Early avian risk assessments both in Florida (Jurczyk 1993; PTI 1995; Exponent 1998, 2000; Rumbold 1999, 2000) and nationally (USEPA 1997) relied on effects data from one of two studies: either from an observational (correlative) field study of common loons, *Gavia immer* (Barr 1986) or from a transgenerational, controlled dosing study of mallard ducks, *Anas platyrhynchos* (Heinz 1979). USEPA (1997), for example, calculated a lowest-observed-adverse-effect level (LOAEL) from the study by Heinz (1979) and, using an uncertainty factor, proposed 0.026 mg/kg body weight (bw)/day as an avian reference dose (RfD). To assess mammalian risk, USEPA (1997) used results of a dietary exposure study where mink (*Mustela vison*) were fed Hg-contaminated fish (Wobeser 1973 as cited by USEPA 1997) and, in this case, was able to report a no-observed-effect-level (NOAEL); they proposed 0.018 mg/kg bw/day as a mammalian RfD (USEPA 1997). These are analogous to the more recently proposed human RfD for MeHg (i.e., 0.0001 mg/kg BW/day; USEPA 2001), except endpoints differed as did uncertainty factors; hence, the more conservative value for humans. Subsequently, these wildlife reference doses were corroborated and improved upon by results from several controlled feeding experiments done in Florida using food spiked with MeHg (typically as MeHgCl) based on environmentally relevant concentrations measured in wild fish (Bouton et al. 1999; Frederick et al. 1997; Spalding et al. 2000; Frederick and Jayasena 2010).

10.3.2.2 Media-Based TRVs (i.e., Based on Hg Concentrations in Prey)

The principal pathway for MeHg exposure to wildlife is dietary consumption of prey and not from drinking water or ingesting sediment (Wiener et al. 2003; Rumbold 2005). Because of the propensity for methylation to occur in aquatic sediments, prey items linked to aquatic food webs particularly, but not limited, to fish are often the focus of these media-based TRVs. For example, USEPA (1997) estimated that tissue concentrations (i.e., a media-based TRV) of 0.077 mg/kg and 0.346 mg/kg in fish at trophic levels 3 and 4, respectively, would be protective of piscivorous wildlife based on the mammalian reference dose mentioned above (they relied on the mammalian RfD because it resulted in a more conservative value than the avian RfD). Depew et al. (2012a) recently reviewed the literature focused on sublethal Hg effects to the common loon and proposed similar thresholds based on [Hg] in prey fish. More specifically, they proposed three TRVs: 0.1 mg/kg, 0.18 mg/kg, and 0.4 mg/kg (wet wt.) in prey fish; the lowest being a threshold for behavioral impacts, with the other two representing thresholds for significant reproductive impairment and reproductive failure in loons, respectively (Depew et al. 2012a). While these focus on loons and, thus, could serve as an avian TRV, it is noteworthy that neuroanatomy changes (N-methyl-D-aspartic acid, NMDA, receptor levels) have also been reported in mink at dietary concentrations as low as 0.1 mg/kg (Basu et al. 2007). Thus, independent studies on different species are generating consistent results.

These TRVs based on fish-tissue residues are analogous to USEPA national water quality criterion for the protection of human health based on a fish-tissue concentration (0.3 mg/kg) that was derived from the previously mentioned human health RfD, and a much lower fish-consumption rate by humans (USEPA 2001).

While most of these early studies traditionally focused on toxicity to top wildlife predators (e.g., USEPA 1997), and considered fish primarily as a biovector to piscivorous wildlife and humans, recent studies have shown toxicity from Hg exposure in fish themselves (for review, see Sandheinrich et al. 2011; Depew et al. 2012b). Wiener et al. (2003) report the dead fish found floating in Minamata Bay, the site of industrial pollution in Japan, had tissue concentrations ranging from 8.4 to 24 mg/kg (and averaged 15 mg/kg). As was the case for wildlife, diet is the principal pathway for exposure to MeHg for fish (Hall et al. 1997; Wiener et al. 2003). Accordingly, while a number of studies of Hg toxicity to fish have been based on aqueous exposures (for review, see Depew et al. 2012b), the most appropriate exposure metrics are based on dietary exposure. From their literature review that focused on toxicity in fish, Depew et al. (2012b) proposed a dietary concentration of 0.5 mg/kg in prey as a toxicity threshold for fish behavior and an even lower concentration of <0.2 mg/kg in prey based on reproduction and other subclinical endpoints expected in the predator fish (the latter was, however, based on a limited number of studies).

10.3.2.3 Tissue-Residue Based TRVs

For Protection of Fish

Results of dosing fish have been used to also derive TRVs based on the tissue residue-effects approach (TRA). For example golden shiners (*Notemigonus crysoleucas*) fed a diet incorporated with MeHg were hyperactive and had altered shoaling behavior when their whole-body [Hg] averaged 0.52 μg/g ww, relative to fish with lower [Hg] (Webber and Haines 2003). Beckvar et al. (2005) proposed a tissue-residue TRV of 0.2 mg/kg ww in whole bodies of juvenile or adult fish such that below this level adverse effects were unlikely to occur in the fish. Based on a comprehensive review of the literature, Sandheinrich et al. (2011) concluded that sublethal effects including changes in reproductive health begin to occur in fish, in both the laboratory and field, when their whole-body [Hg] reaches about 0.3 mg/kg (0.5 mg/kg in fillets). Based on 11 studies that focused on lethality-equivalent endpoints, Dillon et al. (2010) developed a dose-response curve and estimated injury in 50% of the exposed fish when their tissue concentrations reached about 3 mg/kg (whole body). Interestingly, their model supported Beckvar et al. (2005) and predicted only 5.5% injury rate at a tissue concentration of 0.2 mg/kg.

For Protection of Birds and Mammals

Ackerman et al. (2016) recently summarized available information on avian TRVs based on residues measured in their various tissues, including eggs and feathers, and provided conversion factors to translate them into a common, blood-equivalent concentration. They found lowest documented effects in birds (i.e., LOAEL) in terms of: (1) altered breeding behaviors occurred at a 0.2 mg/kg blood-equivalent concentration; (2) reduced egg hatchability at 1.0 mg/kg blood-equivalent concentration and; (3) reduced breeding success in sensitive species at 2.0 mg/kg blood-equivalent concentration (with this becoming increasingly widespread in less sensitive species at 3.0 mg/kg; Ackerman et al. 2016). This extensive review of the literature by Ackerman et al. (2016) included a Florida study by Frederick and Jayasena (2010) that dosed captive white ibises (*Eudocimus albus*) and found altered courtship behavior, with higher proportion of same sex nest pairs, and depressed reproduction even in the low-dose group, suggesting these birds were showing significant reproductive impairment at the lower TRV.

Recent studies of Hg toxicity in mammals have focused on [Hg] in the brain (3–5 mg/kg dry wt) as a TRV (Basu 2012; Dornbos et al. 2013; for review, see Scheuhammer et al. 2015). Yet, many also recognized the need to interpret levels in tissues that can be sampled less invasively in routine monitoring programs, such as fur or hair (Evans et al. 2016; Eccles et al. 2017). Risk estimation was based on correlations observed between [Hg] in fur with [Hg] in brains (Evans et al. 2016; Eccles et al. 2017). Nonetheless, it should not be forgotten that high [Hg] in fur/hair is an effective way for mammals to excrete Hg from the body and that it is removed from the site of action (just as growing feathers are for birds). Based on a literature review, Basu (2012) recommended that 30 mg/kg in fur should be considered the lowest observed adverse effect level (LOAEL) for terrestrial mammalian wildlife.

Yet more recently Eccles (2017) argues that the TRV should be lowered to 15 mg/kg in fur based on their models relating brain-[Hg] to fur-[Hg]. It is interesting to note that in humans the NOAEL is purported to be 6.0 mg/kg in hair (based on data from the Faroe Islands) and that USEPA recommends 1.0 mg/kg as the TRV for humans (as compared to the 15 mg/kg or 30 mg/kg for wildlife; for review, see Basu 2012).

10.4 Ecosystem and Receptor Characteristics That Influence (or Modify) Exposure or Response

As mentioned above, a number of ecosystem characteristics can influence Hg methylation and biomagnification and, thus, exposure and risk (Fig. 10.1). Many of these factors related to biogeochemistry are discussed in detail in other chapters of this volume. However, risk assessors must also consider receptor characteristics that may influence its exposure or its response and, thus, make it more or less susceptible. Many obvious factors influence a receptors susceptibility such as genetics, previous exposure, species-specific (and ontogenetic-specific) trophic position and increased dependence on aquatic food webs (as opposed to terrestrial). Because the rate of Hg methylation is greatest in aquatic sediments, an ecological receptor more heavily dependent or linked to an aquatic food web will have increased chance of exposure. This dependence can vary and depend on hydroperiod or season in the Everglades (for more details, see Chaps. 7 and 8, this volume). For example, some birds may rely more heavily on aquatic insects during the nesting season. As another example, depending on water level, panthers in some areas may feed on raccoons and armadillos and even alligators, which are tied to the aquatic food web, rather than deer or hog (see, Chap. 8, this volume). Those species that prefer wetlands with shorter hydroperiods, which may have higher net methylation rates producing bioavailable MeHg (Snodgrass et al. 2000; Chen et al. 2012), may be at greater risk of exposure. As another example, species with large foraging ranges may be able to dilute their exposure as compared to species with a small home range who happen to find themselves in a hotspot. Foraging range may change during the reproductive season in both location and size.

Sensitivity to Hg can be related to the life stage of an organism. For example, early life stages, particularly embryos *in utero* or *in ova*, are thought to be more vulnerable to chemical stressors because of the number of processes that must occur in the proper sequence for successful development (Wiener et al. 2003). Age-dependent sensitivity may also be related to an organism's inability to compensate behaviorally, e.g., "naive chicks" or panther cubs being less adept at foraging or hunting. Alternatively, an organism may be less vulnerable during times when tissues that fortuitously serve as sites to sequester Hg away from active sites are being replaced and, thus, are able to serve as sinks (e.g., actively growing feathers, hair). Alternatively, when these tissues are not actively growing (i.e., between seasonal molts) and are not being supplied with blood they cannot serve as a sink and the biologically effective dose may increase dramatically. Organisms undergoing rapid growth may also biodilute the assimilated Hg and, thus, lesson the effective dose; however, this will also be dependent on food-consumption rates, food-conversion efficiency and how the energy budget is allocated. Organisms may be more sensitive to Hg when under stress. For example, based on their knowledge of [Hg] that were lethal to grey herons (*Ardea cinerea*), derived from dosing studies, Van der Molen et al. (1982) postulated that the death of a large number of grey herons in the Netherlands after a severe winter resulted from the sublethal effects of Hg combined with the stress of cold weather and undernourishment. In south Florida, Spalding et al. (1994) speculated that Hg may have been a contributing factor in the death of great white herons (*Ardea herodias occidentalis*) that were emaciated and suffering from chronic, often multiple, diseases. Emaciation is a common observation in necropsies of salvaged dead birds and other animals with elevated Hg (Van der Molen et al. 1982; Roelke et al. 1991; Spalding et al. 1994; Scheuhammer et al. 1998; Stone and Okoniewski 2001; Depew et al. 2012a, b). The primary cause of death in many of these cases was often attributed to the proximate factor, e.g., cold stress, parasitic infections, chronic disease (Depew et al. 2012b; Scheuhammer et al. 2007). Immunotoxicological effects from Hg exposure have been postulated for birds and mammals (Scheuhammer et al. 2007).

These context dependencies that modify exposure or response across a range of conditions can be sources of considerable variability and uncertainty in risk estimates.

10.5 Exposure Characterization

Predicting the rate of exposure (e.g., intensity and duration) across the highly variable south Florida landscape is complicated for the reasons discussed above. To model daily dietary dose for a receptor requires knowledge of its diet, the proportion of specific prey items in the diet, [Hg] in those prey items and daily consumption rate. The resulting daily dose of Hg is then compared to the dietary-based reference doses discussed above. While diet information is known for many important ecological receptors in the Everglades (see Chaps. 7 and 8, this volume), we are often lacking recent information on [Hg] in specific prey species (i.e., that are not targets of biomonitoring programs). Previous risk assessments in the Everglades that modeled daily dietary exposure often resorted to making simplifying assumptions (which, as described below, sometimes turned out to be erroneous) and limited the assessment to populations in relatively small, well-defined areas to minimize spatial heterogeneity of Hg exposures (e.g., Roelke et al. 1991; Jurczk 1993, PTI 1995; Exponent 1998, 2000; Rumbold 1999, 2000, 2005; Duvall and Barron 2000; Barron et al. 2004; Rumbold et al. 2008). Some of these assessments used Monte Carlo simulations to express the uncertainties inherent in predicting exposure (Rumbold 2000; Exponent 2000; Barron et al. 2004, Rumbold 2005; Rumbold et al. 2008). Another approach to address the spatial variability involves spatially explicit modeling that allows the ecological receptor to move across the landscape. One early risk assessment in Florida used an individual based model that was spatially explicit (at least for foraging range around a nesting colony; Fleming et al. 1994). That modeling effort was undertaken to test a number of hypotheses regarding declining wood stork populations in the Everglades including Hg impact; however, it was not a robust assessment of Hg risks.

Alternatively, to avoid having to depend on data-intensive, spatially-explicit dietary models, many of the previous studies were media-based. For example, following the severe drought in 1999, Rumbold and Rawlik (2000) reported that most fishes collected under SFWMD's biomonitoring program, e.g., mosquitofish (*Gambusia spp.*), sunfish (*Lepomis spp.*) and bass (*Micropterus salmoides*), contained [Hg] that exceeded USEPA (1997) predator protection criteria for both TL 3 and 4 fish. Other previous ERAs in the Everglades have avoided the spatial heterogeneity issue by characterizing exposure based on [Hg] measured in the receptor's tissues (i.e., tissue-residue based TRV; Sundlof et al. 1994; Rumbold et al. 2001; Brandon 2011). Although the discussion above focused on spatial variability, it is crucial to understand that MeHg availability to food web across the Everglades landscape also varies through time, i.e., old hotspots may cool while new hotspots pop up. As summarized in Chaps. 1–5 (in Volume III), [Hg] have changed over time in a variety of Everglades media, including fish and wildlife, due to a number of different drivers. After several papers reported declines in [Hg] in fish and wildlife from certain areas (Lange et al. 2000, 2005; Rumbold et al. 2001, 2002; Frederick et al. 2002) some began to conclude that the Hg problem had been resolved (for review, see Rumbold et al. 2008). This misconception was further strengthen by the results of a risk assessment in 2004 that was based on the erroneous premise that [Hg] in fish had decreased by 70–90% across the entire system (Barron et al. 2004). However, because MeHg biomagnification is so highly variable across the large expanse of the Everglades Protection Area, caution must be exercised when making system-wide generalizations. At that time, a number of hotspots outside the WCAs were not exhibiting Hg declines (e.g., ENP and Florida Bay) and there were new hotspots north of the WCAs (for review, see Rumbold et al. 2008). Therefore, when assessing Hg risk in south Florida it is crucial to consider both the geographical area and the time frame during which the exposure assessment was conducted. When making decisions on the type of data to look for to characterize exposure or effects it is important to recognize that when it comes time to integrate the two, they must be combined in a way that makes sense, i.e., they must be concordant (Suter 2007). In other words, the metric used to characterize the dose-response (e.g., modeled dietary dose, fish tissue concentrations or toxicity residues in the receptor) will determine the choice of metric to characterize exposure or, inversely, the exposure metric will dictate the effects metrics.

10.6 Characterization of Current Risk

The aim of the current analysis was to use a multiple lines of evidence (LOE) approach to assess current Hg risk for a number of ecological receptors across the Everglades and south Florida.

10.6.1 LOE 1: Risk Based on Fish-Tissue [Hg] as the Media of Exposure

The first line of evidence was based on recent results from the state's program to monitoring [Hg] in fish within Everglades Protection Area. The Florida Fish and Wildlife Conservation Commission (FFWCC; under contract to SFWMD) collect fishes, via electroshocking, at 13 sites within the Everglades each year (and in some of the STAs which are not considered here). They target Largemouth bass (n = 20), sunfish (n = 20), and mosquitofish (a composite sample of about 100 fish) and while bass are processed as fillets, sunfish and mosquitofish are processed as whole-body homogenates (for details, see Julian et al. 2016). These data were augmented with data (published and unpublished) collected by other governmental and non-governmental entities (Fig. 10.2). Recall from above that USEPA (1997) derived their predator protection criteria for [Hg] in trophic level 3 and 4 fish based on controlled dosing studies involving mallards and mink; however, similar thresholds had been proposed by Barr (1986) based on his field studies and reconfirmed recently by Depew et al. (2012a) based on their review of more than 16 separate reports from both field and captive feeding studies. As shown in Fig. 10.2, the distribution of [Hg] in recently collected fishes at most of the freshwater and estuarine sites exceeded one or both of USEPA's thresholds, either entirely or in part; the only exception being mosquitofish, where only fish from WCAU3 had [Hg] in excess of the TL 3 criterion. Overall, 62% of the individual sunfish collected from these sites exceeded the TL 3 criterion. But clearly a greater proportion of fish at certain sites exceeded the criterion; sunfish from over half the sites had a median [Hg] in excess of the criterion.

Because bass feed at TL 4 only when reaching certain age and size and because these [Hg] were for all sampled fish regardless of age or size, observed distributions were compared to both benchmarks. Across all sites, 96% of the bass had [Hg] in excess of the TL 3 criterion whereas 29% exceeded the TL 4 criterion; bass had a median [Hg] that exceeded the TL 4 criterion at 1/3 of the sites. It is worthwhile to mention here that FFWCC measured [Hg] in pig frogs (*Rana grylio*), another important prey item of birds, collected from the WCAs in 2006 and 2008. While [Hg] were lower than levels found in pig frogs from ENP a few years ago (Ugarte et al. 2005), average [Hg] in frogs from five of six of the WCA sites (all but WCA2A) exceeded the TL 3 criterion for protection of predators (T. Lange, FFWCC, pers. comm.). With regard to estuarine fishes, 66% of the TL 3 fish (e.g.,

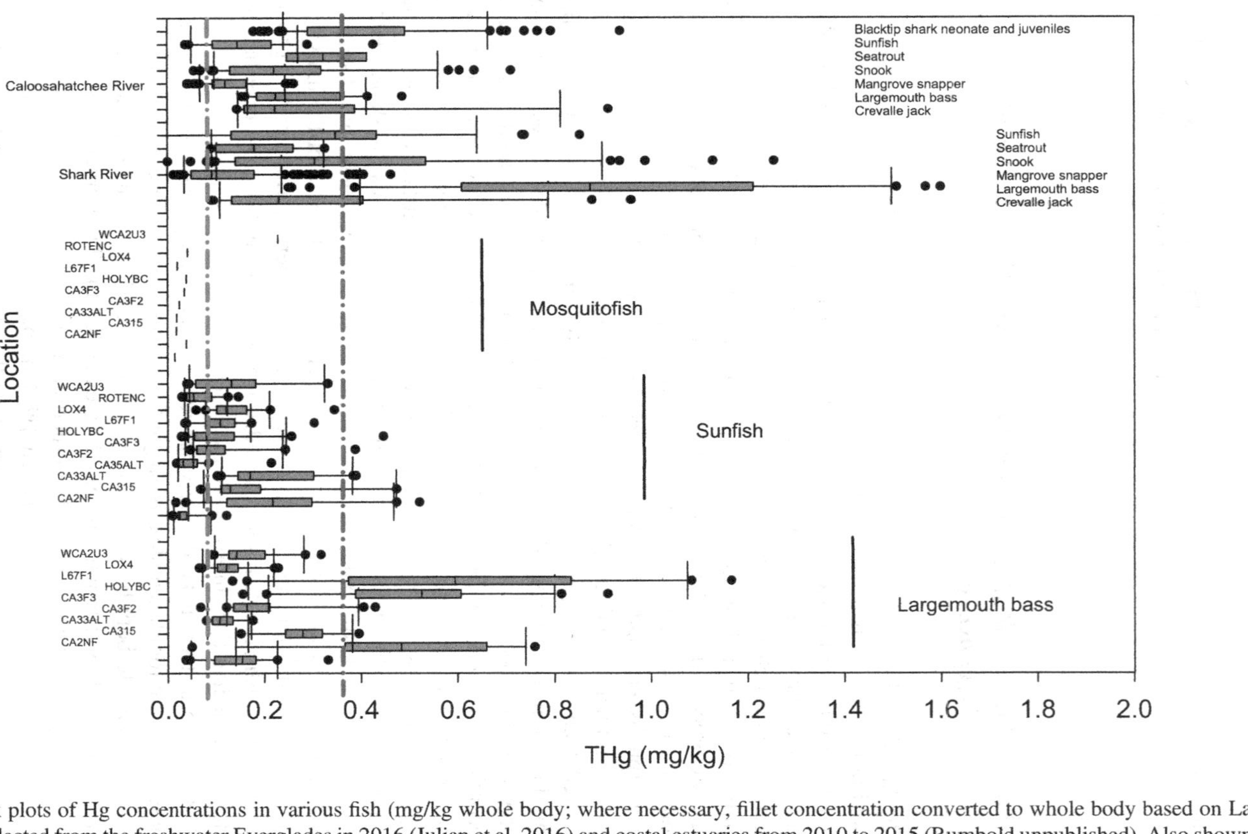

Fig. 10.2 Box plots of Hg concentrations in various fish (mg/kg whole body; where necessary, fillet concentration converted to whole body based on Lange et al. 1998) collected from the freshwater Everglades in 2016 (Julian et al. 2016) and costal estuaries from 2010 to 2015 (Rumbold unpublished). Also shown are USEPA (1997) TRVs (0.077 mg/kg and 0.346 mg/kg for fish at trophic levels 3 and 4, respectively) that would be protective of piscivorous wildlife (*cf.* Depew et al. 2012b)

sunfish and snapper) and 42% of the TL 4 fish (e.g., bass, snook, trout, and jack) collected from the two estuaries exceeded the risk criteria (Fig. 10.2).

Thus, the observed [Hg] in these fish (and frogs) indicate piscivorous predators in south Florida are at risk and this risk is greater in certain areas due to increased prevalence of fish with [Hg] above the predator-prey thresholds. As will be discussed in the next two sections, [Hg] measured in several wildlife species and results of limited eco-epidemiological wildlife surveys both support this conclusion.

10.6.2 LOE 2: Risk Based on Tissue Residues Within Ecological Receptors

The second line of evidence to characterize the risk to fish and wildlife receptors was based on results of surveys of [Hg] in the tissues of a limited number of fish and wildlife species (e.g., alligators, birds, panthers and dolphins) by government and other entities compiled from the published and the unpublished literature. These results were then compared to TRVs derived from tissue-residue effects approach as discussed above.

10.6.2.1 Risk to Fish

The data on fish-tissue [Hg] presented in the previous section can also be assessed in terms of risk to the fish themselves. As shown in Fig. 10.3, the minimum Hg toxicity threshold in fish, based on the review by Sandheinrich et al. (2011), is frequently exceeded in fish of south Florida. In 2016, a portion of Largemouth bass collected at seven of nine sites, where bass were found, contained [Hg] that exceeded the fish-TRV; at three sites almost all the collected fish had levels that exceeded the toxic threshold value for fish (Fig. 10.3). When pooled across sites this represented 32% of individual bass. These fish were therefore at some risk of sublethal effects. Fewer sunfish at each of seven of ten sites had [Hg] that exceeded this criterion (only 8% of individuals) and, thus, fewer of these fish were at risk. None of the mosquitofish populations sampled under this program had mean tissue-[Hg] above the toxic threshold (Fig. 10.3); however, because these are composite samples that physically average the concentration in the sampled fish, there may have been individual fish within the composite with [Hg] > the threshold value of 0.3 mg/kg, particularly at WCA2U3. Several fish species from two south Florida estuaries had individuals with [Hg] exceeding this fish-TRV (Fig. 10.3); when pooled across species, [Hg] in 31% of the individual fish exceeded the toxic threshold suggesting those fish were at risk. In particular, the majority of the distributions of [Hg] in both Largemouth bass from upper Shark River and young blacktip sharks (*Carcharhinus limbatus*) from Caloosahatchee River exceeded the benchmark. While there is some uncertainty in translating [Hg] in muscle to whole fish in these marine species using the formula

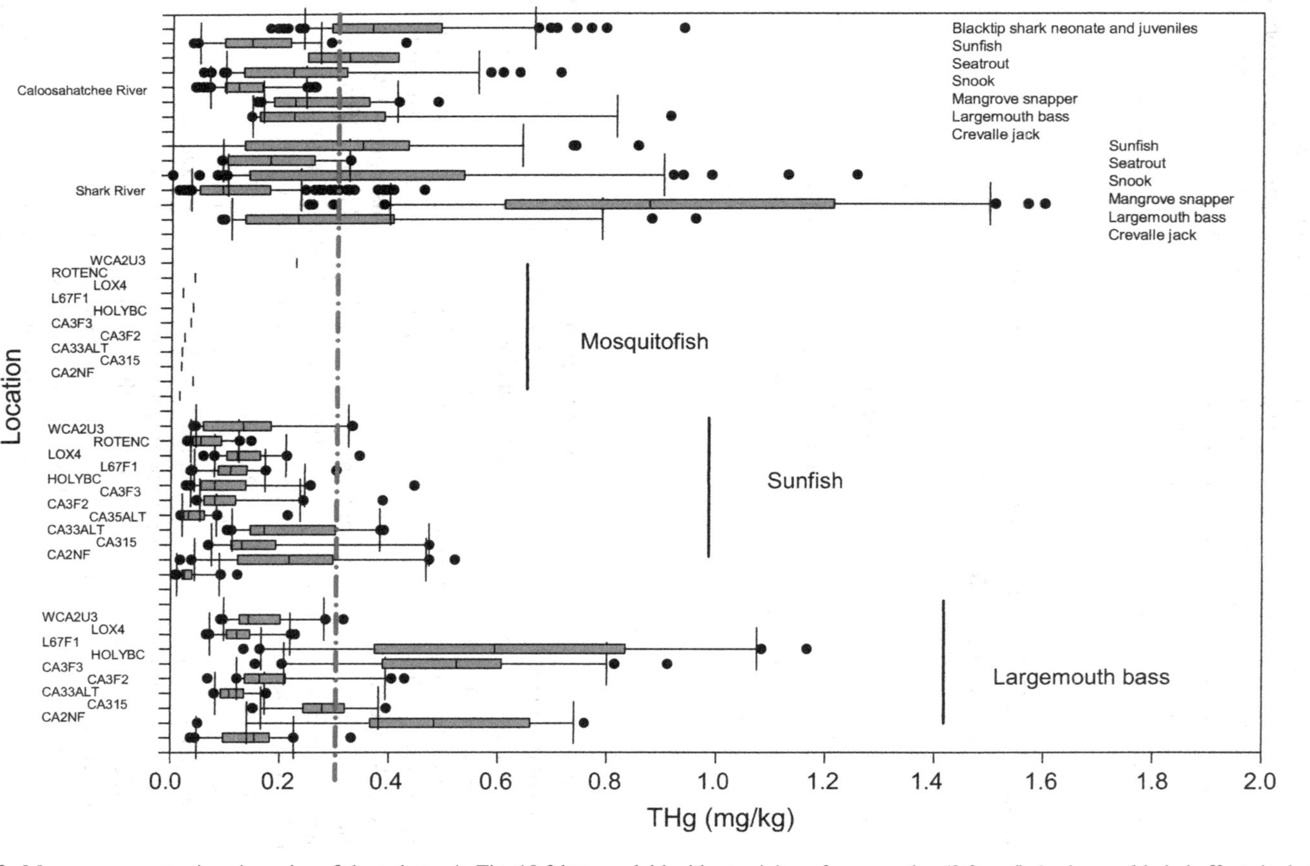

Fig. 10.3 Mercury concentrations in various fish as shown in Fig. 10.2 but overlaid with a toxicity reference value (0.3 mg/kg) where sublethal effects including reproductive effects begin to occur in fish (Sandheinrich et al. 2011)

developed by Lange et al. (1998); the results would be very similar based on others formulas (Peterson et al. 2005). It is also not known to what extent the threshold for toxic effects in finfish are applicable to sharks. Nonetheless, currently available data on Hg in fish does indicate a risk of toxicity for a variety of fish species from several locations. As discussed below, risk to fish themselves was supported also by the results of a limited number of eco-epidemiological surveys done on fish and sharks in south Florida.

10.6.2.2 Risk to Alligators

A number of surveys have been done of [Hg] in south Florida alligators (*Alligator mississippiensis*) with higher levels typically found in animals from ENP (for review, see Rumbold et al. 2002; Nilsen et al. 2017). The FFWCC continues to monitor [Hg] in alligators from the WCAs and reports finding elevated levels in gators particularly from WCA3A (average level of 2.18 mg/kg in muscle in 2011; T. Lange, FFWCC pers. comm.). When Rumbold et al. (2002) compared their findings to possible thresholds based on a captive dosing study by Peters (1983) and a field study by Heaton-Jones et al. (1997), they concluded levels were not a threat to the alligators themselves but they could continue to pose a threat to predators and scavengers of alligators. In the Everglades, even the endangered Florida panther has been reported to kill and consume small to intermediate-size alligators (Roelke et al. 1991). From their review of the literature, Grillitsch and Schiesari (2010) concluded that reptiles appear to be particularly robust against metal intoxication; however, they cautioned that this was based on a relatively small number of studies. More recently, Nilsen et al. (2016) reported that global DNA methylation, which is an epigenetic modification that is sensitive to environmental cues, was associated with increased [Hg] in alligators, particular in animals sampled from the WCAs.

10.6.2.3 Risk to Birds

Hg has been monitored in tissues of south Florida wading birds as far back as 1987, first by UF researchers (Spalding and Forrester 1991; Frederick et al. 2002) and later by both UF researchers (Heath and Frederick 2005) and SFWMD (Rumbold et al. 2001, 2008) and then by UF researchers under contract to SFWMD (Frederick as referenced in Julian et al. 2015). While it remains part of the permit-mandated monitoring program of SFWMD, the most recent reported data is from 2013 (Julian et al. 2015). While early surveys included a variety of tissues, later surveys choose to monitor [Hg] in feathers as a non-intrusive, non-lethal sample; however, assessing risk based on feather-Hg levels is more difficult because it is so removed from the site of action (Rumbold et al. 2001; Ackerman et al. 2016). Nonetheless, as mentioned above, Ackerman et al. (2016) has developed equations by which [Hg] in various tissues, including feather-[Hg], may be translated to blood-equivalent

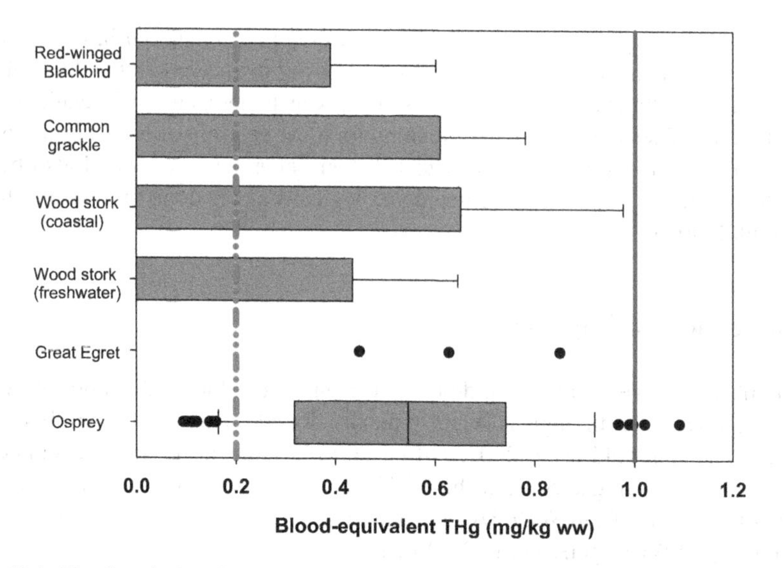

Fig. 10.4 Blood-equivalent Hg concentrations (converted from feather-Hg based on Ackerman et al. 2016) in various birds sampled in south Florida from 2010 to 2014. Also shown are toxicity reference values from Ackerman et al. (2016: 0.2 mg/kg and 1.0 mg/kg where altered behaviors or reduced egg hatchability and breeding success, respectively, begin to occur in sensitive species). Bar for blackbirds and grackles represents mean ± 1 SD measured in blood of adults not feather (Borrego de Pezon 2011); bar for wood storks and egrets are colony means (±1 SD) for nestlings (P. Frederick in Julian et al. 2015); box plot for osprey represents distribution in individual nestlings (Rumbold et al. 2014)

concentration. As also discussed above, they proposed TRVs based on blood-equivalent concentration (lowest documented effects occurred at 0.2 mg/kg; altered breeding behaviors at 1.0 mg/kg; Ackerman et al. 2016).

As evident in Fig. 10.4, recent surveys of [Hg] in feathers of wading birds in south Florida, when translated to blood-equivalent, show levels in most birds fall between the two thresholds (i.e., above the LOAEL but below what is considered moderate risk based on blood-equivalents); however, these data represent colony means (i.e., grand mean for several wood stork, *Mycteria Americana*, colonies and means of three egret colonies) and, thus, individual birds may have levels more closely approaching the upper threshold.

Furthermore, as discussed above, Frederick and Jayasena (2010) found significant reproductive impairment in captive white ibises dosed near the lower TRV of Ackerman et al. (2016).

Rumbold et al. (2016) recently reported feather-[Hg] in 95 osprey (*Pandion haliaetus*) nestlings from across south Florida. When converted to blood equivalent concentration, levels in most birds again fell between the two toxic thresholds, [Hg] in 87% of the nestlings exceeded the lower threshold that represented the LOAEL, while 6%, mostly from Florida Bay, exceeded the upper toxic threshold and, thus, were at moderate risk. Comparing observed feather-[Hg] to values in the published literature, Rumbold et al. (2016) reported that levels were much higher in

Florida birds than in osprey recently sampled in Western Canada, Chesapeake Bay, Delaware Bay, Coastal South Carolina and the Great Lakes but did not exceed historical levels in osprey from two heavily contaminated areas. Because birds at these two contaminated areas showed no evidence of poor nesting success, Rumbold et al. (2016) concluded that Hg did not appear to be a critical stressor increasing the risk of nest failure in Florida ospreys. However, they cautioned that they could not rule out potential organismal-level effects to highly exposed nestlings after fledging. Because [Hg] has been found to dramatically increase in blood and other tissues after feather growth is complete (Spalding et al. 2000; Kenow et al. 2003; Ackerman et al. 2011), several authors argue that risk increases immediately after fledging (Kenow et al. 2003; Rumbold 2005; Ackerman et al. 2011), particularly for naive chicks attempting to forage on their own for the first time.

A small study that sampled blood (and feathers) from red-winged blackbirds (*Agelaius phoeniceus*; n = 5) and common grackles (*Quiscalus quiscula; n = 6*) along Interstate 75 within the WCA and BCNP found average [Hg] in blood (ranged as high as 0.82 mg/kg) to fall between the two of benchmarks (Borrego de Pezon 2011), suggesting some risk, if they are a sensitive species.

10.6.2.4 Risk to Florida Panthers

From the background discussion above, we know that there has been a great deal of interest in the potential risk that Hg poses to the endangered Florida panther ever since high levels were reported. FFWCC began collecting various tissue from panthers in 1978 for different reasons (these samples were subsequently subsampled and a portion analyzed for Hg by Roelke et al. 1991), and continue to collect hair for Hg determination (M. Cunningham, FFWCC, pers. comm.). As evident from Fig. 10.5, several panthers from 2008 to 2015 had hair-[Hg] in excess of one of the two criteria; overall, 8% of the individuals exceeded the lower threshold while 3% exceeded the upper criterion. Before assessing any possible temporal trends, we must also consider sampling locations. The highest Hg value from 2008 to 2015 shown in Fig. 10.5 (i.e., 100 mg/kg) was in an uncollared panther killed by a car near Francis Taylor Wildlife Management Area (WCA3B). This concentration was similar to values reported by Roelke et al. (1991) who sampled several panthers from ENP (highest individual measurement was 130 mg/kg). However, few cats have been sampled recently from the SE or NE regions (Fig. 10.5) with no samples coming from ENP during 2008–2015. There are various reasons for this spatial bias in recent sampling of panthers, including poor habitat in NE region (i.e., WCA3, due to high water levels) and increased focus on western populations (and of course, budget constraints). As described elsewhere (Chap. 8, this volume), panthers occupy different habitats, which affect their diet and Hg exposure. As with any biomonitoring program, care must be taken to understand where samples were collected across a spatially variable landscape. This spatial bias in monitoring panthers may skew our risk perception.

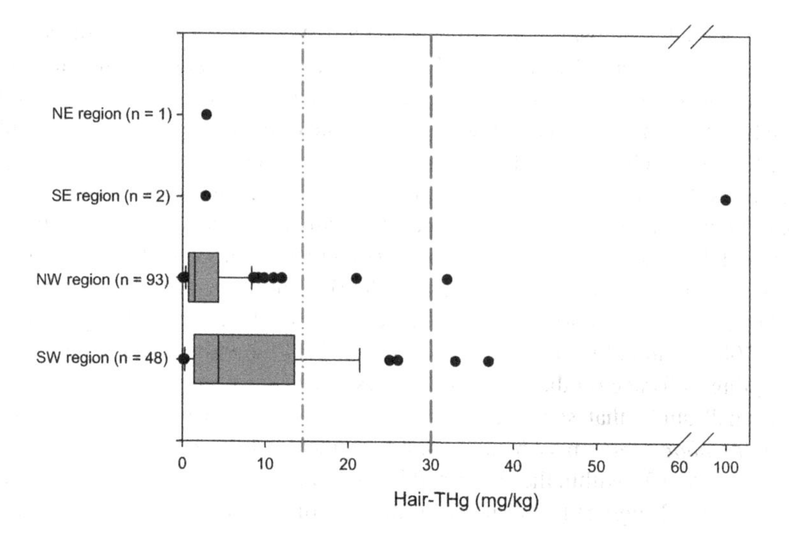

Fig. 10.5 Mercury concentrations in hair (mg/kg) collected from Florida panthers in different regions of south Florida from 2008 to 2015 (Mark Cunningham, FFWCC pers. comm.). Also shown are TRVs for terrestrial mammals based on results of different literature reviews or models (Eccles et al. 2017; Basu 2012)

Yet it is clear that individual panthers in some areas of south Florida are still highly exposed to Hg. As discussed below, this has greater ecological significance considering the endangered status of the Florida panther.

10.6.2.5 Risk to Sharks and Bottlenose Dolphins

As mentioned above, young blacktip sharks caught in Caloosahatchee River Estuary contained elevated [Hg] (much of which was believed to be derived from maternal offloading *in utero*) in excess of the fish-TRV (Fig. 10.3). Two recent surveys have reported even higher [Hg] in adult sharks of several species from south Florida waters (Rumbold et al. 2014; Matulik et al. 2017). It is not surprising therefore that a number of surveys have also found elevated [Hg] in bottlenose dolphins in south Florida estuaries (Woshner et al. 2008; Schaefer et al. 2011; Damseaux et al. 2017). The most recent survey by Damseaux et al. (2017) reported that dolphins from Florida coastal Everglades (e.g., Shark River and Whitewater Bay) had higher [Hg] (in skin biopsy samples) than dolphins from other estuaries of Florida and, to their knowledge the rest of the world. When these skin-[Hg] were translated the average blood-equivalent concentration (mean = 860 μg/l; based on data in Bryan et al. 2007) they greatly exceeded a mammalian-TRV (100 μg/l based on Clarkson and Magos 2006 as cited by Dietz et al. 2013). This mammalian-TRV was developed from observations of humans following the Minamata Bay incident (Dietz et al. 2013). It is noteworthy that except for a very small amount hair in dolphin calves that quickly falls out, these mammals do not have hair as a tissue to dump Hg.

10.6.3 LOE 3: Risk Based on Eco-epidemiological Wildlife Surveys in South Florida

Results from eco-epidemiological wildlife surveys that have looked for possible correlation between abnormalities in biota with higher [Hg] was the third line of evidence used in this ERA; however, these types of surveys are limited in number.

As mentioned earlier, previous surveys have found reproductive health abnormalities in various Everglades wildlife species that were associated with higher tissue Hg residues (e.g., Roelke et al. 1991; Spalding et al. 1994; Heath and Frederick 2005). Roelke et al. (1991), for example, reported that female panthers with blood Hg values >0.5 ppm had significantly fewer surviving offspring than females with blood Hg values <0.25 ppm. Heath and Frederick (2005) found feather-[Hg] in pre-breeding white ibises to be correlated with hormone levels; negatively with estradiol in females and positively with testosterone in males. They felt that these subacute effects could lead fewer birds to nest or more birds to abandon nests (Heath and Frederick 2005). While two other studies reported no significant relation between [Hg] and health metrics or reproductive success in wading birds (Rumbold et al. 2001; Herring et al. 2009), these two surveys were conducted in areas of the WCAs that, at that time, had relatively lower [Hg] (Rumbold et al. 2001; Herring et al. 2009). Rumbold et al. (2001), for example, reported that [Hg] in eggs and nestling feathers for the period of their study were below toxic thresholds and so it should not have been a surprise when they found no impact in brood size or fledging success. More recent surveys in areas of south Florida with high [Hg] have reported statistical associations between [Hg] and health metrics in wading birds. As presented above, average blood-equivalent [Hg] for several wading bird colonies recently exceeded the lower threshold (for altered behaviors) but did not exceed the higher threshold (for reduced egg hatchability and breeding success). Findings of a recent study by Rodriguez et al. (2016) that assessed possible Hg effect on great egret nesting success in the Everglades was in agreement with this prediction of altered behavior. They found that nests with average [Hg] in egg albumin >0.75 mg/kg had survival probabilities of only 42–57% as compared to survival probabilities as high as 90–95% in nests with average egg albumin [Hg] < 0.4 mg/kg. They concluded that Hg was not having an effect through teratogenesis or reduced hatchability but instead primarily through deficits in parental behavior (Rodriguez et al. 2016).

With regard to the health of fishes in south Florida estuaries, Adams et al. (2010) reported on an association of altered neuroanatomy (i.e., reduced N-methyl-D-aspartate receptor levels) and liver histopathology with higher [Hg] in spotted seatrout (*Cynoscion nebulosus*) from Shark River Estuary, ENP. These fish had a mean [Hg] of 0.39 mg/kg (whole body translated from 0.56 mg/kg in muscle) and so their finding effects would be consistent with predictions made above based on the fish-TRV suggested by Sandheinrich et al. (2011, i.e., 0.3 mg/kg). Similarly, the recent survey of young blacktips sharks in Calooshatachee River Estuary that found elevated [Hg] in excess of the fish-TRV (Fig. 10.3) also found an increased number of macrophage centers in livers of juveniles and altered adipose deposition in livers of neonates (Bridenbaugh 2017). As mentioned above, Damseaux et al. (2017) found bottlenose dolphins from Florida coastal Everglades to have the highest [Hg] ever reported in dolphins. While they did not assess health metrics in these dolphins, [Hg] were higher in these dolphins than dolphins from other estuaries where effects have been reported (Woshner et al. 2008; Schaefer et al. 2011). Schaefer et al. (2011), for example, reported a significant inverse relationship between blood [Hg] and several markers of endocrine function in dolphins, particularly populations sampled along the east coast of Florida.

10.7 Discussion

This ERA used three lines of evidence to characterize risk: LOE 1: Risk based on Fish-tissue [Hg] as the media of exposure; LOE 2: Risk based on Tissue Residues within Ecological Receptors and, LOE 3: Risk based on Eco-epidemiological Wildlife Surveys. Together these results suggest that biota in certain regions of south Florida continue to be exposed to levels of Hg that pose a threat. The LOE that was the most direct assessment and therefore had the least uncertainty was the results from eco-epidemiological wildlife surveys (i.e., LOE 3) showing statistical association between health metrics and [Hg]. These observed effects were in agreement with predictions based on tissue-residue effects thresholds for fish and wildlife reported in the literature (i.e., LOE 2). As suggested above, the tissue-residue effects approach is preferable to methods used in previous risk assessments that compared modeled dietary dose to established reference doses because of its lower uncertainty in the estimated risk (e.g., variability in bioavailability, integration of exposure over time and space) and the fact that it is less data intensive. However, because [Hg] cannot be biomonitored in every receptor, it was significant that the tissue-[Hg] residues observed in the fish and wildlife in south Florida were also consistent with predictions based on the USEPA's predator protection criteria (i.e., LOE 1). Clearly, the predator protection criteria based on [Hg] measured in fish as the media of exposure have the broadest application to make inferences about risk to a variety of ecological receptors across the Everglades and south Florida. Recall that over 60% of individual fish (at TL 3) collected from the freshwater Everglades and estuaries in south Florida have [Hg] that put predators at risk. So, while the focus above was on a handful of fish and wildlife species (e.g., wading birds, osprey, eagles, panthers, sharks, dolphins, etc.) that had been surveyed or included in models, they should be viewed as sentinels for other ecological receptors in south Florida that may also be at risk. These include other mammals common to south Florida such as raccoons (*Procyon lotor*); bobcat (*Lynx rufus floridanus*), river otter (*Lontra Canadensis*); the Everglades Mink (*Mustela vison evergladensis*) and several species of bats (Chiroptera). Roelke et al. (1991) reported high [Hg] in raccoons, otters and bobcats sampled in the later 1980s–early 1990s supporting this risk hypothesis. Clearly, other piscivorous bird species in addition to those already reverenced above (Table 10.1; Fig. 10.4) could be at risk. In the past, insectivores were overlooked when monitoring wildlife for Hg exposure. It was assumed that the low trophic level of insects would pose little risk for bioaccumulation in insectivorous predators. However, as reviewed in Chap. 8 (this volume), Loftus' (2000) work in ENP was one of the first studies that showed aquatic insects could accumulate high levels of Hg. More recent surveys of insectivorous bats and birds done in other regions have demonstrated that they can be highly exposed (i.e., from high consumption rates of this low-caloric density prey) and, thus, may be at risk (Nam et al. 2012; Yates et al. 2014; Becker et al. 2017). This risk hypothesis was supported in south Florida by the limited study done on red-winged blackbirds and common grackles (Borrego de Pezon 2011). This risk hypothesis was raised previously regarding the Cape Sable seaside sparrow (*Ammodramus maritimus mirabilis*) nesting in ENP (Rumbold et al. 2008). Subsequently, moderate levels of Hg have been reported in the feathers of these sparrows (particularly population E; D. Krabbenhoft, US Geological Survey, pers. comm.).

To provide some context, [Hg] in south Florida fish and wildlife remain as some of the highest in North America, if not the world (Rumbold et al. 2016; also, see Chap. 9 of this volume; *cf.* Sandheinrich et al. 2011; Dietz et al. 2013; Ackerman et al. 2016; Damseaux et al. 2017).

10.7.1 Ecological Significance

There are several criteria that need be considered when evaluating the ecological significance of these risks. Societal important attributes include economically important species, endangered species or those recognized as playing key roles within the ecosystem. Several of the above species are listed either at the Federal or state level as endangered, threatened or species of special concern. The risk from Hg may have added ecological significance because many of these species (both fish and wildlife) have been recognized as important to the structure, function, and/or health of the Everglades ecosystem to the point that their populations were selected as a restoration target for CERP (Jacobs 2005).

Some may argue that this risk estimation was based largely on non-lethal endpoints, e.g., neurochemical changes, cellular or tissue pathology, altered behavior, etc.) that are not necessarily ecological significant. However, as previously argued, these sub-lethal effects can lead to other keys events in the adverse outcome pathway (Fig. 10.1) leaving the organism at a competitive disadvantage (e.g., unable to find food causing malnutrition, increased susceptibility to disease, depressed reproduction) or at an increased risk of predation. The problem is how to distinguish the impact of this ecological mortality from the noise of natural variability and reductions in populations caused by the numerous other stressors in the Everglades.

Because Hg exposure is so highly variable spatially and temporally in the Everglades and south Florida, so is risk. As previously discussed, risk estimation cannot be simplified to a single system-wide hazard quotient. One of the objectives of this regional risk assessment was to assess the spatial distribution of current risk between sub-regions. Currently, the highest potential ecological risk area for Hg is within the southern freshwater region of the Everglades, particularly ENP (for discussion of added ecological significance of risk in this national park, see Rumbold et al. 2008), NE Florida bay, southwest Florida estuaries and is limited to isolated biological hotspots in the north (e.g., HOLYBC; CA3F3). Basically, it comes down to whether current risk of ecological mortality within local populations in those areas of south Florida is acceptable. Perhaps it is more appropriate, therefore, to consider the risk in terms of meta-populations consisting of source and sink patches. Source patches (in high quality, low-Hg habitat) would have surplus offspring that disperse to other patches; whereas, sink patches (lower quality, high-Hg habitat) are maintained to some degree by immigration of individuals from source patches. Raimondo and Barron (2008) conducted a modeling exercise to assess population-level impacts and meta-population extinction probability of Florida panthers under various Hg exposure scenarios. Their model predicted the meta-population remained stable; however, local populations went extinct even under relatively low Hg stress (Raimondo and Barron 2008). Furthermore, their scenarios involved exposure to only two of the four population centers (ENP and BCNP) and assumed that no land management change altered available habitat during the 100-year simulation period. Clearly, the ecological significance of Hg risk in south Florida must be viewed within the context of cumulative risks (and possible interactions of stressors as discussed above). Without similar modeling exercises assessing cumulative impacts on local populations and stability of meta-populations of more receptors, it is difficult to characterize current risk accurately.

We need to also consider future risk under CERP. It is reasonable to envisage two future environmental scenarios under CERP: (1) where net MeHg production remains status quo and only the distribution of some animals have changed as a result of ecological restoration and, (2) where net MeHg production and bioavailability has increased over a larger area due to some activity associated with CERP. As discussed above, restoration targets under CERP include shifting the distribution of high-density pools of fishes south and, as a result of habitat improvement, reestablishing wading bird nesting colonies along the coastal regions of the southern Everglades. Because Hg risk is currently greatest in Shark River Slough and estuary, reestablishing wading bird super-colonies in this area could represent an "ecological trap" (Dwernychuk and Boag 1972), i.e., created habitat appears suitable but when selected results in reduced fitness. Of potentially greater concern is the possibility that net methylation increases over a larger area due to some management action of CERP. As shown in the conceptual model (Fig. 10.1), a number of environmental factors influence net methylation. Management actions under CERP include changing quantity, quality, timing and distribution of water (and loads of sulfate) and, thus, have a high probability of affecting net methylation rates in restored areas. The potential for net methylation rate and MeHg biomagnification to increase dramatically ($>10\times$ fold increase) over a very short period time has previously been documented in south Florida following changes in water management (e.g., Rumbold et al. 2001; Krabbenhoft and Fink 2001; Rumbold and Fink 2006).

10.7.2 Risk Management

This ERA was developed to provide information to resource managers about current risks and potential unintended consequences from management actions that could increase this risk in the future. State and federal agencies have long been aware of the Hg risks in south Florida to both wildlife and humans. Many of the previous risk assessments were done by government scientists or researchers under contract to those agencies. State agencies continue to biomonitor Hg levels in fish and evaluate results using the USEPA predator protection criteria (Julian et al. 2016) and, so, are fully aware of at least one line of evidence indicating risk (i.e., [Hg] in fish exceeding predator protection criteria). An analysis of previous or potential future management actions to reduce or eliminate Hg risks in Florida are beyond the scope of this assessment; instead, the reader is referred to Chap. 7 (in Volume III). As a water management district, SFWMD has always taken the position that fixing the Hg problem was not in their mandate (i.e., that it was the role of FDEP or federal government to reduce Hg emissions); however, in the past they recognized their management actions could worsen the problem. In at least one case, SFWMD took corrective action to reduce net methylation and MeHg biomagnification in one of their Stormwater Treatment Areas after monitoring revealed significant risk (Rumbold and Fink 2006). Initially the CERP RECOVER team (i.e., REstoration COordination & VERification) recognized that CERP could unintentionally worsen the Hg problem and, accordingly, developed a system-wide performance measure (Jacobs 2005) and monitoring plan to allow for future adaptive management in case it worsened. They contracted with NOAA and FFWCC from 2006 to 2008 to survey [Hg] in fish across south Florida, including estuaries downstream of areas targeted for restoration (Evans 2008), but have discontinued this system-wide monitoring. As demonstrated above, while eco-epidemiological wildlife surveys of Hg impact on health or reproductive metrics are the most direct method to assess risks, they are rare in south Florida due to the difficulty in obtaining funding. While government agencies frequently monitor [Hg] in various media including fish (often in support of consumption advisories for human health protection), annual monitoring programs traditionally do not explicitly look for effects. Unfortunately, without improved estimates of the frequency of these sublethal responses (i.e., biomarkers of effect) and probability of ecological mortality, risks may go underappreciated by resource managers who may view any uncertainty as justification for taking no action. Whether state and federal agencies are doing enough to reduce or eliminate these risks or at least not worsen them is a question for interested parties and the public.

10.8 Conclusions

Using multiple lines of evidence, this regional risk assessment provided a more robust picture of the nature and extent of risk to Everglades biota than could be obtained using any single line of evidence. This risk analysis clearly shows that Hg remains a risk to a variety of ecological receptors in certain sub-regions or hotspots in south Florida. The existing risk is of sublethal effects from Hg toxicity (e.g., lethargy, impaired escape and avoidance behavior, altered spawning behavior, altered brooding behavior, histopathology, altered motor function, etc.), but which could lead to reduced fitness and ecological mortality. In the future, following ecological restoration under CERP, it is anticipated that more animals will be drawn to some of these Hg hotspots; thus, increasing their risk. Furthermore, because numerous environmental factors related to water quality and hydrology are known to influence Hg methylation and bioavailability in south Florida, there is a strong possibility that management actions under CERP could re-distribute or increase the areal extent of the Hg hotspots.

References

Ackerman JT, Eagles-Smith CA, Herzog MP (2011) Bird mercury concentrations change rapidly as chicks age: toxicological risk is highest at hatching and fledging. Environ Sci Technol 45 (12):5418–5425

Ackerman JT, Eagles-Smith CA, Herzog MP, Hartman CA, Peterson SH, Evers DC et al (2016) Avian mercury exposure and toxicological risk across western North America: a synthesis. Sci Total Environ 568:749–769

Adams DH, Sonne C, Basu N, Dietz R, Nam D-H, Leifsson PS, Jensen AL (2010) Mercury contamination in spotted seatrout, *Cynoscion nebulosus*: an assessment of liver, kidney, blood, and nervous system health. Sci Total Environ 408(23):5808–5816

Ankley GT, Bennett RS, Erickson RJ, Hoff DJ, Hornung MW, Johnson RD, Mount DR, Nichols JW, Russom CL, Schmieder PK (2010) Adverse outcome pathways: a conceptual framework to support ecotoxicology research and risk assessment. Environ Toxicol Chem 29:730–741

Barr JF (1986) Population dynamics of the common loon (*Gavia immer*) associated with mercury-contaminated waters in northwestern Ontario. Canadian Wildlife Service Occasional Papers No. 56, 23 pp

Barron MG, Duvall SE, Barron KJ (2004) Retrospective and current risks of mercury to panthers in the Florida Everglades. Ecotoxicology 13(3):223–229

Barst BD, Rosabal M, Campbell PG, Muir DG, Wang X, Köck G, Drevnick PE (2016) Subcellular distribution of trace elements and liver histology of landlocked Arctic char (Salvelinus alpinus) sampled along a mercury contamination gradient. Environ Pollut 212:574–583

Basu N (2012) Piscivorous mammalian wildlife as sentinels of methylmercury exposure and neurotoxicity in humans. Methylmercury Neurotox:357–370

Basu N (2015) Applications and implications of neurochemical biomarkers in environmental toxicology. Environ Toxicol Chem 34(1):22–29

Basu N, Scheuhammer AM, Rouvinen-Watt K, Grochowina N, Evans RD, O'Brien M, Chan HM (2007) Decreased N-methyl-D-aspartic acid (NMDA) receptor levels are associated with mercury exposure in wild and captive mink. Neurotoxicology 28(3):587–593

Becker DJ, Chumchal MM, Bentz AB, Platt SG, Czirják GÁ, Rainwater TR et al (2017) Predictors and immunological correlates of sublethal mercury exposure in vampire bats. R Soc Open Sci 4 (4). https://doi.org/10.1098/rsos.170073

Beckvar N, Dillon TM, Read LB (2005) Approaches for linking whole-body fish tissue residues of mercury or DDT to biological effects thresholds. Environ Toxicol Chem 24(8):2094–2105

Berntssen MH, Aatland A, Handy RD (2003) Chronic dietary mercury exposure causes oxidative stress, brain lesions, and altered behaviour in Atlantic salmon (*Salmo salar*) parr. Aquat Toxicol 65(1):55–72

Borrego de Pezon ND (2011) Mercury accumulation in south Florida blackbirds family: Icteridae. Unpublished MS Thesis, Florida Gulf Coast University, Ft. Myers, FL

Bouton SN, Frederick PC, Spalding MG, McGill H (1999) Effects of chronic, low concentrations of dietary methylmercury on the behavior of juvenile great egrets. Environ Toxicol Chem 18:1934–1939

Brandon AL (2011) Spatial and temporal trends in mercury concentrations in the blood and hair of Florida Panthers (*Puma concolor coryi*). Unpublished MS Thesis, Florida Gulf Coast University, Ft. Myers, FL

Bridenbaugh SN (2017) The hematological and histological effects of maternal offloading of mercury in neonatal and Juvenile Blacktip Sharks (*Carcharhinus limbatus*). Unpublished MS Thesis, Florida Gulf Coast University, Ft. Myers, FL

Bryan CE, Christopher SJ, Balmer BC, Wells RS (2007) Establishing baseline levels of trace elements in blood and skin of bottlenose dolphins in Sarasota Bay, Florida: implications for non-invasive monitoring. Sci Total Environ 388(1–3):325–342

Chen CY, Driscoll CT, Kamman NC (2012) Mercury hotspots in freshwater ecosystems: drivers, processes, and patterns. In: Bank MS (ed) Mercury in the environment. University of California Press, Berkeley, pp 143–166

Corrales J, Naja GM, Dziuba C, Rivero RG, Orem W (2011) Sulfate threshold target to control methylmercury levels in wetland ecosystems. Sci Total Environ 409(11):2156–2162

Damseaux F, Kiszka JJ, Heithaus MR, Scholl G, Eppe G, Thomé J-P et al (2017) Spatial variation in the accumulation of POPs and mercury in bottlenose dolphins of the Lower Florida Keys and the coastal Everglades (South Florida). Environ Pollut 220:577–587

Depew DC, Basu N, Burgess NM, Campbell LM, Devlin EW, Drevnick PE et al (2012a) Toxicity of dietary methylmercury to fish: derivation of ecologically meaningful threshold concentrations. Environ Toxicol Chem 31(7):1536–1547

Depew DC, Basu N, Burgess NM, Campbell LM, Evers DC, Grasman KA, Scheuhammer AM (2012b) Derivation of screening benchmarks for dietary methylmercury exposure for the common loon (Gavia immer): rationale for use in ecological risk assessment. Environ Toxicol Chem 31(10):2399–2407

Dietz R, Sonne C, Basu N, Braune B, O'Hara T, Letcher RJ et al (2013) What are the toxicological effects of mercury in Arctic biota? Sci Total Environ 443:775–790

Dillon T, Beckvar N, Kern J (2010) Residue-based mercury dose–response in fish: an analysis using lethality-equivalent test endpoints. Environ Toxicol Chem 29(11):2559–2565

Dornbos P, Strom S, Basu N (2013) Mercury exposure and neurochemical biomarkers in multiple brain regions of Wisconsin river otters (Lontra canadensis). Ecotoxicology 22(3):469–475

Duvall SE, Barron MG (2000) A screening level probabilistic risk assessment of mercury in Florida Everglades food webs. Ecotoxicol Environ Saf 47(3):298–305

Dwernychuk L, Boag D (1972) Ducks nesting in association with gulls—An ecological trap? Can J Zool 50(5):559–563

Eccles KM, Thomas PJ, Chan HM (2017) Predictive meta-regressions relating mercury tissue concentrations of freshwater piscivorous mammals. Environ Toxicol Chem. https://doi.org/10.1002/etc.3775

Ercal N, Gurer-Orhan H, Aykin-Burns N (2001) Toxic metals and oxidative stress part I: mechanisms involved in metal-induced oxidative damage. Curr Top Med Chem 1(6):529–539

Escher BI, Ashauer R, Dyer S, Hermens JL, Lee JH, Leslie HA et al (2011) Crucial role of mechanisms and modes of toxic action for understanding tissue residue toxicity and internal effect concentrations of organic chemicals. Integr Environ Assess Manag 7(1):28–49

Evans DW (2008) Annual report: Assessment of mercury bioaccumulation in sentinel fish in south Florida. NOAA Center for Coastal Fisheries and Habitat Research. NOS Agreement Code MOA-204-191/6945. Beaufort, NC

Evans RD, Hickie B, Rouvinen-Watt K, Wang W (2016) Partitioning and kinetics of methylmercury among organs in captive mink (*Neovison vison*): a stable isotope tracer study. Environ Toxicol Pharmacol 42:163–169

Evers DC, Han Y-J, Driscoll CT, Kamman NC, Goodale MW, Lambert KF et al (2007) Biological mercury hotspots in the northeastern United States and southeastern Canada. Bioscience 57 (1):29–43

Exponent (1998) Ecological risks to wading birds of the Everglades in relation to phosphorus reductions in water and mercury bioaccumulation in fishes. Prepared for the Sugar Cane Growers Cooperative. Exponent, Bellevue, WA

Exponent (2000) An evaluation of population risks to avian and mammalian wildlife in the Northern Everglades. Technical memorandum prepared for the Sugar Cane Growers Cooperative of Florida. Exponent, Bellevue, WA. Reprinted in 2001 Everglades Consolidated Report, Appendix 1–2d. http://my.sfwmd.gov/portal

Fay KA, Villeneuve DL, LaLone CA, Song Y, Tollefsen KE, Ankley GT (2017) Practical approaches to adverse outcome pathway development and weight-of-evidence evaluation as illustrated by ecotoxicological case studies. Environ Toxicol Chem 36(6):1429–1449

Fleming DM, Wolff WF, DeAngelis DL (1994) Importance of landscape heterogeneity to wood storks in Florida Everglades. Environ Manag 18(5):743–757

Frederick P, Jayasena N (2010) Altered pairing behaviour and reproductive success in white ibises exposed to environmentally relevant concentrations of methylmercury. Proc R Soc Lond B Biol Sci 278:1851–1857

Frederick P, Spalding MG, Sepulveda MS, Williams GE Jr, Bouton S, Lynch H, . . . Hoffman D (1997) Effects of environmental mercury exposure on reproduction, health and sur vival of wading birds in Florida Everglades. Report to Florida Department of Environmental Protection, Gainesville, FL

Frederick PC, Spalding MG, Dusek R (2002) Wading birds as bioindicators of mercury contamination in Florida: annual and geographic variation. Environ Toxicol Chem 21:262–264

Grillitsch B, Schiesari L (2010) The ecotoxicology of metals in reptiles. In: Sparling DW, Linder G, Bishop CA, Krest SK (eds) Ecotoxicology of amphibians and reptiles, 2nd edn. Society of Environmental Toxicology and Chemistry (SETAC), Pensacola, pp 337–448

Hall BD, Bodaly RA, Fudge RJP, Rudd JWM, Rosenberg DM (1997) Food as the dominant pathway of methylmercury uptake by fish. Water Air Soil Pollut 100:13–24

Heath JA, Frederick PC (2005) Relationships among mercury concentrations, hormones, and nesting effort of white ibises (*Eudocimus albus*) in the Florida Everglades. Auk 122(1):255–267

Heaton-Jones TG, Homer BL, Heaton-Jones DL, Sundlof SF (1997) Mercury distribution in American alligators (*Alligator mississippiensis*) in Florida. J Zoo Wildl Med 28:62–70

Heinz GH (1979) Methylmercury: reproductive and behavioral effects on three generations of Mallard ducks. J Wildl Manag 43:394–401

Henry KA, Cristol DA, Varian-Ramos CW, Bradley EL (2015) Oxidative stress in songbirds exposed to dietary methylmercury. Ecotoxicology 24(3):520–526

Herring G, Gawlik DE, Rumbold DG (2009) Feather mercury concentrations and physiological condition of great egret and white ibis nestlings in the Florida Everglades. Sci Total Environ 407 (8):2641–2649

Hightower JM (2008) Diagnosis mercury: money, politics, and poison. Island Press, Washington, DC

Hoffman DJ (1995) Wildlife toxicity testing. Handb Ecotoxicol:356–391

Hoffman DJ, Spalding MG, Frederick PC (2005) Subchronic effects of methylmercury on plasma and organ biochemistries in great egret nestlings. Environ Toxicol Chem 24(12):3078–3084

Hoffman DJ, Eagles-Smith CA, Ackerman JT, Adelsbach TL, Stebbins KR (2011) Oxidative stress response of Forster's terns (*Sterna forsteri*) and Caspian terns (*Hydroprogne caspia*) to mercury and selenium bioaccumulation in liver, kidney, and brain. Environ Toxicol Chem 30 (4):920–929

Jacobs K (2005) Appendix 7-1: Summary of CERP systemwide assessment performance measures. In: 2005 South Florida Environmental Report. South Florida Water Management District, West Palm Beach, FL. http://www.sfwmd.gov/sfer/

Julian P, Gu B, Redfield G, Weaver K, Lange T, Frederick P, McCray JM, Wright AL, Dierberg FE, DeBusk TA, Jerauld M, DeBusk WF, Bae H, Ogram A (2015) Chapter 3B: Mercury and sulfur environmental assessment for the Everglades. In: 2015 South Florida Environmental Report. South Florida Water Management District, West Palm Beach, FL

Julian P, Gu B, Redfield G, Weaver K (2016) Chapter 3B: Mercury and sulfur environmental assessment for the Everglades. In: 2016 South Florida Environmental Report. South Florida Water Management District, West Palm Beach, FL

Jurczk NU (1993) An ecological risk assessment of the impact of mercury contamination in the Florida Everglades. Unpublished MS Thesis, Univrsity of Florida, Gainesville, FL

Karagas MR, Choi AL, Oken E, Horvat M, Schoeny R, Kamai E et al (2012) Evidence on the human health effects of low-level methylmercury exposure. Environ Health Perspect 120 (6):799

Kenow KP, Gutreuter S, Hines RK, Meyer MW, Fournier F, Karasov WH (2003) Effects of methyl mercury exposure on the growth of juvenile common loons. Ecotoxicology 12(1–4):171–181

Kenow KP, Hoffman DJ, Hines RK, Meyer MW, Bickham JW, Matson CW et al (2008) Effects of methylmercury exposure on glutathione metabolism, oxidative stress, and chromosomal damage in captive-reared common loon (*Gavia immer*) chicks. Environ Pollut 156(3):732–738

Kirkpatrick M, Benoit J, Everett W, Gibson J, Rist M, Fredette N (2015) The effects of methylmercury exposure on behavior and biomarkers of oxidative stress in adult mice. Neurotoxicology 50:170–178

Krabbenhoft DP, Fink L (2001) Appendix 7-8: The effect of dry down and natural fires on mercury methylation in the Florida Everglades. In: 2001 Everglades Consolidated Report. South Florida Water Management District, West Palm Beach, FL

Lange TR, Richard DA, Royals HE (1998) Trophic relationships of mercury bioaccumulation in fish from the Florida Everglades. Annual Report. Florida Game and Fresh Water Fish Commission, Fisheries Research Laboratory, Eustis, FL. Prepared for the Florida Department of Environmental Protection, Tallahassee, FL

Lange TR, Richard DA, Royals HE (1999) Trophic relationships of mercury bioaccumulation in fish from the Florida Everglades. Annual Report. Florida Game and Fresh Water Fish Commission, Fisheries Research Laboratory, Eustis, FL. Prepared for the Florida Department of Environmental Protection, Tallahassee, FL

Lange TR, Richard DA, Royals HE (2000) Long-term trends of mercury bioaccumulation in Florida's Largemouth bass. Abstracts of the annual meeting of the South Florida mercury science program, Tarpon Springs, FL, 8–11 May 2000

Lange TR, Richard DA, Sargent B (2005) Annual fish mercury monitoring report, August 2005. Long-term monitoring of Mercury in largemouth bass from the Everglades and Peninsular Florida. Florida Fish and Wildlife Conservation Commission, Eustis, FL

Loftus WF (2000) Accumulation and fate of mercury in an Everglades aquatic food web. PhD dissertation, Florida International University, Miami, FL

Matulik AG, Kerstetter DW, Hammerschlag N, Divoll T, Hammerschmidt CR, Evers DC (2017) Bioaccumulation and biomagnification of mercury and methylmercury in four sympatric coastal sharks in a protected subtropical lagoon. Mar Pollut Bull 116:357–364

McCarty L, Landrum P, Luoma S, Meador J, Merten A, Shephard B, Van Wezel A (2011) Advancing environmental toxicology through chemical dosimetry: external exposures versus tissue residues. Integr Environ Assess Manag 7(1):7–27

McLean AR, Bush E (1999) Chapter 10: The C&SF restudy. In: Everglades Interim Report. South Florida Water Management District, West Palm Beach, FL. http://my.sfwmd.gov/portal/page/portal

Meador JP, Adams WJ, Escher BI, McCarty LS, McElroy AE, Sappington KG (2011) The tissue residue approach for toxicity assessment: findings and critical reviews from a Society of Environmental Toxicology and Chemistry Pellston Workshop. Integr Environ Assess Manag 7(1):2–6

Mela M, Randi M, Ventura D, Carvalho C, Pelletier E, Ribeiro CO (2007) Effects of dietary methylmercury on liver and kidney histology in the neotropical fish Hoplias malabaricus. Ecotoxicol Environ Saf 68(3):426–435

Mieiro C, Ahmad I, Pereira M, Duarte A, Pacheco M (2010) Antioxidant system breakdown in brain of feral golden grey mullet (Liza aurata) as an effect of mercury exposure. Ecotoxicology 19(6):1034–1045

Nam D-H, Yates D, Ardapple P, Evers DC, Schmerfeld J, Basu N (2012) Elevated mercury exposure and neurochemical alterations in little brown bats (Myotis lucifugus) from a site with historical mercury contamination. Ecotoxicology 21(4):1094–1101

Newman MC (2015) Fundamentals of Ecotoxicology, 4th edn. CRC Press, Boca Raton, FL, p 654

Nilsen FM, Parrott BB, Bowden JA, Kassim BL, Somerville SE, Bryan TA et al (2016) Global DNA methylation loss associated with mercury contamination and aging in the American alligator (Alligator mississippiensis). Sci Total Environ 545:389–397

Nilsen FM, Kassim BL, Delaney JP, Lange TR, Brunell AM, Guillette LJ et al (2017) Trace element biodistribution in the American alligator (*Alligator mississippiensis*). Chemosphere 181:343–351

Peters LJ (1983) Mercury accumulation in the American alligator, *Alligator mississippiensis*. Unpublished MS thesis, University of Florida, Gainesville, FL

Peterson SA, Van Sickle J, Hughes RM, Schacher JA, Echols SF (2005) A biopsy procedure for determining filet and predicting whole-fish mercury concentration. Arch Environ Contam Toxicol 48(1):99–107

PTI (1995) Ecological risks to wading birds of the Everglades in relation to phosphorus reductions in water and mercury bioaccumulation in fishes. Prepared for Sugar Cane Growers Cooperative. PTI Environmental Services, Bellevue, WA

Raimondo S, Barron MG (2008) Population-level modeling of mercury stress in the Florida panther (Puma concolor coryi) metapopulation. In: Akcakaya HR (ed) Demographic toxicity: methods in ecological risk assessment. Oxford University Press, New York, NY, pp 40–53

Rodriguez IA, Vitale NE, Orzechowski S, Frederick P (2016) From egg to fledge: partitioning effects of mercury exposure on reproduction in Great Egrets (*Ardea alba*). Presented at the 37th Annual Meeting and 7th World Congress of Society of Environmental Toxicology and Chemistry, Orlando, FL, 6–10 November 2016

Roelke M, Schultz D, Facemire C, Sundlof S, Royals H (1991) Mercury contamination in Florida panthers. Prepared by the Technical Subcommittee of the Florida Panther Interagency Committee

Rumbold DG (1999) Appendix 7.2: Ecological risk assessment of mercury in the Florida Everglades: risk to wading birds. In: Everglades Interim Report. South Florida Water Management District, West Palm Beach, FL. http://my.sfwmd.gov/portal/page/portal

Rumbold DG (2000) Appendix 7.3b: Methylmercury risk to Everglades wading birds: a probabilistic ecological risk assessment. In: 2000 Everglades Consolidated Report. South Florida Water Management District, West Palm Beach, FL. http://my.sfwmd.gov/portal/

Rumbold DG (2005) A probabilistic risk assessment of the effects of methylmercury on great egrets and bald eagles foraging at a constructed wetland in South Florida relative to the Everglades. Hum Ecol Risk Assess 11(2):365–388

Rumbold DG, Fink LE (2006) Extreme spatial variability and unprecedented methylmercury concentrations within a constructed wetland. Environ Monit Assess 112(1-3):115–135

Rumbold DG, Rawlik P (2000) Annual permit compliance monitoring report for mercury in stormwater treatment areas and downstream receiving waters. Appendix 7-2 in Everglades Consolidated Report. South Florida Water Management District, West Palm Beach, FL

Rumbold DG, Rawlik P, Fink L (1999) Appendix 7-2: Ecological risk assessment of mercury in the Florida Everglades: risk to wading birds. In: The Everglades Interim Report to the Florida Legislature. South Florida Water Management District, West Palm Beach, FL

Rumbold DG, Fink L, Laine K, Matson F, Niemczyk S, Rawlik P (2001) Appendix 7-9: Annual permit compliance monitoring report for mercury in stormwater treatment areas and downstream receiving waters of the Everglades Protection Area. In: 2001 Everglades Consolidated Report. South Florida Water Management District, West Palm Beach, FL. http://www.sfwmd.gov/sfer/

Rumbold DG, Fink LE, Laine KA, Niemczyk SL, Chandrasekhar T, Wankel SD, Kendall C (2002) Levels of mercury in alligators (*Alligator mississippiensis*) collected along a transect through the Florida Everglades. [Article]. Sci Total Environ 297(1–3):239–252. https://doi.org/10.1016/s0048-9697(02)00132-8

Rumbold DG, Lange TR, Axelrad DM, Atkeson TD (2008) Ecological risk of methylmercury in Everglades National Park, Florida, USA. Ecotoxicology 17:632–641

Rumbold DG, Wasno R, Hammerschlag N, Volety A (2014) Mercury accumulation in sharks from the coastal waters of Southwest Florida. Arch Environ Contam Toxicol 67:402–412

Rumbold DG, Miller KE, Dellinger TA, Haas N (2016) Mercury concentrations in feathers of adult and nestling osprey (Pandion haliaetus) from coastal and freshwater environments of Florida. Arch Environ Contam Toxicol 72:31–38

Sandheinrich MB, Bhavsar SP, Bodaly R, Drevnick PE, Paul EA (2011) Ecological risk of methylmercury to piscivorous fish of the Great Lakes region. Ecotoxicology:1–11

Sappington KG, Bridges TS, Bradbury SP, Erickson RJ, Hendriks AJ, Lanno RP et al (2011) Application of the tissue residue approach in ecological risk assessment. Integr Environ Assess Manag 7(1):116–140

Schaefer A, Stavros H-C, Bossart G, Fair P, Goldstein J, Reif J (2011) Associations between mercury and hepatic, renal, endocrine, and hematological parameters in Atlantic Bottlenose Dolphins (*Tursiops truncatus*) along the eastern coast of Florida and South Carolina. Arch Environ Contam Toxicol 61(4):688–695

Scheuhammer AM, Wong AH, Bond D (1998) Mercury and selenium accumulation in common loons (Gavia immer) and common mergansers (Mergus merganser) from eastern Canada. Environ Toxicol Chem 17(2):197–201

Scheuhammer AM, Meyer MW, Sandheinrich MB, Murray MW (2007) Effects of environmental methylmercury on the health of wild birds, mammals, and fish. Ambio 36(1):12–18

Scheuhammer A, Braune B, Chan HM, Frouin H, Krey A, Letcher R et al (2015) Recent progress on our understanding of the biological effects of mercury in fish and wildlife in the Canadian Arctic. Sci Total Environ 509:91–103

Schueneman T (2001) Characterization of sulfur sources in the EAA. Soil Crop Sci Soc Fl Proc 60:49–52

Snodgrass JW, Jagoe CH, Bryan ALJ, Brant HA, Burger J (2000) Effects of trophic status and wetland morphology, hydroperiod, and water chemistry on mercury concentrations in fish. Can J Fish Aquat Sci 57:171–180

Spalding MG, Forrester DJ (1991) Effects of parasitism and disease on the nesting success of colonial wading birds (Ciconiiformes) in southern Florida. Report to the Florida Game and Fresh Water Fish Commission. Report no. NG88-008. Tallahassee, FL

Spalding MG, Bjork RD, Powell GV, Sundlof SF (1994) Mercury and cause of death in great white herons. J Wildl Manag 58:735–739

Spalding MG, Frederick PC, McGill HC, Bouton SN, McDowell LR (2000) Methylmercury accumulation in tissues and its effects on growth and appetite in captive great egrets. J Wildl Dis 36(3):411–422
Stober J, Scheidt D, Jones R, Thornton K, Gandy L, Stevens D, Trexler J, Rathbun S (1998) South Florida ecosystem assessment: monitoring for ecosystem restoration. Final Technical Report – Phase I. EPA 904-R-98-002. USEPA Region 4 Science and Ecosystem Support Division and Office of Research and Development, Athens, GA, p 285 plus appendices
Stone WB, Okoniewski JC (2001) Necropsy findings and environmental contaminants in common loons from New York. J Wildl Dis 37(1):178–184
Sundlof SF, Spalding MG, Wentworth JD, Steible CK (1994) Mercury in livers of wading birds (ciconiiformes) in southern Florida. Arch Environ Contam Toxicol 27(3):299–305
Suter GW (2007) Ecological risk assessment, 2nd edn. CRC Press, Boca Raton, FL, p 643
Ugarte CA, Rice KG, Donnelly MA (2005) Variation of total mercury concentrations in pig frogs (Rana grylio) across the Florida Everglades, USA. Sci Total Environ 345:51–59
USEPA (1997) Mercury study report to Congress, vol VI: an ecological assessment for anthropogenic mercury emissions in the United States. U.S. Environmental Protection Agency, Washington, DC. EPA-452/R-97-008
USEPA (1998) Guidelines for ecological risk assessment. U.S. Environmental Protection Agency, Washington, DC. EPA/630/R-95/002F
USEPA (2001) Water quality criterion for the protection of human health: methylmercury. Washington, DC. EPA/823/R-01-001
Valavanidis A, Vlahogianni T, Dassenakis M, Scoullos M (2006) Molecular biomarkers of oxidative stress in aquatic organisms in relation to toxic environmental pollutants. Ecotoxicol Environ Saf 64(2):178–189
Van der Molen EJ, Blok AA, de Graaf GJ (1982) Winter starvation and mercury intoxication in grey herons (*Ardea cinerea*) in the Netherlands. Ardea 70:173–184
Ware FJ, Royals H, Lange T (1990) Mercury contamination in Florida largemouth bass. Paper presented at the Proceedings of the Annual Conference of the Southeast Association of Fish and Wildlife Agencies
Webber HM, Haines TA (2003) Mercury effects on predator avoidance behavior of a forage fish, Golden Shiner (*Notemigonus Crysoleucas*). Environ Toxicol Chem 22(7):1556–1561
Wiener JG, Krabbenhoft DP, Heinz GH, Scheuhammer AM (2003) Ecotoxicology of mercury. Handb Ecotoxicol 2:409–463
Woshner V, Knott K, Wells R, Willetto C, Swor R, O'Hara T (2008) Mercury and selenium in blood and epidermis of bottlenose dolphins (*Tursiops truncatus*) from Sarasota Bay, FL: interaction and relevance to life history and hematologic parameters. EcoHealth 5(3):360–370
Yates DE, Adams EM, Angelo SE, Evers DC, Schmerfeld J, Moore MS et al (2014) Mercury in bats from the northeastern United States. Ecotoxicology 23(1):45–55

Chapter 11
Everglades Mercury: Human Health Risk

Donald M. Axelrad, Charles Jagoe, and Alan Becker

Abstract Human exposure to mercury, a potent neurotoxicant, results primarily from consumption of fish contaminated with methylmercury. The Everglades is a mercury-in-fish hotspot by reason of its high deposition rate of atmospheric mercury, agricultural inputs of sulfate, and the biogeochemistry of the ecosystem. Anglers and hunters and their families who eat their catch from the Everglades are a subpopulation potentially at risk of excessive mercury exposure. Current fish consumption advisories for the Everglades recommend that anglers and hunters limit and, in some cases refrain, from consuming freshwater, marine and estuarine fish species, as well as pig frogs and alligators. These advisories however may not be sufficiently protective, particularly for those that consume higher than average amounts of fish and game. Here, we develop and apply a probabilistic risk assessment to examine risks to sport and subsistence hunters and anglers consuming Everglades fish and wildlife, discuss advantages of this methodology, and examine the premise that current methods and advisories are sufficiently protective.

Keywords Mercury · Methylmercury · Fish consumption advisories · Risk assessment · Subsistence anglers

11.1 Introduction

For many the Everglades brings to mind America's "River of Grass" (Douglas 1947); a special subtropical ecosystem in south Florida, expansive, with slow moving waters, dominated by sawgrass prairies separated by open-water channels

D. M. Axelrad (✉) · A. Becker
College of Pharmacy, Institute of Public Health, Florida A & M University, Tallahassee, FL, USA
e-mail: donald.axelrad@famu.edu; alan.becker@famu.edu

C. Jagoe
School of the Environment, Florida A & M University, Tallahassee, FL, USA
e-mail: charles.jagoe@famu.edu

© Springer Nature Switzerland AG 2019

D. G. Rumbold et al. (eds.), *Mercury and the Everglades. A Synthesis and Model for Complex Ecosystem Restoration*, https://doi.org/10.1007/978-3-030-32057-7_11

and with occasional tree islands. It is known for its high diversity of plants and animals, including a number of threatened or endangered animal species, and an abundance of wading birds and alligators (NPS 2019).

The Everglades is nationally and internationally recognized as an important ecosystem, and portions have been designated as a National Park, an International Biosphere Reserve, a World Heritage Site, and a Ramsar Wetland of International Importance (NPS 2015). These national and international designations offer protections to the Everglades, limiting some human impacts, so that the Everglades may offer esthetic and recreational benefits for present and future generations. However, it cannot be overlooked that humans are part of this ecosystem—indigenous peoples arrived in the Everglades region about 15,000 years ago, (McCally 1999), and today Florida is the U.S.'s third most populous State with 21 million people.

Human impacts on the ecosystem include fishing and hunting (FFWCC 2018a, b), and these activities are actively encouraged, and managed by the State government (FFWCC 2018c). Recreational hunters and anglers in the Everglades who consume predatory fish and wildlife risk exposure to methylmercury (MeHg), an organic mercury form that is both a potent toxicant and readily biomagnified through aquatic food webs (Mergler et al. 2007). Wet deposition of mercury (Hg) from the atmosphere is high in south Florida despite relatively low emissions of industrial Hg to the atmosphere in the State (Chaps. 3–6, Volume I). This high rate of wet deposition results in part from convective thunderstorms common to the area scavenging the elevated concentrations of inorganic mercury [Hg(II)] present in the free troposphere, and also by increasing Hg(II) concentrations in non-convective rain by mixing Hg(II) down to lower altitudes (Chap. 3, Vol. I).

This high rate of atmospheric wet deposition of Hg, when combined with biogeochemical factors in the Everglades including: sulfate inputs, high concentrations of dissolved organic matter (DOM) and changes in DOM quantity, low sediment redox in much of the area, and high average annual temperatures, produce nearly optimal conditions to promote the production of bioaccumulative and toxic MeHg (Chap. 2, Volume I; Chap. 2, this volume), and which result in very high MeHg concentrations in predator fish (Chap. 9, this volume). MeHg is a particular challenge as regards nutritional health because this toxicant is mainly contained in fish, a very nutritious food high in omega-3 fatty acids, providing known cognitive and cardiovascular benefits (Mergler et al. 2007).

Fish consumption advisories are a means of promoting consumption of nutritionally beneficial fish, while informing the public of fish species and fishing areas that pose risks due to high contaminant concentrations. The US EPA (2013) notes that "A fish consumption advisory is not a regulation, but rather guidance issued to help protect public health. These advisories may include recommendations to limit or avoid eating certain fish and wildlife species caught from specific waterbodies or from waterbody types (e.g., all lakes) due to chemical contamination. An advisory may be issued for the general public, including recreational and subsistence anglers, or it may be issued specifically for sensitive subpopulations, such as pregnant women, nursing mothers, and children."

Nationally, fish consumption advisories cover 42% of the Nation's total lake acreage, 36% of the nation's total river miles, and 79% of the Nation's coastal waters. Advisories are in force for all 50 US states to protect the public from health risks associated with eating contaminated wild-caught fish (US EPA 2013). For the U.S., fish consumption advisories have been issued for 34 different contaminants, mostly bioaccumulative, persistent toxicants including Hg, PCBs, chlordane, dioxins, and DDT. There are more Hg advisories in effect than for all other contaminants in fish and wildlife combined. Of all advisories issued nationally, 81% are based at least partially on Hg, and Hg advisories are in effect for all 50 States (US EPA 2013).

At the national level, the FDA guidance document "What You Need to Know About Mercury in Fish and Shellfish" (US FDA 2004) and as updated (US FDA 2018), recommends that women who might become pregnant, pregnant women, nursing mothers, and children do not eat six listed types or species of marine fish. They also recommend eating up to 12 ounces a week of (cooked weight) fish that are lower in Hg, and checking state advisories about the safety of consuming fish from local lakes, rivers, and coastal areas.

In Florida, high levels of MeHg in Everglades fish and wildlife were first reported in Everglades National Park (ENP) as early as 1974 by Ogden et al. (1974) who reported that "levels of Hg in freshwater vertebrates... are great enough to deserve more intensive study." Ware et al. (1990) also found high levels of Hg in some Everglades fish and wildlife species (sampling 1982–1987). Consequently, in 1989 the State of Florida developed "Exposure Guidelines for Mercury in Fish", and the Florida Department of Health and Rehabilitative Services, Florida Game and Fresh Water Fish Commission, and Florida Department of Environmental Regulation issued health advisories urging limited or no consumption of fish from many waters in Florida, including the Everglades, because of high levels of Hg in fish.

Florida fish consumption advice in 1989 was guided by the then World Health Organization Provisional Tolerable Weekly Intake (PTWI) of 3.3 micrograms MeHg per kilogram of human body weight (μg MeHg/kg-body weight), equivalent to an RfD (reference dose) of 0.43 μg MeHg/kg body-weight/day. From 2003 to the present, Florida Fish Consumption Advisories have been based on the US EPA MeHg RfD of 0.1 μg MeHg/kg body-weight/day (NRC 2000).

While data on human exposure to MeHg from consuming Everglades fish and wildlife are scant, in 1992–1993 Fleming et al. (1995) approached Everglades-area anglers and hunters to participate in a Hg exposure survey. These subjects were sport fishermen, residents who ate local fish, and, subsistence fishermen, all of whom who had eaten Everglades fish or wildlife at least once/month for the prior 3 months. Over 93% of subjects were subsistence fishermen, Everglades residents, or both. Of the 350 subjects, 119 (34%) had Hg in hair above the then detection limit: 1.26 ppm total Hg (THg) in human hair. For these 119 subjects, mean Hg concentration in hair was 3.62 ppm $\pm$ 3.0 ppm (standard deviation); range 1.28–15.57 ppm. The US EPA RfD (reference dose) translated to a human hair THg concentration is 1.1 ppm, indicating that >1/3rd of Everglades area subjects in 1992–1993 were above the MeHg RfD.

Just as the Federal National Listing of Fish Advisories indicates the scope of Hg as a contaminant in fish in US waterbodies (US EPA 2013), the Florida Department of Environmental Protection (FDEP) determines the number of "impairments" of waterbodies (lakes, streams, estuaries, and coastal waters) in Florida for not meeting the water quality standard for Hg levels in fish as an indicator of the statewide scope of the Hg problem (FDEP 2018). Currently, Hg in fish is the second most frequently

Table 11.1 Present basis for Florida Department of Health (FDOH 2019) fish consumption advisories regarding mercury levels in fish (deterministic calculation) for women of childbearing age

Toxicological value or basis of advisory
Reference dose (RfD) = 0.1 μg MeHg/kg body-wt/day (US EPA)
One meal = 8 oz. uncooked fish (~6 oz. cooked)
Women's body weight = 64 kg
2 meals/week < 0.1 ppm total mercury in fresh fish tissue
1 meal/week ≤ 0.2 ppm total mercury in fresh fish tissue
1 meal/month < 0.85 ppm total mercury in fresh fish tissue
No consumption ≥ 0.85 ppm total mercury in fresh fish tissue

identified cause of waterbody impairments across all surface water types in Florida, after dissolved oxygen in water; however, dissolved oxygen is not a contaminant in fish (FDEP 2018).

Determination of water quality impairment for Hg in Florida is based on whether Hg concentrations in fish tissue in a waterbody exceed Florida Department of Health (FDOH) advisory limits. The FDOH sets two limits for fish consumption advisories: (1) A general population advisory, where the total Hg in edible fish muscle equals or exceeds 0.3 ppm, and (2) An advisory for women of childbearing age and young children (under 10 years), where the total Hg in edible fish muscle equals or exceeds 0.1 ppm (FDEP 2013; FDOH 2019; Table 11.1).

When waterbodies are identified as impaired by a pollutant such as Hg, mandatory Hg source reductions under Section 303(d)(1)(C) of the U.S. Clean Water Act are triggered, the so-called Total Maximum Daily Load (TMDL) provision. The US EPA requires development of a TMDL to identify the maximum amount of a pollutant that a body of water can receive while still meeting water quality standards. The remediation goal of a TMDL is to meet that pollution-source reduction target. Florida's statewide Hg TMDL identified that an 86% reduction of anthropogenic Hg emissions was required from both within and outside of Florida and the U.S. (FDEP 2013).

Florida surface waters remain frequently listed as impaired because of elevated Hg in fish concentrations even though recently, waterbodies so listed have been deleted from the state's impaired waters list on administrative grounds, not because Hg levels in fish tissue have declined. It is anticipated that by 2020, all 863 Florida waterbody segments currently remaining as listed as impaired for Hg levels in fish (there were 1199 listed in 2014), will be placed on US EPA's Category 4a list, *"Impaired for one or more designated uses and a TMDL (total maximum daily load) has been completed"*. This administrative delisting of Florida waterbodies results from Florida having completed a Hg TMDL in 2012 (FDEP 2013), and such removal of waterbodies from the impaired waters list is as per US EPA guidelines.

Irrespective of the delisting of Florida's waterbodies from the impaired waters list, there is no evidence that Hg in fish tissue has declined in many of these waterbodies. In fact, it is expected that the dominant source of Hg to Florida's waterbodies—atmospheric deposition of Hg from Hg sourced outside of the US—will not decline for decades to centuries (Chap. 6, Volume I; Chap. 3, Volume III).

As discussed below, the State is also not currently pursuing the oft-recommended approach of determining a sulfur input budget for the Everglades, and assessing whether reductions in sulfur inputs to the ecosystem are feasible and can reduce bacterial MeHg production, MeHg bioaccumulation, and MeHg levels in fish.

FDOH advisories for fish consumption will therefore remain, and may represent Florida's only means of reducing the public's exposure to toxic MeHg in fish for the foreseeable future. As such, it is important that these fish consumption advisories be based on valid assumptions regarding human exposure to MeHg from consuming fish, and that they be widely disseminated.

Current FDOH fish consumption advisories for the Everglades recommend that anglers and hunters limit and, in some cases refrain, from consuming 13 freshwater species, over 60 marine and estuarine fish species, pig frogs (*Lithobates grylio*) and alligators (*Alligator mississippiensis*; FDOH 2008, 2019). Guidance for consumption of a popular Everglades sport fish, the Florida Largemouth bass, for women of childbearing age children younger than 10, and all other individuals, is displayed in Table 11.2.

11.2 Methylmercury Toxicology

Grandjean et al. (2010), wrote: "Since 1980, when the term "methylmercury compounds" was introduced as a medical subject heading, the U.S. National Library of Medicine has listed > 1,000 publications on experimental toxicology of this substance. At present, MeHg is one of the environmental pollutants with the most extensive toxicology documentation." This trend has continued, with over 1200 publications in the last 5 years containing the keywords "mercury" and "human health".

When Hg contaminated fish are consumed by humans, MeHg is mostly absorbed in the digestive tract (up to 95%) and is distributed to all tissues equally (Heckey et al. 1991; Aberg et al. 1969; Miettinen 1973; Clarkson 1997). In the bloodstream about 90% of the MeHg is covalently bound to the SH group of hemoglobin (Kershaw et al. 1980). Smith et al. (1994) reported that the half-life of MeHg in blood is approximately 48–53 days (Cox et al. 1986), and the whole body half-life is 70–74 days. About 10% of the body burden of MeHg is found in the brain. MeHg transport from the blood to the brain occurs rapidly, within 2 days, ultimately leading to a brain to blood MeHg ratio of approximately 10–1 in humans and 5–1 in primates respectively (ATSDR 1999). MeHg is demethylated to inorganic Hg in the body and is excreted mostly via the kidney and feces (Lovejoy et al. 1974). However, inorganic Hg accumulates in brain tissue (Hursh et al. 1989; WHO 1990; Clarkson et al. 1988; Aschner and Aschner 1990), as demethylation may occur after MeHg crosses the blood-brain-barrier, and the inorganic species is not readily eliminated

Table 11.2 Fish consumption guidance issued by the Florida Department of Health for consumption of Everglades Largemouth bass for women of childbearing age, children younger than 10 years, and the general population, as regards Hg exposure

Location	Species	Women of childbearing age, children Number of meals	All other individuals Number of meals
Water Conservation Area 1 (Loxahatchee National Wildlife Refuge—LNWR) (Including the L-7 and L-40 Canals)	Largemouth bass	One per month	One per week
Water Conservation Area (WCA) 2A (Including the L-35B and L-38E Canals)	Largemouth bass	One per month	One per week
Water Conservation Area (WCA) 2B	Largemouth bass	One per month	One per week
Water Conservation Area (WCA) 3A (Including Alligator Alley Canals, Tamiami Canal, L-38W Canal, and the L-67 Canal in Miami Dade County)	Largemouth bass	DO NOT EAT	One per week
L-67A Canal in Broward County	Largemouth bass	One per month	One per month
Water Conservation Area (WCA) 3B (including L-67C and Miami Canals)	Largemouth bass	DO NOT EAT	One per week
Big Cypress Preserve	Largemouth bass	DO NOT EAT	One per month
Everglades National Park: All drainages (coming out of Big Cypress National Preserve) North and West of Shark Slough to Park boundary (includes Broad River, Rogers River, Lostmans River, New River, Turner River, Halfway Creek, and Barron River)	Largemouth bass	One per month	One per week
Everglades National Park: Shark River Slough south of Tamiami Trail, including Harney and Shark Rivers and Tarpon Bay and its tributaries (Squawk, Rookery, North Prong, & Otter Creeks)	Largemouth bass	DO NOT EAT	One per month

For locations, see Fig. 11.1

from brain tissues; this may account for the persistence of Hg in the brain (Zheng 2002) as well as the resultant brain damage.

Fetal and infant Hg exposure results from Hg exposure in utero and via ingestion of breast milk. MeHg crosses the blood-brain-barrier via the MeHg-L-cysteine complex (Kerper et al. 1992) and crosses the placenta to the fetus and fetal brain. Higher concentrations of Hg occur in the fetus than in maternal tissues (Stern and Smith 2003; Lippmann 2009; Nordberg et al. 2007), and fetal brain and kidney MeHg concentrations are 6–10 times higher than in other fetal organs. Biomonitoring of maternal blood indicates the fetal Hg blood levels will on average be 70% higher than maternal levels (Mergler et al. 2007).

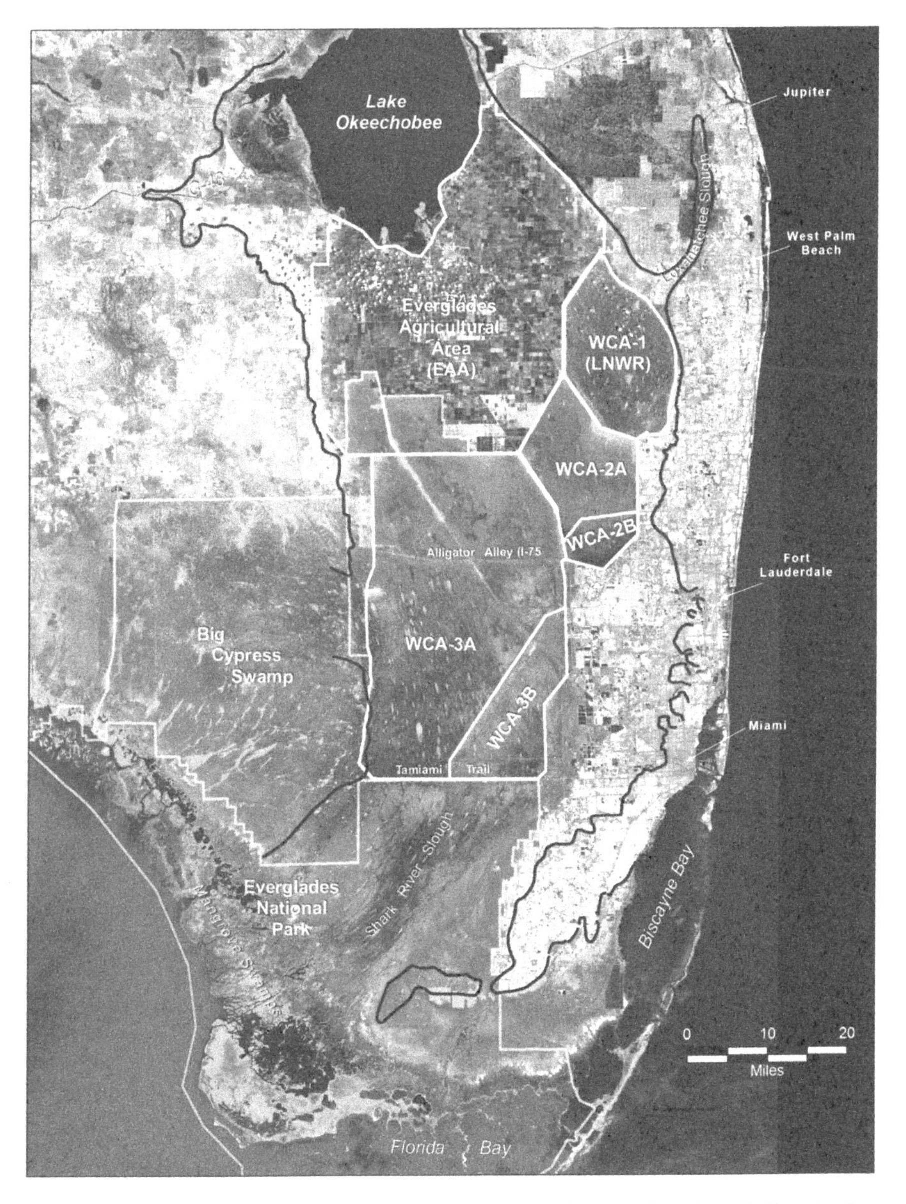

Fig. 11.1 The Everglades Protection Area includes Water Conservation Area 1 (Loxahatchee National Wildlife Refuge—LNWR), Water Conservation Areas (WCA) 2A, 2B, 3A, 3B, the Everglades National Park, and Florida Bay and SW Florida estuaries (e.g. Shark River Slough). Relative areas: WCA-1, 588 km^2; WCA-2A + WCA-2B, 544 km^2; WCA-3A + WCA-3B, 2370 km^2; Everglades National Park, 6105 km^2. Image courtesy of Frank S. Razem, GISP, PMP

MeHg accumulates in growing hair in humans, presumably by entering the active hair follicle and with incorporation into hair strands (Clarkson 1997). Hair contains keratins rich in cysteine, and Hg has a high affinity for the –SH groups on cysteine. The concentration ratio of MeHg between newly formed hair and the simultaneous blood Hg level is about 250 to 1 in humans (Cernichiari et al. 1995), and THg in hair correlates well with blood Hg levels (Suzuki et al. 1993). THg concentrations in maternal hair grown during pregnancy correlates well with fetal blood THg levels and is a good predictor of MeHg and THg levels in the fetal brain (Cernichiari et al. 1995). Maternal Hg concentrations in scalp hair exceeding 10–20 μg/g (IPCS 1990) are associated with greater fetal neurological damage.

11.2.1 Health Effects of High Mercury Exposure: Nervous System

Effects of Hg on the nervous system are well documented from human, animal and cell culture studies. Acute neural health effects are illustrated by incidents of exposure to high levels of dietary MeHg in Iraq (1970s) and Minamata, Japan (beginning in the 1950s). Minamata disease, arising from consumption of fish highly contaminated with MeHg from an industrial discharge, affected cortical and cerebellar function. Adults with Minamata disease exhibited peripheral neuropathy, dysarthria, tremor, cerebellar ataxia, gait disturbance, ophthalmological impairment (e.g. visual field constriction and disturbed ocular movements), audiological impairments (e.g. hearing loss), disturbance of equilibrium, and subjective symptoms such as headache, muscle and joint pain, forgetfulness, and fatigue (Harada 1997; Uchino et al. 1995; Tsubaki and Takahashi 1986). In the Iraq episode, which resulted from consumption of grain treated with a MeHg-based fungicide, affected individuals had similar symptoms, including an inability to walk, cerebellar ataxia, speech difficulties, paraplegia, spasticity, abnormal reflexes, restriction of visual fields or blindness, tremors, paresthesia, insomnia, confusion, hallucinations, excitement, and loss of consciousness (NRC 2000).

11.2.2 Low-Level MeHg Exposure from Maternal Seafood Consumption and Outcomes of Offspring

While severe neurotoxic effects resulting from acute exposure to very high concentrations of MeHg are well documented, the effects of chronic exposure through consumption of seafood containing environmentally-realistic concentrations of MeHg are less clear. Two major studies in the 1980s of populations whose diets consisted of a high proportion of seafood produced different results, with one linking

MeHg in the diet to developmental effects in children, and the other finding no such relationship.

The Faroe Islands study in 1986–1987 involved a large cohort (N = 1022) of consumers of a diet rich in fish and whale meat. Metrics of MeHg exposure used for the Faroe Islands Study were hair and blood Hg concentrations, which taken together are more predictive of exposure (Karagas et al. 2012). Children from the Faroes were tested with neurobehavioral assessments of I.Q. attention, fine motor function, language, visual-spatial abilities at different ages. Potential effects were first studied in 7 year olds (Grandjean et al. 1997) and again at age 14, with cognitive effects observed in both (Debes et al. 2006). Clinical exams with 22 year old subjects from the Faroe Island study conducted in 2008–2009 found a correlation between deficits in the Boston naming test and other verbal performance tests with cord blood Hg levels, and a 2.2 I.Q. point deficit associated with a 10-fold increase in prenatal MeHg exposure (Debes et al. 2016). Seafood in the Faroes also contained PCBs, but exposure to PCBs had minimal effects on the inverse relationship between Hg exposure and IQ (Grandjean et al. 1997, 2012; Jacobson et al. 2015).

The Seychelles Islands study, originally conducted in 1989–1990, was another large study (N = 779) but where the population consumed only ocean fish (Grandjean et al. 1998). In the Seychelles study, MeHg exposure levels were approximately 10 times higher than in the U.S. (van Wijngaarden et al. 2013). MeHg exposure was not associated with adverse developmental outcomes (Davidson et al. 1998), nor did it cause detectable neurological effects (Debes et al. 2016), suggesting that prenatal and postnatal exposure to MeHg in fish at these exposure levels did not adversely affect neurobehavioral development (Davidson et al. 1998; Myers et al. 2003). While the Seychelles Islands study did not demonstrate any adverse neurobehavioral effects (Myers and Davidson 1998), it was suggested that effects of prenatal exposure to Hg can be obscured by the beneficial effects of docosahexaenoic acid (DHA). In the Seychelles study, DHA had beneficial effects that were reduced at high MeHg exposure levels (Lynch et al. 2011). DHA supplementation was also found benefit children's cognition by Kuratko et al. (2013).

In a U.S. cohort of 3 year olds, higher fish intake was associated with better child cognitive test performance, and higher Hg levels were associated with poorer test scores (Oken et al. 2008). Jacobson et al. (2015) found an association of prenatal Hg exposure with poorer performance of school-age children on IQ tests, and determined that accounting for DHA consumption strengthened the association of IQ decline with Hg, thus supporting the hypothesis that benefits from DHA intake can confound the relationship between IQ and Hg exposure.

A study in New Zealand, originally reported in 1986, involved a smaller number of 6–7 year olds (N = 237), and determined that Hg levels in mothers' hair correlated with neurotoxic effects (Crump et al. 1998). The New Zealand population showed a consistent relationship between high prenatal MeHg exposure and decreased performance of 6–7 year olds on psychological and scholastic tests (Kjellstrom et al. 1989).

11.2.3 Neurological Health Effects of Low Level Mercury Exposure

High MeHg exposure in the Faroes Islands study correlated with poor scores on the Finger Tapping Test, the Boston Naming Test and the California Verbal Learning Test (Grandjean et al. 1997), as well as effects on attention, fine-motor function, confrontational naming, visual-spacial abilities, and verbal memory (NRC 2000). The Faroes Islands and New Zealand studies found evidence that low-level MeHg exposure in pregnant women is associated with language, attention, visual-motor and memory deficits in offspring (Grandjean et al. 1999). A decline in I.Q. per log increase in cord blood MeHg was observed in 4 year olds by Lederman et al. (2008). Based on cord blood as a biomarker, the lowest benchmark dose for a neurobehavioral endpoint for the Boston Naming Test was 58 ppb of MeHg. A benchmark dose level is the lowest dose estimated from modeled data expected to be associated with a small increase (1–10%) in adverse health effects (NRC 2000).

Other studies have documented neurological effects from MeHg at low levels of exposure, as reviewed by Karagas et al. (2012). For example, prenatal exposure to MeHg affected neonatal motor function (Suzuki et al. 2010) and was correlated with behavioral problems (Gao et al. 2007). MeHg exposure has been associated with decrements in infant cognition (Oken et al. 2005), poorer performance on the psychomotor development index (Jedrychowski et al. 2006, 2007), decreased vocabulary and visual motor ability at age 3 (Oken et al. 2008), and decreased cognition, memory and verbal skills at 4 years (Freire et al. 2010). Based on such findings, risk-based reference doses have been developed (NRC 2000) that were used to develop fish consumption advisories in Florida and elsewhere.

11.3 Methods: Probabilistic Risk Assessment

Current fish consumption advisories in Florida are based on deterministic calculations. Under this paradigm, the concentration of Hg in fish tissue that can be safely consumed over some time period (per week or month) without exceeding the current US EPA RfD of 0.1 μg MeHg/kg body-weight/day is calculated, based on the average body weight of the consumer population. A specified meal size, typically 8 ounces fresh weight of fish, is assumed. The average Hg concentration of fish in a particular water body is then used to generate an advisory, typically (1) eat no more than one meal per week, (2) eat no more than one meal per month, (3) do not consume, or (4) no restriction for consumption.

An alternative approach to deterministic calculations of risk is to develop and apply a probabilistic risk assessment. Using this method, a probability distribution is generated for a consumer population that eats fish of a particular species from a specific location, based on the actual Hg concentration distribution measured in fish sampled from that location. The probability distribution can then be examined for minimum, median, and maximum doses (as well as quartiles, 5th and 95th percentiles, etc.), and the percentage of times a consumer would exceed the US EPA RfD under various fish consumption scenarios is directly determined.

To apply this method to assess risks to fish and wildlife consumers in the Everglades, we obtained datasets of Hg concentrations in Everglades fish and wildlife from the Florida Fish and Wildlife Conservation Commission (FFWCC) (Chap. 9, this volume)—sport-fish sample collections are routinely conducted by FFWCC to assess public health risks. In addition to fish muscle Hg concentrations, data for each fish included species, collection location, year, and fish size and weight. This dataset contained muscle Hg concentrations for over 1700 fish from 6 species from 2009 thru 2017 collected at locations throughout the Everglades system. Of these, 56% were fish sampled in the last 5 years, providing a contemporary picture of the fish populations taken and consumed from the Everglades. All fish were species taken by recreational and subsistence anglers, in size ranges representative of those consumed by these groups.

We focused our calculations on risks to women of childbearing age, reasoning that, because this group is known to be particularly sensitive to Hg exposure, results obtained would be conservative and conclusions protective of lower risk segments of the population. We obtained data on body weights of women of child bearing age from the US Centers for Disease Control (CDC) National Health and Nutrition Examination Survey (NHANES; https://www.cdc.gov/nchs/nhanes/nhanes_products.htm). We used data from 2013 to 2014, as this was the most complete and recent data set available when we began calculations. Body weights (n=1829) ranged from 33.9 to 201.6 kg, with a median of 68.5 kg.

To calculate the actual Hg dose received by an individual, we assumed an 8 oz. (226 g) (uncooked weight) meal eaten once per week or once per month. We randomly selected fish Hg concentrations for a particular fish and location from the FFWCC dataset, and a body weight from the NHANES data set, and calculated the dose to this individual as:

$$\left(\left(\text{Hg concentration } \mu\text{g g}^{-1} \times 226\text{ g}\right)/\text{body weight kg}\right)/\text{time between meals in days}$$

This calculation was repeated 10,000 times, randomly selecting (with replacement) a fish Hg concentration and body weight each time, using Resampling Stats for Excel v.4.0 (http://www.resample.com/excel/). Initial trials involving 5000–20,000 iterations yielded similar results, so 10,000 iterations was selected for consistency. Probability distributions of the values generated were examined using JMP 13 (SAS, Cary NC).

We used this same approach to estimate probabilities of exceeding the US EPA MeHg RfD for consumers of pig frogs and alligator meat. Both frogs and alligators are harvested and consumed by recreational and subsistence hunters, and indeed may be regarded as a delicacy by some groups. Data on pig frog (n = 135) and alligator meat (n = 137) Hg concentrations from the Everglades were also obtained from the FFWCC and used to generate dose calculations as described above.

11.4 Results: Probabilistic Risk Assessment

Different areas within the Everglades currently have different fish consumption advisories (Table 11.2). Initially, we focused on the L67A canal in Broward County, FL, located in Water Conservation Area 3. This canal is popular with both recreational and subsistence anglers, and FDOH currently recommends consumption of no more than one 8 oz. meal per month of Largemouth bass for children and women of childbearing age. Figure 11.2 shows the distribution of doses calculated for women of childbearing age consuming Largemouth bass from this canal. Graphic displays of the distribution of doses for a particular population such as these are useful in visualizing the probabilities of meeting or exceeding specific dose limits. Here, if this group consumed one 8 oz. meal per week, the MeHg RfD would be exceeded over 92% of the time, two times the RfD would be exceeded 67% of the time, and 17% of weekly fish consumers would have MeHg exposure exceeding five times the RfD (bars shown in red). MeHg exposure is lower when women of childbearing age consume one meal per month of Largemouth bass. However, one meal per month still results in a 33% probability of exceeding the RfD. This raises the question of whether the current advisory is sufficiently protective, given that the reference dose will be exceeded by about 1/3 of women of child bearing age who consume bass from this canal at the maximum rate currently recommended by FDOH (one 8 oz. meal per month) .

The risk is less for those consuming sunfish from the L67A canal (Fig. 11.3). Those consuming one meal per week of these species will exceed the RfD about 44% of the time, while those consuming one meal per month have only a 0.5% probability of exceeding the RfD—the maximum sunfish consumption rate currently recommended by FDOH for the L67A Canal is one 8 oz. meal per month.

In addition to L67A canal, there are a number of other areas in the Everglades extensively used by recreational and subsistence anglers. These include the Alligator Alley Canal, Shark River Slough in the Everglades National park, the Loxahatchee National Wildlife Refuge, the larger WCA-3A area and the L35B canal in WCA-2A. We performed separate analyses for each of these areas, assuming one 8 oz. meal per month of Largemouth bass consumed by women of childbearing age. Table 11.3 summarizes these results. Median Hg doses (as μg Hg/kg body-weight/day) were highest in the southern portions of the Everglades, particularly the Alligator Alley Canal and Shark River Slough, where the probability of exceeding the US EPA MeHg RfD was 42–66%, depending on the size of the bass consumed. FDOH recommends not consuming bass from these areas, but in fact, they are still popular with anglers. Median Hg exposure from consuming fish was lowest in WCA-1, the Loxahatchee National Wildlife Refuge, where the probability of exceeding the US EPA MeHg reference dose was roughly 1:50 to 1:100 when consuming one 8 oz. meal per month. These results demonstrate the necessity of considering spatial variation in fish Hg concentrations when developing fish consumption advisories. They also demonstrate the utility of a probabilistic approach, as risk managers can obtain specific probabilities of exceeding a particular dose, and hopefully make better informed decisions about what constitutes an acceptable risk.

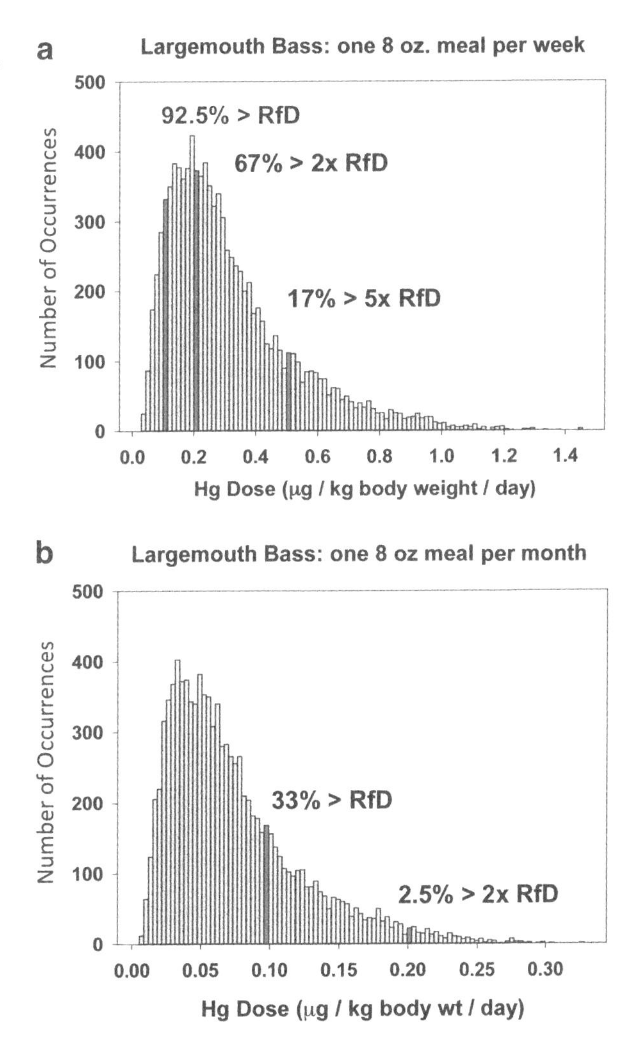

Fig. 11.2 Distribution of mercury doses for women of childbearing age consuming (**a**) one 8 oz. meal per week and (**b**) one 8 oz. meal per month of Largemouth bass taken from the L67A Canal, Water Conservation Area 3. Shown are the probabilities of exceeding the RfD by factors of 2×, or 5× the RfD under each fish consumption scenario—the maximum bass consumption rate currently recommended by FDOH for the L67A Canal is one 8 oz. meal per month

One method to reduce Hg exposure to fish consumers is to encourage the taking of smaller fish, as fish size and age are strongly correlated with Hg concentrations in edible tissues. In 2016, FFWCC eliminated the statewide minimum length limit for bass while also allowing anglers to harvest only one bass 16 in. or longer per day; therefore, providing the opportunity to harvest smaller bass for consumption (www.myfwc.com). To assess the efficacy of this approach, we calculated Hg doses for our target population, assuming that fish of different size classes were consumed. We ran models including all size classes, and for Largemouth bass <16, <14, and <12 in. in total length (Table 11.3), assuming one 8 oz. meal per month for women of childbearing age. In some locations, reducing the size of fish taken and consumed made a noticeable difference in the percentage of times a consumer would exceed the

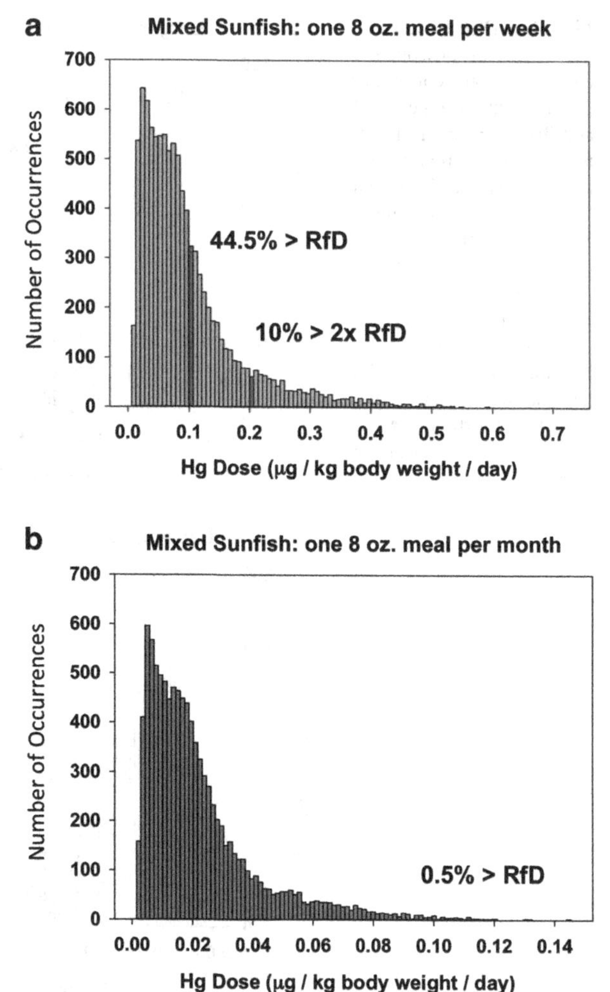

Fig. 11.3 Distribution of mercury doses for women of childbearing age consuming (**a**) one 8 oz. meal per week and (**b**) one 8 oz. meal per month of sunfish (mixed species, including bluegill, redspotted sunfish, spotted sunfish and warmouth) taken from the L67A Canal, Water Conservation Area 3. Shown are the probabilities of exceeding the RfD, or 2× the RfD under each fish consumption scenario

reference dose. For example, exclusive harvest of bass <12 in. would reduce the probability of exceeding the US EPA MeHg RfD to about 1 in 10 in the L67A and L35 canals (probability of exceedance 11% and 13%, respectively). In comparison, consumption of bass of all sizes from these locations results in probabilities of about 1 in 3 to 1 in 4 of exceeding the US EPA RfD (27% and 29% for L67A and L35 canals, respectively). Reducing the probability of exceeding the US EPA RfD is clearly desirable, and the magnitude of the reductions in these cases might make recommendations to anglers to consume smaller size fish worthwhile.

In other geographic locations, harvest of smaller size fish makes little difference (Loxahatchee, Shark River Slough). These are areas where fish Hg concentrations are either relatively high or relatively low. Harvest of only bass <12 in. in Shark River Slough reduces the probability of women of childbearing age exceeding the

Table 11.3 Range, 5th and 95th percentiles, quartiles and median calculated Hg doses (μg/kg body-weight/day) for women of childbearing age consuming one 8 oz. meal per month of Everglades Largemouth bass, by location

Size	Maximum	95th percentile	75th percentile	Median	25th percentile	5th percentile	Minimum	% exceeding RfD
WCA 3A								
<12 in.	0.394	0.151	0.088	0.061	0.043	0.025	0.009	18
<14 in.	0.305	0.163	0.097	0.066	0.045	0.025	0.007	23
<16 in.	0.483	0.184	0.108	0.072	0.048	0.029	0.008	29
All sizes	0.395	0.206	0.124	0.081	0.052	0.028	0.006	37
L67A Canal								
<12 in.	0.312	0.132	0.076	0.053	0.034	0.012	0.002	11
<14 in.	0.351	0.133	0.078	0.055	0.036	0.014	0.002	12
<16 in.	0.336	0.162	0.082	0.060	0.039	0.014	0.002	18
All sizes	0.324	0.182	0.104	0.068	0.044	0.016	0.003	27
L35 Canal								
<12 in.	0.300	0.137	0.076	0.046	0.025	0.011	0.003	13
<14 in.	0.316	0.14	0.083	0.051	0.028	0.012	0.004	16
<16 in.	0.394	0.165	0.094	0.058	0.031	0.012	0.004	22
All sizes	0.391	0.181	0.108	0.066	0.034	0.013	0.003	29
Loxahatchee								
<12 in.	0.146	0.068	0.042	0.029	0.019	0.009	0.003	1
<14 in.	0.168	0.08	0.047	0.031	0.021	0.01	0.002	2
<16 in.	0.163	0.083	0.049	0.032	0.021	0.011	0.002	2
All sizes	0.298	0.11	0.057	0.036	0.023	0.012	0.003	7
Shark River Slough								
<12 in.	0.354	0.182	0.125	0.091	0.067	0.041	0.012	42
<14 in.	0.567	0.286	0.181	0.126	0.086	0.048	0.017	66
<16 in.	0.56	0.286	0.178	0.124	0.085	0.05	0.014	66
All sizes	0.36	0.276	0.179	0.126	0.086	0.049	0.015	66

(continued)

Table 11.3 (continued)

Size	Maximum	95th percentile	75th percentile	Median	25th percentile	5th percentile	Minimum	% exceeding RfD
Alligator Alley								
<12 in.	0.293	0.181	0.131	0.098	0.074	0.049	0.022	48
<14 in.	0.308	0.196	0.141	0.108	0.082	0.054	0.024	58
<16 in.	0.509	0.218	0.148	0.12	0.085	0.055	0.023	61
All sizes	0.512	0.255	0.165	0.121	0.088	0.057	0.028	65

Doses were calculated for a population consuming all sizes of bass, and for those consuming fish that are <16, <14 and <12 in. in total length. Percent of times a hypothetical consumer would exceed the US EPA RfD under these scenarios are also shown for each location and size class

US EPA RfD when consuming one meal per month from 66% to 42%; clearly, the likelihood of exceeding the RfD is substantial in either case. In contrast, women of childbearing age consuming fish from WCA-1, the Loxahatchee National Wildlife Refuge have only a 7 in 100 chance of exceeding the RfD when consuming bass of any size; reducing the size to <12 in. produces a modest improvement to 1 in 100.

Sunfish generally feed at lower trophic levels than Largemouth bass, and tend to have shorter life-spans, resulting in Hg concentrations that are generally lower than those in bass. Accordingly, there is less risk associated with consuming these species. Median Hg doses to women of childbearing age consuming one meal per week of mixed sunfish species range from 0.237 to 0.086 μg Hg/kg body-wt/day (RfD = 0.1), with the highest doses from Shark River Slough and the lowest from WCA-1, the Loxahatchee National Wildlife Refuge. In these locations, women consuming one meal per week have probabilities of 92% and 43%, respectively, of exceeding the RfD. By reducing consumption to one meal per month, median doses fall to 0.053 and 0.020 μg Hg/kg-body weight/day for Shark River Slough and Loxahatchee, respectively. In the former, there is a 13% chance of exceeding the RfD; in the latter, only a 0.5% chance (Table 11.4).

Mayan cichlids (*Mayaheros urophthalmus*) are an invasive species, tolerant of a wide range of salinities, that first appeared in Florida Bay in 1983 and are now an abundant and popular gamefish in south Florida as far north as Lake Okeechobee (Florida Museum of Natural History, 2018). For probabilistic risk assessment, we only had sufficient data on Hg concentrations in this species from one area, WCA-3A. Compared to Largemouth bass, this species has lower Hg concentrations, resulting in less risk to consumers. Calculated median doses for women of child-bearing age consuming one 8 oz. meal were 0.072 and 0.016 μg Hg/kg body-weight/day for one meal per week and one meal per month, respectively (Table 11.5). Both are below the US EPA MeHg RfD. In fact, consumers of one meal per week have a 37% chance, and those consuming one meal per month a 0.5% chance, of exceeding the RfD. Thus, a recommendation of restricting consumption to one meal per month appears sufficiently protective for this species.

In addition to fish, alligators and frogs are both popular game species harvested and consumed from the waters of the Everglades. The public is advised not to consume alligator meat harvested in the Everglades, although such consumption does occur, and in recognition of this fact, FFWCC provides guidance on how to butcher an alligator, store alligator meat, and where to find alligator recipes (FFWCC 2019).

The risk of exposure to high levels of MeHg from consuming alligator is however high. From consuming one 8 oz. meal of alligator meat per week, women of childbearing age have a 95% chance of exceeding the MeHg reference dose, a 23% chance of exceeding 10 times the RfD, and nearly a 1 in 50 chance of exceeding 20 times the RfD (Fig. 11.4). If consumption is reduced to one meal per month, there is a still a 66% chance of exceeding the RfD. Given that, consuming only one meal per month, there is still a two out of three chance of exceeding the reference dose, alligator meat from the Everglades contains too much Hg to risk any consumption.

Table 11.4 Range, 5th and 95th percentiles, quartiles and median calculated Hg doses (μg/kg body-weight/day) for women of childbearing age consuming Everglades sunfish (mixed species), by location

Frequency	Maximum	95th percentile	75th percentile	Median	25th percentile	5th percentile	Minimum	% exceeding RfD
WCA 3A								
1 meal/week	1.059	0.501	0.291	0.182	0.099	0.042	0.009	75
1 meal/month	0.237	0.114	0.067	0.042	0.023	0.01	0.002	9
L67A Canal								
1 meal/week	0.724	0.277	0.123	0.075	0.042	0.018	0.006	44
1 meal/month	0.145	0.064	0.028	0.017	0.009	0.004	0.001	0.5
L35 Canal								
1 meal/week	0.74	0.408	0.238	0.132	0.069	0.03	0.004	61
1 meal/month	0.168	0.093	0.054	0.03	0.016	0.007	0.001	3
Loxahatchee								
1 meal/week	0.704	0.292	0.148	0.087	0.044	0.02	0.006	43
1 meal/month	0.136	0.068	0.033	0.02	0.01	0.005	0.001	0.5
Shark River Slough								
1 meal/week	1.221	0.617	0.364	0.258	0.17	0.046	0.0155	88
1 meal/month	0.216	0.129	0.08	0.057	0.035	0.009	0.003	12
Alligator Alley								
1 meal/week	1.113	0.125	0.345	0.236	0.16	0.087	0.025	92
1 meal/month	0.217	0.054	0.078	0.053	0.036	0.02	0.006	13

Doses were calculated for a population consuming one meal per week and one meal per month. Percent of times a hypothetical consumer would exceed the US EPA RfD are also shown for each location and size class

Table 11.5 Range, 5th and 95th percentiles, quartiles and median calculated Hg doses (μg/kg body-weight/day) for women of childbearing age consuming Mayan cichlids

Frequency	Maximum	95th percentile	75th percentile	Median	25th percentile	5th percentile	Minimum	% exceeding RfD
WCA 3A								
1 meal/week	0.497	0.295	0.152	0.072	0.042	0.018	0.006	37
1 meal/month	0.123	0.067	0.035	0.016	0.009	0.004	0.001	0.5

Doses were calculated for a population consuming one meal per week and one meal per month. Percent of times a hypothetical consumer would exceed the US EPA RfD are also shown for each location and size class

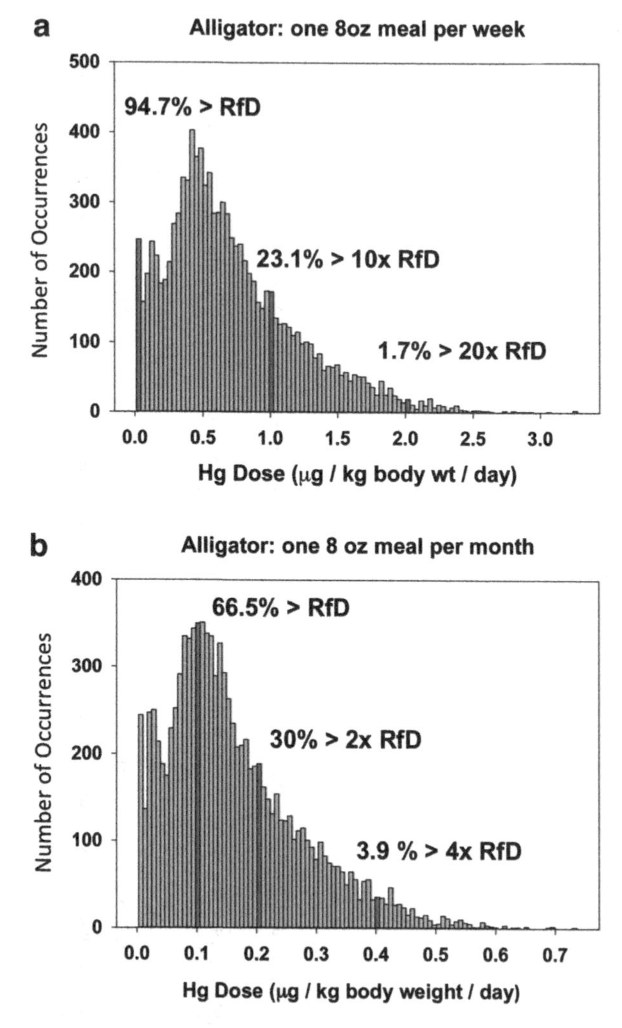

Fig. 11.4 Distribution of mercury doses for women of childbearing age consuming (**a**) one 8 oz. meal per week and (**b**) one 8 oz. meal per month of alligator meat from the Everglades. Shown are the probabilities of exceeding the RfD, or up to 2×, 4×, 10× or 20× the RfD under each alligator meat consumption scenario

Pig frogs contain lower concentrations of Hg, and therefore the odds of exceeding the RfD with weekly or monthly consumption is lower (Fig. 11.5). When consuming one meal per month, there is about a 1 in 500 chance of exceeding the RfD. More frequent consumption poses greater risk; one meal per week increases the chances of RfD exceedance to 28%

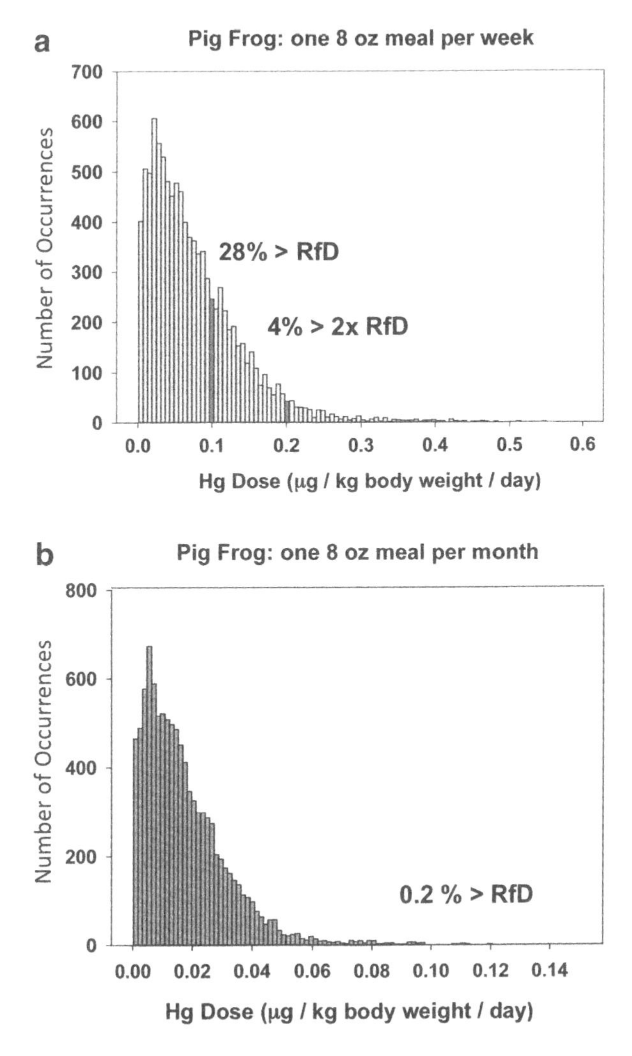

Fig. 11.5 Distribution of mercury doses for women of childbearing age consuming (**a**) one 8 oz. meal per week and (**b**) one 8 oz. meal per month of pig frog meat from the Everglades. Shown are the probabilities of exceeding the RfD, or 2× the RfD under each pig frog consumption scenario

11.5 Discussion

The portion of the Everglades Protection Area comprising Water Conservation Area 1 (Loxahatchee National Wildlife Refuge—LNWR), Water Conservation Areas (WCA) 2A, 2B, 3A, 3B, and the Everglades National Park, has very high overall distribution of mean Largemouth bass total Hg concentrations with 96% of all sites exceeding the US EPA 0.3 mg/kg Hg in fish human health criterion (Chap. 9, this volume). Hg levels are also elevated in sunfish, alligators and pig frogs—species for which we have assessed human MeHg exposures for various human consumption rates of these fish and wildlife.

When healthful rates of human consumption of fish (USDHHS and USDA, 2015) taken by anglers from waterbodies are advised against by States because of fish contaminant levels, there is a formal State/Federal process for reducing these contaminant levels (US EPA 2018b). Waters—lakes, rivers, coastal waters—are deemed "Impaired" when they fail to meet water quality standards, as in the case for the Everglades where Hg levels in fish are elevated to the extent that the State has issued both "no consumption" and "limited consumption" advisories (Table 11.1).

For these "Impaired Waters", under the Clean Water Act Section 303(d), States are required to develop a Total Maximum Daily Load (TMDL). The TMDL identifies the maximum amount of a pollutant that a waterbody can receive while still meeting water quality standards. For Hg for the Everglades, meeting standards means that Everglades waters are not listed as impaired by the state, due to high mercury levels in fish, with consequent FDOH fish consumption advisories. Having identified impaired waters, waterbody restoration is then to be conducted to reduce pollutant loads—in the case of the Everglades to reduce Hg inputs from various sources identified in the TMDL in order that the waterbody meets water quality standards (US EPA 2018a).

The State of Florida's Hg TMDL (FDEP 2013) as accepted by the US EPA will however not result in toxicologically acceptable Hg levels in Everglades fish for many decades or centuries, and will therefore not make fish safe for human consumption for that term. This is because while atmospheric Hg deposition to the Everglades is more than 95% of the total Hg input to this ecosystem (Chap. 4, Volume I), the sources of Hg contributing to atmospheric deposition to the Everglades currently are dominated by sources outside of Florida and the U.S.—from regional and larger scale Hg emissions rather than from local and statewide Hg sources. These local plus statewide Hg sources now likely contribute only about 3% of the wet plus dry Hg deposition to the Everglades (Chap. 6, Volume I). While Florida emissions of Hg presently are a minor contributor to atmospheric Hg deposited in the Everglades, atmospheric deposition of Hg to the Everglades remains very high (Chap. 6, Volume I; Chap. 3, Volume III).

The explanation for this is that natural Hg emissions together with re-emissions of legacy anthropogenic Hg stored in terrestrial and oceanic reservoirs presently contribute two-thirds or more to the global Hg budget, with direct anthropogenic Hg emissions comprising the remainder (Chap. 6, Volume I). Modeling in fact indicates that atmospheric deposition of Hg globally will likely increase up to year 2050, and legacy anthropogenic Hg will then maintain atmospheric Hg deposition at elevated levels for centuries. All things being equal, Hg levels in fish will not reach safe levels for recommended fish consumption rates (USDHHS and USDA, 2015), for that term.

It would seem reasonable that in this case where input reductions in the "primary" pollutant (Hg for the Everglades) cannot be achieved, that the State/Federal process would evaluate the likely efficacy of other potential solutions to the Hg problem. The Everglades receives tens of thousands of metric tons per year of sulfate runoff as a result of agricultural practices and operations in the Everglades Agricultural Area north of the Everglades, and from agriculture north of Lake Okeechobee, (Landing 2013). Control of sulfur inputs to reduce Hg methylation by naturally-occurring sulfate-reducing bacteria—which convert inorganic Hg to its strongly bioaccumulated form, MeHg—could represent a means of reducing MeHg in Everglades fish (Bates et al. 2002; Gilmour et al. 2007; Orem et al. 2011; Axelrad et al.

2013; Chap. 2, this volume). To date, this approach has not been pursued by the State.

Thus in Florida, limitation of human exposure to MeHg through consumption of fish is managed mainly via consumption advisories which cover 31 freshwater and over 60 marine species as well as other game species within the Everglades (FDOH 2019) (Chap. 9, this volume).

The deterministic method presently utilized by FDOH to determine safe frequency of fish consumption as the basis for fish consumption advisories does not take into account variation in body weight in the consumer population, nor variation in Hg content among fish of a particular species within a water body. As well, this approach does not allow evaluation of the probability of exceeding the MeHg reference dose when following the fish consumption advisory.

These shortcomings make it difficult to estimate the true risk to a particular consumer population. This consideration also raises the question of what is an acceptable probability of exceeding the RfD if consumption advisories are followed. In other words, would one be satisfied that the consumption advisory was sufficiently protective if the reference dose was only exceeded 50% of the time? Would a 95% probability of not exceeding the reference dose be desirable, and what fish Hg concentration would be necessary to achieve this? While deterministic calculations cannot address such considerations, a probabilistic approach provides a more robust framework to evaluate various protective actions.

Here we conducted probabilistic risk assessments for MeHg exposure for women of childbearing age regarding Everglades fish, pig frogs and alligators; fish and wildlife consumed by those who fish and hunt in the Everglades, these fish and wildlife species having been determined to have elevated Hg concentrations in the freshwater Everglades (Chap. 9, this volume). These probabilistic risk assessments indicate greater risk to women of childbearing age than do the deterministic calculations presently used by FDOH.

Risks to those who consume fish from the Everglades remain high, and communication of risk (fish consumption advisories) is complicated by a high degree of spatial and species-level variations in Hg in Everglades fish (Chap. 9, this volume). Nonetheless, if the State does not intend to investigate alternative means of reducing MeHg levels in Everglades fish apart from atmospheric Hg deposition loading reduction (which will be unachievable for many decades to centuries), the remaining best option to protect public health re MeHg exposure from consuming Everglades fish and wildlife, may be fish consumption advisories. It must be noted that efforts more active than fish consumption advisories must also be considered. It is unrealistic to expect that subsistence anglers will limit fish consumption to one meal per month simply because a fish consumption advisory so advises.

On the basis of our findings, we would recommend that:

- FDOH replace deterministic MeHg exposure risk assessment with a probabilistic approach;
- FDOH in consultation with FFWCC consider advising anglers to consume smaller size Largemouth bass in some areas of the Everglades;

- FDOH/FFWCC emphasize that any (Everglades wild) alligator consumption is hazardous to women of childbearing age and children;
- FDOH/FFWCC/FDEP consider signposting the Everglades with fish and wildlife consumption advice, and further develop programs for public education;
- FDOH list fish and wildlife consumption advisories from all Florida agencies in its annual "Guide To Eating Fish (and Wildlife) Caught In Florida" (FDOH 2019).

We further encourage for the future that:

- FDEP investigate alternative means of reducing MeHg in Everglades fish and wildlife apart from atmospheric Hg deposition loading reduction, such as sulfur loading reduction to the Everglades; and,
- The US EPA through its TMDL process require States to pursue alternative means of reducing contaminants in fish apart from reducing inputs of the primary pollutant (in the Everglades case, Hg) via investigating feasible alternative approaches.

References

Aberg B, Ekman L, Falk R, Greitz U, Persson G (1969) Metabolism of methyl mercury (^{203}Hg) compounds in man. Arch Environ Health 19:478–484

Aschner M, Aschner JL (1990) Mercury neurotoxicity: mechanisms of blood-brain barrier transport. Neurosci Biobehav Rev 14:169–176

ATSDR (Agency for Toxic Substances and Disease Registry) (1999) Toxicological profile for mercury. (Update) US Department of Health and Human Services Public, Agency for Toxic Substances and Disease Registry, Atlanta, GA

Axelrad DM, Pollman C, Gu B, Lange T (2013) Mercury and sulfur environmental assessment for the Everglades. South Florida Environmental Report, vol 1, Chapter 3B, South Florida Water Management District, West Palm Beach, FL, 57 pp. http://my.sfwmd.gov/portal/page/portal/pg_grp_sfwmd_sfer/portlet_prevreport/2013_sfer/v1/chapters/v1_ch3b.pdf

Bates AL, Orem WH, Harvey JW, Spiker EC (2002) Tracing sources of sulfur in the Florida Everglades. J Environ Qual 31:287–299

Cernichiari E, Brewer R, Myers GJ, Marsh DO, Lapham LW, Cox C, Shamlaye CF, Berlin M, Davidson PW, Clarkson TW (1995) Monitoring methylmercury during pregnancy: maternal hair predicts fetal brain exposure. Neurotoxicology 16:705–710

Clarkson TW (1997) The toxicology of mercury. Crit Rev Clin Lab Sci 34:369–403

Clarkson TW, Hursh JB, Sager PR, Syversen TLM (1988) Mercury. In: Clarkson TW, Friberg L, Norgberg GF, Sager PR (eds) Biological monitoring of toxic metals. Plenum Press, New York, pp 199–246

Cox C, Clarkson TW, Marsh DO, Amin-Kaki L, Tikneti S, Myers GG (1986) Dose-response analysis of infants prenatally exposed to methylmercury: an application of a single compartment model to single-strand hair analysis. Environ Res 49:193–214

Crump KS, Kjellstrom T, Shipp AM, Silvers A, Stewart A (1998) Influence of prenatal mercury exposure upon scholastic and psychological test performance: benchmarked analysis of a New Zealand Cohort. Risk Anal 18(6):701–713

Davidson PW, Myers GJ, Cos C, Axtell C, Shamlaye C, Sloane-Reeves J, Cernichiari E, Needham L, Choi A, Wang Y, Berlin M, Clarkson TW (1998) Effects of prenatal and postnatal

methylmercury exposure from fish consumption on neurodevelopment: outcomes at 66 months of age in the Seychelles Child Development Study. J Am Med Assoc 280(8):701–707
Debes DC, Dramer JH, Kaplan E, Weuhe P, White RF, Grandjean P (2006) Impact of prenatal methylmercury exposure on neurobehavioral function at age 14 years. Nerotoxicol Teratol 28:536–547
Debes F, Weihe P, Grandjean P (2016) Cognitive deficits as age 22 years associated with prenatal exposure to methymercury. Cortex 74:358–369
Douglas MS (1947) The Everglades: River of Grass. Rinehart, New York. Reprinted (1988) by Pineapple Press, Sarasota, FL
FDEP (Florida Department of Environmental Protection) (2013) Final Report. Mercury TMDL for the State of Florida. Division of Environmental Assessment and Restoration. Bureau of Watershed Restoration. Watershed Evaluation and TMDL Section. 24 October 2013. https://floridadep.gov/sites/default/files/Mercury-TMDL.pdf
FDEP (Florida Department of Environmental Protection) (2018) Final Integrated Water Quality Assessment for Florida: 2018 Sections 303(d), 305(b), and 314 Report and Listing Update. Division of Environmental Assessment and Restoration. Florida Department of Environmental Protection. June 2018. https://floridadep.gov/dear/bioassessment/documents/integrated-303d305b-report-2018
FDOH (Florida Department of Health) (2008) News Release: consumption advisory for pig frog legs for Everglades and Francis S. Taylor Wildlife management area (Water Conservation Areas 2 and 3) in Palm Beach, Broward and Miami-Dade Counties. 14 May 2008. Florida Department of Health, Tallahassee, FL
FDOH (Florida Department of Health) (2019) Your guide to eating fish caught in Florida. http://www.floridahealth.gov/programs-and-services/prevention/healthy-weight/nutrition/seafood-consumption/_documents/fish-advisory-big-book2019.pdf
FFWCC (Florida Fish and Wildlife Conservation Commission Florida) (2018a) Freshwater Fishing Regulations 2017–2018. http://myfwc.com/media/4234225/2017FLFWRegulations.pdf
FFWCC (Florida Fish and Wildlife Conservation Commission Florida) (2018b) Florida Hunting Regulations 2017–2018. http://www.eregulations.com/wp-content/uploads/2017/06/17FLHD_LR.pdf
FFWCC (Florida Fish and Wildlife Conservation Commission Florida). (2018c) Everglades and Francis S. Taylor Wildlife management areas. http://myfwc.com/viewing/recreation/wmas/lead/everglades/
FFWCC (Florida Fish and Wildlife Conservation Commission Florida) (2019) Guide to alligator hunting in Florida. https://myfwc.com/media/16675/alligator-hunting-guide.pdf. Accessed 10 May 2019
Florida Museum of Natural History (2018) Discover fishes. Mayaheros urophthalmus. https://www.floridamuseum.ufl.edu/discover-fish/species-profiles/mayaheros-urophthalmus/
Fleming LE, Watkins S, Kaderman R, Levin B, Ayyar DR, Bizzio M, Stephens D, Bean JA (1995) Mercury exposure in humans through food consumption from the Everglades in Florida. Water Air Soil Pollut 80:41–48
Freire C, Ramos R, Lopez-Espinosa MJ, Diez S, Vioque J, Ballester F, Fernandez MF (2010) Hair mercury levels, fish consumptions, and cognitive development in preschool children from Granada, Spain. Environ Res 110(1):96–104
Gao Y, Yan CH, Tian Y, Wang Y, Xie HF, Zhou C, Yu XD, Yu XG, Tong S, Zhou QX, Shen XM (2007) Prenatal exposure to mercury and neurobehavioral development of neonates in Zhoushan City, China. Environ Res 105(3):390–399
Gilmour CC, Krabbenhoft D, Orem W, Aiken G, Roden D (2007) Appendix 3B-2. Status Report on ACME Studies on the control of mercury methylation and bioaccumulation in the Everglades. In: Redfield G (ed) 2007 South Florida Environmental Report. South Florida Water Management District, West Palm Beach, FL. http://my.sfwmd.gov/portal/page/portal/pg_grp_sfwmd_sfer/portlet_prevreport/volume1/appendices/v1_app_3b-2.pdf

Grandjean P, Weihe P, White RF, Debes F, Araki S, Yokoyama K, Murata K, Sorensen N, Dahl R, Jorgensen PJ (1997) Cognitive deficits associated with 7-year-old children with prenatal exposure to methylmercury. Neurotoxicol Teratol 19(6):417–428

Grandjean P, Weihe P, White RF (1998) Cognitive performance of children prenatal exposed to safe levels of methylmercury. Environ Res 77:165–172

Grandjean P, Butdtz-Jorgensen E, White R, Jorgensen P, Weihe P, Debes F, Keiding N (1999) Methylmercury exposure biomarkers as indicators of neurotoxicity in children aged 7 years. Am J Epidemiol 150:301–305

Grandjean P, Satoh H, Katsuyuki Murata K, Eto K (2010) Adverse effects of methylmercury: environmental health research implications. Environ Health Perspect 118:1137–1145

Grandjean P, Weihe P, Nielsen F, Heinzow B, Debes F, Budtz-Jorgensen E (2012) Neurobehavioral deficits at age 7 years associated with prenatal exposure to toxicants from maternal seafood diet. Neurotoxicol Teratol 34:466–472

Harada M (1997) Neurotoxicity of methylmercury: Minamata and the Amazon. In: Yasui M, Strong MJ, Ota K, Verity MA (eds) Mineral and metal neurotoxicology. CRC Press, Boca Raton, FL, pp 177–188

Hecky RE, Ramsey DJ, Bodaly RA, Strange NE (1991) In: Suzuki T (ed) Advances in mercury toxicology. Plenum Press, New York, pp 33–52

Hursh JB, Clarkson TW, Miles EF (1989) Percutaneous absorption of mercury vapor by man. Arch Environ Health 44:120–127

IPCS (International Program on Chemical Safety) (1990) Environmental Health Criteria Document 101: Methylmercury. World Health Organization, Geneva

Jacobson JL, Muckle G, Ayotte P, Dewailly E, Jacobson SW (2015) Relation of prenatal methylmercury exposure from environmental sources to childhood IQ. Environ Health Perspect 123 (8):827–833

Jedrychowski W, Jankowski J, Flak E, Skarupa A, Mroz E, Sochacka-Tatara E, Lisowska-Miszczyk I, Szparowska-Wohn A, Rauh B, Stolicki Z, Kaim I, Perera F (2006) Effects of prenatal exposure to mercury on cognitive and psychomotor function in one-year-old infants: epidemiology cohort study in Poland. Ann Epidemiol 16(6):439–447

Jedrychowski W, Perera F, Jankowski J, Rauh V, Flak E, Caldwell KL, Jones RL, Pac A, Lisowska-Miszczyk I (2007) Fish consumption in pregnancy, cord blood mercury level and cognitive and psychomotor development of infants followed over the first three years of life: Krakow epidemiologic study. Environ Int 33(8):1057–1062

Karagas M, Choi AL, Oken E, Horvat M, Schoeny R, Kamai E, Cowell W, Grandjean P, Korrick S (2012) Evidence on human health effects on low-level methylmercury exposure. Environ Health Perspect 120(6):799–806

Kerper LE, Ballatori N, Clarkson TW (1992) Methylmercury transport across the blood-brain barrier by amino acid carrier. Am J Physiol 262(5):R761–R765

Kershaw TG, Clarkson TW, Dhahir PH (1980) The relationship between blood-brain levels and dose of methylmercury in man. Arch Environ Health 35:28–36

Kjellstrom T, Kennedy P, Wallis S (1989) Physical and mental development of children with prenatal exposure to mercury from fish. Stage II: Interviews and psychological tests at age 6. National Swedish Environmental Protection Board. Report 3642. Solna, Sweden

Kuratko CN, Barrett EC, Nelson EB, Salem N Jr (2013) The relationship of docosahexaenoic acid (DHA) with learning and behavior in healthy children: a review. Nutrients 5(7):2777–2810

Landing WL (2013) Review and comment in writing on two documents: Regional sulfur mass balance centered on the EAA: an outline for investigations and suggested projects for EAA sulfur mass balance study (draft). Report to the Florida Department of Environmental Protection. DEP Contract: SP703. 16 January 2013

Lederman SA, Jones RL, Caldwell JL, Rouh V, Sheets SE, Tang D, Viswanathan S, Becker M, Stein JL, Wang RY, Perera F (2008) Relations between cord blood mercury levels and early child development in World Trade Center Cohort. Environ Health Perspect 113:1085–1096

Lippmann M (ed) (2009) Environmental toxicants: human exposures and their health effects, 3rd edn. Wiley, Hoboken, NJ

Lovejoy HN, Bell ZG, Bizena TR (1974) Mercury exposure evaluations and their correlations with urine mercury excretions. J Occup Med 15:590

Lynch ML, Huang LS, Cox S, Strain JJ, Myers GJ, Bonham MP, Shamlaye CF, Stokes-Riner A, Wallace JMW, Duffy EM, Clarkson TW, Davidson PW (2011) Varying coefficient function models to explore interactions between maternal nutritional status and prenatal methylmercury toxicity in the Seychelles Child Development Nutrition Study. Environ Res 111(1):75–80

McCally D (1999) The Everglades: an environmental history. University Press of Florida. ISBN 0-8130-2302-5

Mergler D, Anderson HA, Chan LHM, Mahaffey KR, Murray M, Sakamoto M, Stern AH (2007) Methylmercury exposure and health effects in humans: a worldwide concern. Ambio 36 (1):3–11

Miettinen JK (1973) Absorption and elimination of dietary (Hg^{++}) and methylmercury in man. In: Miller MW, Clarkson TW (eds) Mercury, mercurial, and mercaptans. C.C. Thomas, Springfield, IL, pp 233–246

Myers GJ, Davidson PW (1998) Low level prenatal methylmercury exposure and children: neurological, developmental and behavioral research. Environ Health Perspect 106:841–847

Myers GJ, Davidson PW, Cox C, Shamlaye CF, Palumbo D, Cernichiari E, Sloane-Reeves J, Wilding GE, Kost J, Huang L-S, Clarkson TW (2003) Prenatal methylmercury exposure from ocean fish consumption in the Seychelles child development study. Lancet 361(9370):1686–1692

Nordberg GF, Fowler NA, Nordberg M, Friberg L (eds) (2007) Handbook on the toxicology of metals, 3rd edn. Academic Press, San Diego, CA

NPS (National Park Service) (2015) Everglades is internationally significant. https://www.nps.gov/ever/learn/news/internationaldesignations.htm

NPS (National Park Service) (2019) Everglades National Park. Threatened and endangered species. https://www.nps.gov/ever/learn/nature/techecklist.htm

NRC (National Research Council) (2000) Toxicological effects of methylmercury. The National Academies Press, Washington, DC. https://doi.org/10.17226/9899

Ogden JC, Robertson WB, Davis GE, Schmidt TW (1974) Pesticides, polychlorinated biphenyls and heavy metals in upper food chain levels, Everglades National Park and vicinity. U.S. Department of the Interior, National Technical Information Service, No. PB-235 359

Oken E, Wright RO, Kleinman KP, Bellinger D, Amarasiriwardena CJ, Hu H, Rich-Edwards JW, Gilman MW (2005) Maternal fish consumption, hair mercury, and infant cognition in the U.S. cohort. Environ Health Perspect 113:1376–1380

Oken E, Radesky JS, Wright RO, Bellinger DC, Anarasiriward CJ, Kleinman KP (2008) Maternal fish intake during pregnancy, blood mercury levels, and child cognition at age 3 years in a U.S. Cohort. Am J Epidemiol 167(10):1171–1181

Orem W, Gilmour C, Axelrad D, Krabbenhoft D, Scheidt D, Kalla P, McCormick P, Gabriel M, Aiken G (2011) Sulfur in the South Florida ecosystem: distribution, sources, biogeochemistry, impacts, and management for restoration. Crit Rev Environ Sci Technol 41(S1):249–288. https://doi.org/10.1080/10643389.2010.531201

Smith JC, Allen P, Turner MD (1994) The kinetics of intravenously administered methylmercury in man. Toxicol Appl Pharmacol 128:251–256

Stern AH, Smith AE (2003) An assessment of the cord blood: maternal blood methylmercury ratio: implications for risk assessment. Environ Health Perspect 111(12):1465–1470

Suzuki T, Hongo T, Yoshinaga J, Imai H, Nakazawa M, Matsuo N, Akagi H (1993) The hair-organ relationship in mercury concentration in contemporary Japanese. Arch Environ Health 48:221–229

Suzuki K, Nakai K, Sugawara T, Nakamura T, Ohba T, Simada M, Hosakawa T, Okamura K, Sakai T, Kurokawa N, Murata K, Satoh C, Satoh H (2010) Neurobehavioral effects of prenatal exposure to methylmercury and PCBs, and seafood intake: neonatal behavioral assessment scale results of Tohoku study of child development. Environ Res 110(7):699–704

Tsubaki T, Takahashi H (1986) Clinical aspects of Minamata disease. Neurological aspects of methylmercury poisoning in Minamata. In: Tsubaki T, Takahashi H (eds) Recent advances in Minamata disease studies. Kodansha, Tokyo, pp 41–57

Uchino M, Okajima T, Eto K, Kumamoto T, Mishima I, Ando M (1995) Neurologic features of chronic Minamata disease (organic mercury poisoning) certified at autopsy. Intern Med 34 (8):744–747

USDA (U.S. Department of Health and Human Services and U.S. Department of Agriculture) (2015) 2015–2020 dietary guidelines for Americans, 8th edn. December 2015. http://health.gov/dietaryguidelines/2015/guidelines/

US EPA (2013) 2011 National Listing of Fish Advisories. EPA-820-F-13-058. USEPA, National Fish Advisory Program, Office of Science and Technology. Washington, DC. http://water.epa.gov/scitech/swguidance/fishshellfish/fishadvisories

US EPA (2018a) Impaired waters and TMDLs. Impaired waters restoration process: planning. https://www.epa.gov/tmdl/impaired-waters-restoration-process-planning

US EPA (2018b) Impaired waters and TMDLs. Overview of total maximum daily loads (TMDLs). https://www.epa.gov/tmdl/overview-total-maximum-daily-loads-tmdls

US FDA (U.S. Food and Drug Administration) (2004) What you need to know about mercury in fish and shellfish. 2004 EPA and FDA advice. https://www.fda.gov/Food/FoodborneIllnessContaminants/Metals/ucm351781.htm

US FDA (U.S. Food and Drug Administration) (2018) Eating fish: what pregnant women and parents should know. https://www.fda.gov/food/consumers/eating-fish-what-pregnant-women-and-parents-should-know

van Wijngaarden E, Thurston SW, Myers GJ, Strain JJ, Weiss B, Zarcone T, Watson GE, Zareba G, McSorley EM, Mulhern MS, Yeates AJ, Henderson J, Gedeon J, Shamlaye CF, Davidon PW (2013) Prenatal methylmercury exposure in relation to neurodevelopment and behavior at 19 years of age in the Seychelles Child development Study. Neurotoxicol Teratol 39:19–25

Ware FJ, Royals H, Lange T (1990) Mercury contamination in Florida largemouth bass. Proc Annu Conf SEAFWA 44:5–12

World Health Organization (1990) Methylmercury. WHO, Geneva. http://inchem.org/documents/ehc/ehc/ehc101.htm

Zheng W (2002) Blood-brain barrier and blood-CSF barrier in metal-induced neurotoxicities. In: Massaro EJ (ed) Handbook of neurotoxicology, vol 1. Humana Press, Totowa, NJ, pp 161–193

Chapter 12
Mercury Biomagnification Through Everglades' Food Webs and the Resulting Risk for Environmental and Human Health: A Synthesis

Darren G. Rumbold

Abstract This chapter attempts to synthesize information presented in Chaps. 7 through 11 of this volume. It examines the influential factors that control the amount of methylmercury that enters and is subsequently biomagnified through the food web to top predators. It begins with a summary of current risk to both wildlife and humans from mercury exposure in south Florida.

Keywords Biomagnification · Food webs · Risk

The concern about mercury (Hg) in Florida is driven by the risk that it poses to top predators including humans. Because methylmercury (MeHg) biomagnifies (Chap. 7, this volume), top predators including humans are at greater risk than the rest of the food web. Chapter 10 of this volume presented an ecological risk analysis based on three separate lines of evidence showing MeHg remains a risk to fish and wildlife in south Florida hotspots. The greatest risk was for sub-lethal impacts such as cellular or tissue pathology, neurochemical changes and altered behavior. It was pointed out that eco-epidemiological surveys in south Florida have documented these effects in several fish and wildlife species at current levels. As reviewed in Chap. 10, while these sub-lethal effects may not lead to acute somatic death, they can lead to other key events in the adverse outcome pathway (Fig. 10.1) leaving the organism at a competitive disadvantage (e.g., unable to find food causing malnutrition, increased susceptibility to disease, depressed reproduction) or at an increased risk of predation; this being equivalent to ecological mortality.

Chapter 11 of this volume presented a human-health probabilistic risk analysis that examined risks to sport and subsistence anglers and hunters consuming Everglades' fish and wildlife. It reviews an early study by Fleming et al. (1995) that

---

D. G. Rumbold (✉)
Florida Gulf Coast University, Fort Myers, FL, USA
e-mail: drumbold@fgcu.edu

© Springer Nature Switzerland AG 2019

D. G. Rumbold et al. (eds.), *Mercury and the Everglades. A Synthesis and Model for Complex Ecosystem Restoration*, https://doi.org/10.1007/978-3-030-32057-7_12

surveyed hair Hg in 350 participants and found over 1/3 of Florida anglers and hunters had levels exceeding US EPA's reference (i.e., safe) dose. A recent survey of Hg in the hair of women of childbearing age (including pregnant women) in Martin County, Florida continued to find a higher mean hair Hg level and a higher percentage of women with ≥ 1 μg/g hair Hg level than those reported at the national level and in other regional studies (Nair et al. 2014). As reviewed in Chap. 11 (and in Chaps. 8–10 of this volume), fishes from the Everglades (as well as Florida Bay and coastal waters of southwest Florida) continue to have Hg levels exceeding US EPA's national water quality criterion for the protection of human health. Not surprisingly, the Florida Department of Health continues to issue advisories for human consumption of fish due to the high levels of Hg in their tissues (now over 60 species; 2018). As discussed in Chap. 11 of this volume, because of the areal extent of the fishes covered under these advisories, FDEP developed a statewide total maximum daily load (TMDL) for mercury in 2012. As stated in the TMDL document, "mercury is unique among impairments for which TMDLs have been produced to date, in that impairments are made based upon potential risks to human health, not upon whether concentrations of a pollutant exceed the state's water quality criteria (Chapter 62-302.530, Florida Administrative Code)." Despite these past administrative actions, the human-health risk analysis presented in Chap. 11 clearly shows continued risk to Floridians who consume too much locally-caught fish, even if they follow those consumption advisories, which raises the question of whether the current advisory is sufficiently protective. Chapter 11 also finds significant risk from eating pig frogs or alligators caught in the Everglades.

These health risks are a direct result of the exceptionally high levels of Hg biomagnified in south Florida's biota. While levels of Hg declined in certain species from some areas in the early 2000s, data contained in Chaps. 8–10 of this volume clearly show biota from the Everglades and wider south Florida (e.g., fishes, sharks, dolphins, birds, panthers) continue to have Hg levels much higher than biota from other regions. The areal extent and the degree of risk to ecological receptors and humans continues to make Florida's Hg problem one of the worst in the world.

Many of the drivers potentially responsible for Florida's Hg problem were reviewed elsewhere in this book. Chapters 7 through 9 of this volume focus on the influence that differences in food webs and the natural history or biology of the ecological entities (i.e., organisms) themselves have on biomagnification. Intraspecific and interspecific variation in the concentrations of biomagnified Hg are influenced by complex interactions of a myriad of biological traits including diet composition, trophic position, bioenergetics, age, size, and growth rate. Surveys in Florida have found individuals of certain species with surprisingly high or low Hg levels in their tissues as a result of unanticipated interactions of these traits (Krabbenhoft et al. 2012; Rumbold et al. 2014, 2018). Chapter 8 presents information on four animal groups: mosquitofish (*Gambusia holbrooki*), Largemouth bass (*Micropterus salmoides*), wading birds and, Florida panthers (*Felis concolor coryi*) to illustrate how these traits, particularly diet and trophic position affect the amount of Hg that is biomagnified. As discussed in Chap. 8, community structure can shift in the Everglades as a result of water management and land use patterns and, consequently, so can diets and Hg exposure. One of the best examples of this in the Everglades is the increased Hg exposure of panthers that are linked more strongly with aquatic food webs in certain habitats (Chap. 8, this volume). Yet, mosquitofish, bass and wading birds all show seasonal, hydrologically-dependent changes in diet and trophic position (Chap. 8).

Loftus (2000, as cited in Chap. 8) comprehensively sampled the biota in Everglades National Park (ENP) examining their gut contents and determining their Hg content; he was one of the first to identify high Hg concentrations in aquatic insects. He also used stable isotopes of carbon and nitrogen as an independent assessment of the trophic position, which supported the findings from the gut content analysis. He found δ15N values of the invertebrates and fishes, including mosquitofish, correlated significantly with mean total Hg levels. Most importantly, mosquitofish, which are known for having a diverse and variable diet, experienced seasonal changes in their diet as marshes began to dry up with trophic position increasing with time following a dry-down (Chap. 8). Using caged, captive-reared mosquitofish, he found that fish at two of the three long-hydroperiod locations had eaten greater volumes of algae and fewer chironomids and, likely as a result, had lower Hg levels than the short-hydroperiod fish. It was theorized that heterotrophic bacteria metabolizing detritus in flocculent sediments was an important food source and Hg pathway for the chironomids (and amphipods). As discussed in Chap. 8, hydrologically-dependent changes in diet and trophic position of mosquitofish was supported by the findings of a survey of 20 sites across the WCAs and ENP by Williams and Trexler (2006). They found trophic position of mosquitofish ranged from 4 to 4.5 in areas remaining inundated for over 8 years and those that recently dried. However, they reported finding no relationship between trophic position and measures of nutrient status or indicators of productivity at a given location (Williams and Trexler 2006). This would be in agreement with an earlier study done on mosquitofish by Kendall et al. (2003) which found no statistically significant difference in δ15N values for fish from high and low nutrient sites. They concluded that there was no evidence for shorter food chains at high nutrient sites. Alternatively, as discussed in Chap. 8, more recent work that modeled δ15N and δ13C measured in mosquitofish from 21 Everglades sites suggested the presence of niche partitioning, which could lead to variation in food-chain length during the wet season and, in turn could affect biomagnification (Abbey-Lee et al. 2013 as cited in Chap. 8). Nonetheless, a much larger study by US EPA (Stober et al. 2001, p. 6-23), which examined the gut contents of 2784 mosquitofish collected from 259 sites across the Everglades, found that "though mosquitofish consume a variety of animal prey, all the animal types had similar trophic scores." More to the point, they (Stober et al. 2001, p. 6-24) concluded that "the hypothesis that trophic position could be used to explain Hg concentration in mosquitofish was not supported."

Interestingly, Williams and Trexler (2006) speculated that the absence of a relationship between mosquitofish trophic position and nutrients might be explained by the importance of the microbial loop (as opposed to phytoplankton or vascular plants) as the likely major route of energy flow in the Everglades. As discussed in Chap. 7 of this volume, the importance of the microbial loop has implications also with regard to the apparent lack of the "biodilution effect" of Hg in nutrient-enriched Everglades' marshes.

Chapter 8 also reviewed how Largemouth bass, with their opportunistic feeding nature, also have marked changes in composition of diet across habitats in response to changes in hydroperiod. This is not surprising given that food webs supporting bass also varied among habitat types. As reviewed in that chapter, slough and wet prairie habitats in the WCAs offer the most suitable depths for bass and are characterized by thick growth of submerged aquatic vegetation (SAV). Savino and Stein (1982, as cited in Chap. 8) suggested that prey-capture efficiency of bass decreases with increasing SAV which can lead to decreased piscivory within complex vegetated marsh habitat. Food webs varied most between canal and marsh sites probably due to habitat structure affecting prey availability. Bass in a given habitat also tend to shift to greater piscivory as they get larger (Chap. 8); the transition to piscivory came at a substantially smaller size in canal bass. Because marsh bass have a more omnivorous diet, they typically have lower trophic scores than bass from nearby canal sites (Chap. 8). Yet, bass from marsh sites can have much higher Hg than nearby canals. This lead Chap. 8 authors to suggest "among site differences in food webs were less critical to bass Hg concentrations than MeHg availability at the base of the food web." They also point out that this conclusion is consistent with the findings of Bemis et al. (2003 as cited in Chap. 8) who utilized SIA at 12 marsh and canal locations in the WCAs and found δ15N explained little (5%) of the variance in bass Hg across sites.

As reviewed in Chap. 8, wading birds are also considered generalist feeders that will take a wide variety of prey types and sizes in a range of wetland habitats. Exposure of Hg to Everglades' aquatic birds is likely to increase with increasing trophic level of prey, and with increasing size of prey items. Therefore, hydrologically-dependent changes in size or trophic level of prey, particularly fish, as discussed above could lead to changes in Hg exposure to wading birds.

Chapter 8 also reviewed the food habits of panthers which vary across their range within south Florida. North of I-75, their diet is comprised of a larger proportion of ungulates, especially feral swine (Maehr et al. 1990 as cited in Chap. 8). South of I-75, raccoon and deer were the most important prey species, comprising equal proportions of panther diets. Panther diet has also varied temporally. As discussed in Chap. 8, significant declines in deer and feral swine in some areas south of I-75 have been observed. Regardless of the cause, these declines appear to have resulted in a shift in panther prey selection from ungulates to raccoons over the past two decades (Chap. 8 and references therein). Variations in Hg levels among panther populations in south Florida are most often attributed to differences in diet (deer and hog versus raccoon and alligator; Chap. 8) and the degree of linkage with aquatic food webs where methylation occurs in sediments (see Chap. 1, this volume).

Clearly, differences in community structure and preferred prey abundance brought about by variations in hydrology, season, fire or land use can affect diet and trophic position of localized populations within a species. In turn, these changes often affect their Hg exposure and accumulation. For these reasons, care must be taken when a biomonitoring program is based solely on a single species. It is important to recognize that these factors as well as nutrient inputs can also affect methylation potential (for review, see Chap. 1, this volume) which can affect their Hg exposure. Chapter 7 presents a relatively new approach that assesses biomagnification through an entire food web (the Trophic Magnification Slope which is the regression of log-transformed tissue-Hg concentration on δ15N values in taxa from an ecosystem) which has been found useful in untangling factors altering basal MeHg from factors affecting biomagnification efficiency. As reviewed in Chap. 7, although several studies have found TMS as a measure of this efficiency to be affected by various factors, the spatial and temporal differences in basal MeHg entering the food web are often larger and are the dominant factor determining variation in biomagnified Hg in top predators. It is highly likely that this is the case in Everglades's marshes where, as shown above, nutrients do not have the same effect on carbon and Hg transfer as in lakes and where Hg methylation potential is spatially so highly variable (Rumbold and Fink 2006; for review, see Chaps. 1 and 3, this volume).

References

Fleming LE, Watkins S, Kaderman R, Levin B, Ayyar DR, Bizzio M, Stephens D, Bean JA (1995) Mercury exposure in humans through food consumption from the Everglades of Florida. Water Air Soil Pollut 80:41–48

Kendall C, Bemis BE, Trexler J, Lange T, Stober JQ (2003) Is food web structure a main control on mercury concentrations in fish in the Everglades? Greater Everglades Ecosystem Restoration (GEER) Meeting, Palm Harbor, FL, April 2003. Program and Abstracts

Krabbenhoft D, DeWild J, Tate M, Ogorek TJ, Hart K, Snow S, Demopoulos A (2012) Mercury bioaccumulation in pythons from the Florida Everglades region. Presented at the 9th INTECOL International Wetlands Conference, Orlando, FL, 3–8 June 2012

Nair A, Jordan M, Watkins S, Washam R, DuClos C, Jones S et al (2014) Fish consumption and hair mercury levels in women of childbearing age, Martin County, Florida. Matern Child Health J 18(10):2352–2361

Rumbold DG, Fink LE (2006) Extreme spatial variability and unprecedented methylmercury concentrations within a constructed wetland. Environ Monit Assess 112(1–3):115–135

Rumbold D, Wasno R, Hammerschlag N, Volety A (2014) Mercury accumulation in sharks from the coastal waters of southwest Florida. Arch Environ Contam Toxicol 67:402–412

Rumbold DG, Lienhardt CT, Parsons ML (2018) Mercury biomagnification through a coral reef ecosystem. Arch Environ Contam Toxicol 75(1):121–133

Stober Q, Thornton K, Jones R, et al. (2001) South Florida ecosystem assessment: Phase I/II–Everglades stressor interactions: hydropatterns, eutrophication, habitat alteration, and mercury contamination (summary). Water Management Division, and Office of Research and Development EPA-904-R-01-002

Williams AJ, Trexler JC (2006) A preliminary analysis of the correlation of food-web characteristics with hydrology and nutrient gradients in the southern Everglades. Hydrobiologia 569:493–504